From Madman to Crime Fighter

From Madman to Crime Fighter

The Scientist in Western Culture

ROSLYNN D. HAYNES
UNIVERSITY OF NEW SOUTH WALES

Johns Hopkins University Press
Baltimore

© 2017 Johns Hopkins University Press
All rights reserved. Published 2017
Printed in the United States of America on acid-free paper
2 4 6 8 9 7 5 3 1

An earlier version of this book was published as
From Faust to Strangelove: Representations of the Scientist in Western Literature (© 1994 Johns Hopkins University Press).

Johns Hopkins University Press
2715 North Charles Street
Baltimore, Maryland 21218-4363
www.press.jhu.edu

LIBRARY OF CONGRESS CATALOGING-IN-PUBLICATION DATA

Names: Haynes, Roslynn D. (Roslynn Doris), 1940– author.
Title: From madman to crime fighter : the scientist in western culture / Roslynn D. Haynes.
Description: Baltimore : Johns Hopkins University Press, 2017. | Includes bibliographical references and index.
Identifiers: LCCN 2016049245| ISBN 9781421423043 (pbk. : alk. paper) | ISBN 9781421423050 (electronic) | ISBN 1421423049 (pbk. : alk. paper) | ISBN 1421423057 (electronic)
Subjects: LCSH: Scientists in literature. | Literature and science. | Scientists in motion pictures. | Science in motion pictures.
Classification: LCC PN56.S35 H39 2017 | DDC 809/.9336—dc23
LC record available at https://lccn.loc.gov/2016049245

A catalog record for this book is available from the British Library.

Special discounts are available for bulk purchases of this book. For more information, please contact Special Sales at 410-516-6936 or specialsales@press.jhu.edu.

Johns Hopkins University Press uses environmentally friendly book materials, including recycled text paper that is composed of at least 30 percent post-consumer waste, whenever possible.

CONTENTS

List of Illustrations vii
Preface ix
Acknowledgments xi

INTRODUCTION	Myths of Science	1
1	Evil Alchemists and Doctor Faustus	12
2	Bacon's New Scientists	28
3	Foolish Virtuosi	41
4	Newton: A Scientist for God	55
5	Arrogant and Godless: Scientists in Eighteenth-Century Satire	68
6	Inhuman Scientists: The Romantic Perception	75
7	Frankenstein and the Creature	91
8	Victorian Scientists: Doubt and Struggle	105
9	The Scientist as Adventurer	128
10	Efficiency and Power: The Scientist under Scrutiny	142
11	The Scientist as Hero	161
12	Mad, Bad, and Dangerous to Know: Reality Overtakes Fiction	186
13	The Impersonal Scientist	211
14	*Scientia Gratia Scientiae*: The Amoral Scientist	235
15	Robots, Androids, Cyborgs, and Clones: Who Is in Control?	261
16	Pandora's Box	282
17	The Scientist as Woman	302
18	Idealism and Conscience	319
CONCLUSION	New Images of Scientists	337

Appendix: Films and TV Series with Scientist Characters 341
Notes 343
Bibliography 375
Index 401

ILLUSTRATIONS

Figure 1. *The Alchemist*, by Jacques Louis Perée, after David Teniers II — 16
Figure 2. *The Alchemist in Search of the Philosopher's Stone Discovers Phosphorus*, by Joseph Wright — 24
Figure 3. *New Atlantis*, illustration from *The New Atlantis*, by Sir Francis Bacon, 1627 — 32
Figure 4. Frontispiece from Thomas Sprat's *History of the Royal Society of London*, 1667 — 37
Figure 5. Engraving from *Dell'Historia Naturale*, by Ferrante Imperato, Naples, 1599 — 45
Figure 6. *The Astronomer*, by Thomas Rowlandson, 1815 — 49
Figure 7. *Portrait of Sir Isaac Newton*, ca. 1715–20 — 56
Figure 8. *A Philosopher Giving a Lecture on the Orrery*, by Joseph Wright of Derby, 1764–66 — 63
Figure 9. *Bedlam: A Rake's Progress, Plate VIII*, by William Hogarth, 1735 — 74
Figure 10. *Newton*, by William Blake, 1795 — 82
Figure 11. *An Experiment on a Bird in the Air Pump*, by Joseph Wright of Derby, ca. 1767–68 — 83
Figure 12. Frontispiece to *Frankenstein*, by T. Holst (1831) — 97
Figure 13. *Nightmare*, by John Henry Fuseli, 1781 — 101
Figure 14. Detail of a lithograph of *Michael Faraday Delivering a Christmas Lecture in 1856*, by Alexander Blaikley — 107
Figure 15. *Charles Robert Darwin*, from *Punch* (1882), by Linley Sambourne — 120
Figure 16. *Eleven O'Clock P.M.: A Scientific Conversazione*, from *Twice round the Clock; or, The Hour of the Day and Night in London* (1859), by George Augustus Sala and William McConnell — 129
Figure 17. *Nautilus Salon*, from Hetzel edition of *Vingt mille lieues sous les mers*, by Alphonse Marie Adolphe de Neuville and Edouard Riou — 131
Figure 18. *The Strange Case of Dr. Jekyll and Mr. Hyde*, poster from the 1880s — 147
Figure 19. Cover illustration for *Amazing Stories* depicting a scene from H. G. Wells's *War of the Worlds*, by Frank R. Paul, August 1927 — 165
Figure 20. Uncredited illustration in the first issue of *Marvel*, October 1938 — 168

Figure 21. *The Test Tube Baby*, by Albert Robida,
from *Vingtième Siècle*, 1883 — 238

Figure 22. Rotwang and the Robot Maria, from Fritz Lang's
Metropolis (1927) — 265

Figure 23. Pierre and Marie Curie in their laboratory, ca. 1904 — 321

PREFACE

Until *From Faust to Strangelove* was published in 1994, the main body of evidence for the image of scientists in popular culture came from research into the way primary school children depicted "a scientist" (old, male, with a great deal of hair like Einstein, and indications of being mad and dangerous). This research was a sequel to the pilot study carried out by Margaret Mead and Rhoda Métraux in 1957 to assess how scientists were regarded among US high school students. Their responses, while positive about scientists at an official level, were "overwhelmingly negative" when the questions touched on a career in science or a scientist as a marriage partner.

In *From Faust to Strangelove*, I investigated the curious anomaly that such negative concepts and images existed in twentieth-century Western society, where science was central to the social structure at every level and was promoted and rewarded with a significant proportion of national budgets. Surprisingly, this analysis of Western literature depicting scientist characters from the medieval period to the 1990s yielded only seven distinct stereotypes in as many centuries, although they reemerged in particular social and political contexts. The great majority of these were obsessive to the point of madness and morally compromised if not completely evil.

In the two decades and more since *From Faust to Strangelove*, there has been considerable scholarly interest in exploring how far the stereotypes identified in the literature applied to film, and valuable research has been done in this area. As well, there have been research grants, conferences, higher degrees, a flood of journal articles and books, and an interdisciplinary research group, Fiction Meets Science, based at the Universities of Bremen and Oldenburg, Germany, and the Hanse-Wissenschaftskolleg to investigate the sociological context, content, and purpose of recent science novels.

From Madman to Crime Fighter required a new title, since its range extends far beyond Stanley Kubrick's 1964 film *Dr. Strangelove, or: How I Learned to Stop Worrying and Love the Bomb*. This revised and updated work develops the themes of the earlier book into the early 2000s, engaging in particular with a range of films that supplement or counterpoint the literary examples and situating both fiction and films in the sociological context of their time. This book includes two completely new chapters, "Robots, Androids, Cyborgs, and Clones: Who Is in Control?" and "The Scientist as Woman." The latter investigates the growing number of female scientists in fiction and film, the specific risks and difficulties they experience, the way they are depicted, and what this says about the roles and perceptions of women in the profession. It also suggests that women, while often depicted as preoccupied with problems, are not all victims. Some, like Abby Sciuto of *NCIS*, portray a joy in their work and an ability to gain the respect and friendship of their colleagues and to rise above the cliché problems of the professional woman. As well, there is a new introduction and a concluding section, "Watershed: The New Scientists." There are additional notes and an expanded bibliography referencing the large body of recent research and the films discussed in the text.

Many recent films continue to recycle the older literary stereotypes focusing on the "mad, bad scientist" figure as the instigator of dangerous scenarios that threaten societies, the environment, or even the planet. On the other hand, in mainstream fiction there has been a shift toward depicting scientist figures, not as stereotypes but as individuals, whose lives, like those of lawyers, bankers, doctors, or teachers, intersect with those of laypeople, as they struggle with the personal, professional, and moral issues arising from their research.

From Madman to Crime Fighter looks forward to a further volume that focuses exclusively on twenty-first-century novels and films and explores the new perception of scientists in different disciplines as risk monitors and potentially risk averters in the face of environmental dangers and climate change.

ACKNOWLEDGMENTS

From Madman to Crime Fighter has benefited immensely from stimulating conversations with fellow researchers at conferences and within the context of the Fiction Meets Science group* at the Universities of Bremen and Oldenburg and the Hanse-Wissenschaftskolleg, Delmenhorst, Germany, where I have been privileged to be granted two terms as a research fellow.

As was the case when this book was published in its original version, the topic has proved a most effective conversation starter. Everyone, it seems, has a favorite example to share of a scientist character in fiction or film, and such discussions have significantly broadened the scope of this study.

I am greatly indebted to George Roupe for his meticulous and patient copy editing of the typescript through many email interchanges, cross checking all the numerous references for consistency.

Most of all I am grateful to my husband, Raymond, who has watched, reviewed, and discussed with me hundreds of hours of films about scientists, and to my daughters, Nicola and Rowena, whose help, support, and enthusiastic encouragement over many years have been essential for the book's completion.

---

* The Fiction Meets Science project receives funding from the VolkswagenStiftung.

From Madman to Crime Fighter

INTRODUCTION

Myths of Science

Science is a powerful cultural force. Each year Western governments award multimillion-dollar grants to teams of scientists, usually on the basis of the reputation of the principal investigator, to carry out research on a topic that less than 1 percent of the population would be able to understand. Such is the trust that science will deliver great benefits to society. Yet, at the same time, there is a deep-seated fear in the modern "world risk society"[1] that science has brought forth the greatest terrors and ongoing risks to populations unable to control them because only scientists have access to this powerful knowledge.

There are ample historical reasons for both these responses. Science has given us fundamental survival techniques in agriculture, in engineering and medical science, in transport and communications, and in safety criteria. It is our knowledge base for socioeconomic structure and core politics, and for the quality of life that Western society takes as its right. It holds out the best hope we have for monitoring and repairing environmental degradation and possibly, if we are lucky, halting climate change and species extinction. It offers release from endemic poverty, demeaning toil, and infant mortality and promises longevity and well-being for the majority of people. Science has also produced weapons of mass destruction and other inventions that, while less violent, even potentially benign, nevertheless entail possible dangers—physical, moral, and humanitarian—over which we currently have little control: nuclear power stations, genetic engineering, reproductive technology, agribusiness, patenting of genes, cloning, genetically modified foods, and artificial intelligence. With the almost exponential rate of scientific discoveries across all fields of knowledge during the twentieth century and the increasing effect they produce within a short time from their inception, on people and on nature, the potential for the evil reputation of science and scientists has been immense and cumulative, outweighing the damage created by others who more obviously wield short-term power—politicians, judges, economists, generals, corporate giants, or media barons.

Beneath such rational arguments there lies within us a much deeper cultural "knowledge," a cumulative mythological wisdom. Myths are the signature of cultures: they express in enduring form, across many generations, the hopes, fears, values, transgressions, and punishments that underpin the social fabric. In oral cultures myths were created by storytellers, whose powerful narratives spoke to the lived experiences of their listeners. In modern Western society our enduring myths have been created by successive images derived first from oral traditions, reinforced over centuries by the narratives of fiction and, more recently, film and other popular media. One set of deeply embedded myths concerns science as the quest for knowledge, its practitioners, and its imagined impact on our lives.

Far older than the printed word is a fear of too much knowledge and the belief that some things should remain hidden and unknown. Stories from prehistory (Adam, Prometheus, Daedalus, Icarus, Pandora, and the Norse god Odin) warn us to beware of desiring to learn more than the gods have decreed proper for humans to know. In medieval Europe, for a range of social, political, and religious reasons, these fears were readily identified with medieval alchemists, whose closely guarded learning, derived from the Arab world, was regarded by the Catholic Church as doubly dangerous, coming as it did from the "infidels" and all too likely to confirm the Genesis story of Adam's Fall. The alchemists were the cultural progenitors of centuries of fictional scientists who assumed the same characteristics of medieval superstition: secrecy, dangerous knowledge inaccessible to others, arcane symbols and language unintelligible to outsiders, an identifiable otherness that carried both allure and fear. Indeed, the master narrative concerning science and scientists is about fear—fear of specialized knowledge and the authority that powerful knowledge confers on the few, leaving the majority ignorant and impotent.

Such was the cumulative power of this alchemist figure through narrative reiteration that the most powerful creation myth of modern times is not that of Genesis or Darwin but *Frankenstein*. Why does this story, emerging from the "waking dream" of eighteen-year-old Mary Shelley and published in 1818, continue to haunt us, still providing the most universally invoked imagery for science in the twenty-first century? Why is it best read as a myth of modernity?[2] Why do journalists resort to the name "Frankenstein" as verbal and visual shorthand for any potentially dangerous new discoveries and processes such as "Frankenfoods"?

In the belief that popular prejudices are influenced more by images than by demonstrable facts, this book sets out to explore these questions by tracing the representation of the scientist as a character in Western literature and film from the thirteenth century and contextualizing it in the social, cultural, and intellectual climate of successive periods. Very few actual scientists (Isaac Newton, Marie Curie, and Albert Einstein are the only significant exceptions) have contributed to the popular image of "the scientist." On the other hand, the cavalcade of immoral fictional characters from Dr. Faustus through Dr. Frankenstein, Dr. Jekyll, Dr. Moreau, Dr. Caligari, and Dr. Cyclops to Dr. Strangelove and beyond have been immensely influential in the evolution of the unattractive stereotypes that continue in uneasy coexistence with the manifest dependence of Western society on its scientists. And paradoxical images have been communicated more broadly and more strikingly in films. Statistically scientists have provided the cinema with more villains than any other profession. After a survey of over a thousand horror films produced between 1931 and 1984, Andrew Tudor concluded that mad scientists or their creations have provided the villains or monsters of one-third of these and that scientific or psychiatric research produced the greatest number (39 percent) of the threats in all these horror films. For comparison, scientists have been the heroes or saviors in only 11 percent of these films.[3] More recently Peter Weingart and colleagues have analyzed 222 movies and found a somewhat more complex pattern. Even though a large number of the films presented scientists as benevolent, this was an ambivalent quality, since it frequently included naïveté: the initially "good" scientists are often manipulated by powerful evil interests or may themselves become corrupted through ambition and, like their ancestor Dr. Faustus, prepared to sacrifice ethical principles for knowledge.[4]

Yet, despite its prevalence, there is nothing personal in this vilification. The mad, evil scientists are semiotic figures. They are not intended to be realistic or, usually, to depict any particular scientist. On the contrary, they represent, as it were, the colonized view of science. Just as imperial history is written by the colonizers but we now recognize that the colonized retaliated in unofficial or oral histories, in stories, and in parody, so the official history of science records the discoveries of great scientists, the successes, the chain of influence, and the breakthroughs, but there is also an unofficial history, seen from the perspective of ordinary people who fail to understand, or who feel threatened by, the progress of science. In this record, parody becomes a mode of defense, of "writing back" to power, and in this process the mad scientist plays a major part. He exists to protest against the "great men" account of science, which tells us we have nothing to fear because these good and brilliant people are in control and trustworthy. The mad scientist of books and movies confirms our suspicion that this is not so. He enacts our nightmares that new, experimental knowledge, unknown to the rest of society, may misfire or be deliberately misused, impacting on individual lives or humankind.

In modern society this fear is no longer mainly about magic; after all, we live in a knowledge society, specifically science-based knowledge. But the power this knowledge engenders comes with an inherent social risk. In every sphere, whether economics, politics, the military, education, sport, or business, scientific knowledge and science-based technology confer a perceived competitive advantage on those who "own" it and exclude those who do not. With the ever-increasing scientification of modern Western society, the fear of science and scientists emanates from this exclusion process, whereby those who are not privy to such influential knowledge are marginalized and disempowered. Hence the cavalcade of mad and evil scientists from past centuries continues to resonate, however irrationally.

These imaginary scientists are expressions of their creators' response to the role of science and technology in a particular time and social context and are therefore interesting in their own right, but viewed chronologically over seven centuries they achieve an additional historical significance as ideological markers of the changing perception of science over time. Studying the evolution of representations of scientists in Western literature, and more recently in film, allows us to see how clusters of these fictional images have coalesced to produce archetypes that have acquired a cumulative, even mythical, importance. From these centuries of stereotypical images we can draw some surprising conclusions.

1. The stereotypes are all male, mostly middle-aged or old, and all Caucasian.
2. The majority of these stereotypes (as well as the vast majority of scientist characters) reflect a distinctly negative judgment by writers and filmmakers. They are morally compromised if not outright evil, obsessive, dangerous, mad, and uncaring about and dissociated from society.
3. In the whole pageant of fictional scientists, from the medieval alchemist to the twenty-first-century atomic physicist, molecular chemist, geneticist, or artificial intelligence designer, the number of stereotypes

is remarkably small—just seven in as many centuries. We can characterize them as follows:

(i) The morally suspect alchemist, who reappears at critical times as the obsessed or maniacal scientist, has been the most frequently recurring stereotype. Driven to pursue an arcane intellectual goal that carries strong suggestions of ideological evil, this figure, originally medieval, has been reincarnated as the sinister biologist producing new (and hence allegedly unlawful) species through the quasi-magical processes of genetic engineering. Ultimately the perennial fascination of the alchemist narrative is that it tells a story of what we both desire and fear to know—the story of power beyond our dreams but also beyond our control. Paradoxically, no century has had more control over the material world than ours, and yet in many ways we feel as vulnerable and powerless as a medieval peasant. We are confronted with an unpredictable world where we are stalked by terrorism, by fast-spreading pandemics, by a latent and recurrent nuclear threat, by violent weather conditions associated with climate change, by environmental destruction on a massive scale and species extinction.

(ii) The scientist as idealist. First elaborated by Francis Bacon in the seventeenth century, this figure represents the one unambiguously acceptable scientist, sometimes holding out the possibility of a scientifically sustainable utopia with fulfillment and plenty for all, but in recent times more frequently engaged in moral conflict with a technology-based system that fails to provide for individual human values.

(iii) The stupid virtuoso, out of touch with the real world of social intercourse. Initially this figure appears more comic than sinister, but he too carries disturbing implications. Preoccupied with the trivialities of his private hobby world of amateur science, he ignores social obligations. The modern counterpart of this character from the seventeenth-century Restoration stage, the absentminded professor of early twentieth-century films, while less overtly censured, is nevertheless shown to be a potentially dangerous figure, a moral failure by default.

(iv) The unemotional scientist who has reneged on human relationships and suppressed all human affections in the cause of science.

Originally depicted by the nineteenth-century Romantic writers in reaction against the mechanization and reification of the Industrial Revolution, this has been an enduring stereotype, recurring as the most frequent image of the scientist in twentieth-century plays, novels, and films. During the 1950s there was some ambivalence about this figure: while his emotional deficiency was regretted, it was also accepted, even with some qualified admiration, as the price of an altruistic dedication to science.

(v) The heroic adventurer in the physical or the intellectual world. Towering like a superman over his contemporaries, exploring new territories, or engaging with new concepts, this character has emerged at periods of scientific optimism. His particular appeal to adolescent audiences through the implicit promise of transcending boundaries, whether material, social, or intellectual, ensured the popularity of this stereotype in comics and space opera. Deeper analysis of such heroes, however, suggests the danger of their charismatic power; they rarely consider consequences and, in the guise of neoimperialist space travelers, they impose their particular brand of terraforming colonization on the universe.

(vi) The mad, bad, dangerous scientist. The moral characteristics of this figure are not new; his ancestry lies in the alchemist tradition, but twentieth-century writers and filmmakers grafted on a new ruthlessness that soared to the heights of megalomania. This became more feasible with the increasing power of science to produce cataclysmic results on a scale hitherto unimaginable. Nuclear physics, whether in war or peace, raised fundamentally new problems for humanity— questions about moral responsibility for research; the impossibility of unlearning knowledge (the Pandora myth); decisions about how and by whom knowledge should be "owned," controlled, and implemented; the moral consequences of genetic engineering and associated technologies.

(vii) The helpless scientist. This peculiarly twentieth-century character has lost control over his discovery (which, monster-like, has grown beyond his expectations), either physically or, as frequently happens in wartime and in the corporate world, in a managerial sense, being forced to relinquish any say over the direction of its implementation.

These fictional protagonists have been fundamental in the evolution of Western society's ambiguous love-hate attitude toward science, which resurfaces periodically in debates over the use of public money for science research, the benefits and dangers of nuclear power and state-of-the-art weapons, the responsibility of science and technology for environmental pollution, many facets of genetic engineering and artificial intelligence, and the reification of the individual in a postmodern technocracy.

There is also some good news. Although literature has most frequently acted as a mirror, reflecting contemporary attitudes toward science and scientists, it has sometimes been a searchlight, pointing the way to new directions and insights. This was certainly true in the case of Francis Bacon's *New Atlantis* (1626), which was directly responsible for the establishment of the Royal Society. It was arguably true in the case of the English Romantic poets of the early nineteenth century, who, in defiance of the whole Newtonian materialist edifice, opted for an understanding of the world in terms of an interactive relationship, a position not dissimilar to that posited by Heisenberg's uncertainty principle and much of subsequent twentieth-century physics. It could also be true today. Some of our most perceptive contemporary writers have described and analyzed the communication breakdown between the disciplines, not merely between the so-called two cultures that C. P. Snow pinpointed, but between specialized areas within disciplines; a few have dared to propose alternatives. They have pointed out the limitations of the traditional assumption that the scientist must be a detached and wholly rational observer of nature. They have suggested a correlation between this inherently male attitude and the conventionally objective terminology of science, the reports couched in the passive voice to edit out any suggestion of an involved observer, the long-standing model of a passive, female nature laid open to the gaze of (male) scientists. These writers and filmmakers are proposing instead that scientists, who, after all, pride themselves on their mental flexibility, might reexamine their assumptions and their unacknowledged mind-set. They have created female scientist figures such as forensic scientist Abby Sciuto on the television series *NCIS*, as intellectually astute, successful, and imaginative as any of her male colleagues, with skill in communications and an endearing personality that intimidates no one, and humane and quirky interests outside her work. As such, Abby stands as the antithesis of her alchemist forebears.

There are lessons for readers, too. The distancing effects of fiction and film also serve a useful sociological function. Novels and films featuring mad scientists often provide a valuable forum for consideration of ideas taboo in their own time frame. They posit a society in which current values are questioned and new moral concepts and social values explored. The prohibited topics are rendered psychologically safe for discussion because they are the inventions of "mad" scientists.[5] More importantly, when we understand how deeply we have been influenced by the figure of the alchemist, wielding a power we both desire to access and fear to engage, we can discard that fear and become members of the democracy of science, taking responsibility for its governance and outcomes. Science itself has given us the means to do so—free information via the Internet about the latest discoveries in every aspect of research; journalists and scientists who present science programs in accessible terms; social media, which allow us to protest en masse about moral and ethical issues arising from potential breakthroughs and innovations. These facilities are, at one level, the gift of technology; they are also the legacy of the writers discussed in these chapters, who probed, applauded, satirized, demonized, or sought to understand and empathize with the scientists of their day.

In the past science has been a despotic ruler, from whom we have hoped to acquire favors and benevolence. But now we can enter into a democratic relationship with science, for we have some powerful leverage. Scientists are all too aware that their research depends on public support: it must be approved by ethics committees, which include nonscientists, and it must gain financial support—usually a considerable amount. In such cases scientists are only too keen to appear on the media, explaining their research and, especially in medical fields, announcing their latest breakthrough.

We will, of course, continue to visit the alchemists' cave, reinvented as the glittering laboratories of wealthy institutes, but now we may come openly, more alert to deceptions and ready to chat with our alchemists about what they are up to.

A note on science fiction and on the words *science* and *scientist*

This book is about the scientist as a character. Its focus is not science fiction. Insofar as it refers to examples of science fiction in novels, stories, pulp magazines, and films it does so only in relation to a general cultural context or because some of these novels, for example *Frankenstein*, *The Island of Doctor Moreau*, *The Dispossessed*, contain interesting scientist characters. But in most science fiction the characters are not highly developed because the author's main interest is in the intellectual game of ideas, not in characterization. H. G. Wells conceded that his characters were subsidiary to his sociological purpose of exploring science-related issues: "I found that I had to abandon questions of individuation. . . . I had very many things to say."[6] This is true of even classic science fiction writers such as Jack London, Arthur C. Clarke, Gregory Benford, and William Gibson. Science fiction aficionados may be disappointed to find some favorite author or story absent from this discussion, but there are many comprehensive treatments of science fiction already available to supply the deficiency.

The word *science* came into English in the Middle Ages, from the French *science*, which meant knowledge in the broadest sense.[7] Subsequently it acquired the connotation of accurate and systematic knowledge, especially that derived from philosophical demonstration, for example, by a syllogism. The Scholastic philosophers used the word *science* to refer to specialized branches of philosophy; thus the seven sciences of medieval learning were grammar, logic, arithmetic, rhetoric, music, geometry, and astronomy. Francis Bacon extended this definition to include knowledge derived from observation and experiment,[8] and hence the study of the natural world came to be called *natural philosophy*, although this term was still regarded as suspect by the Scholastic purists. Even Isaac Newton took the precaution of setting out his *Principia* in the format of Euclid's *Geometry* in order to elevate it to the status of a science. Although by 1820 the astronomer William Herschel was rejecting the old Scholastic nexus between science and deductive logic and aligning science exclusively with Bacon's experimental method, the terms *science* and *philosophy* were used synonymously until about 1850. After this time, the term *philosophy* was assigned to theological and metaphysical science, and *science* to experimental and physical science. The relation of social and even biological sciences to the physical sciences remained less clear-cut. Ross considers that the growing prestige of the physical sciences in the nineteenth century explains why they could arrogate to themselves the word formerly used to designate all knowledge, but the hierarchical structure within the sciences is still implicit in much discussion. It has frequently been remarked that, in Ross's words, "a higher status is claimed by, and generally accorded to, the physical and biological sciences, and to physical sciences in particular."[9] There often seems to be an implicit assumption, at least by physical scientists, that mathematical content is an index of scientific status.

The word *scientist* appeared much later, and even when the neologism was coined anonymously by William Whewell,[10] it was greeted with acrimony by most of the doyens of the scientific community. By analogy with *dentist*, it was thought to have connotations of specialization and professionalism distasteful to heirs of the amateur tradition of British science. Their ideal was "that of a man liberally educated, whose avocation was science as an intellectual *cum* philanthropic recreation, to which he might indeed devote most of his time without ever surrendering his claim to be a private gentleman of wide culture. In particular, to be thought to be pursuing science for money was distasteful."[11] A character such as Lord Hollingshed in Elizabeth Gaskell's novel *Wives and Daughters* epitomizes this ideal.[12] However, as Ross points out, this superior attitude could not survive the educational reforms of the midcentury, when science became one of the learned professions, along with medicine, theology, and the law. Roger Hamley in the same Gaskell novel exemplifies this new approach to science.

In the twentieth century, as science has acquired professional respectability and, perhaps more importantly, associations of wealth and power, the words *science* and *scientist* have come to bear more positive but scarcely less dangerous connotations of accuracy and even infallibility. Disciplines desirous of demonstrating their intellectual credentials lay claim to the title *science* in much the same way as the terms *philosophy* and *philosopher* were sought-after appellations in the eighteenth century. Terms such as *domestic science, military science, political science,* and *building science* really have meaning only in the original sense of science as knowledge, but their proponents imply in the use of the word a level of rigor that physical and biological scientists would contest.

While this would not, perhaps, matter very greatly if it were merely a dispute over semantics, its implications are more disturbing. Basic to the desire to qualify as a science is public deference to scientific opinion, which assumes an importance in the popular mind not only out of proportion to its likely validity but in violation of the very basis of scientific method. Insofar as scientists exploit this mistaken credulity for their own ends, they are guilty of betraying that tradition of open questioning of all authorities that has allowed science to develop to the position it now holds. Insofar as they themselves believe it, they have returned to the pre-Baconian tradition of the alchemist in search of absolute and unquestionable truth—the philosopher's stone, perpetual motion, and the elixir of life. Although the processes whereby science actually advances have been questioned, Karl Popper's criterion that "the scientific status of a theory is its falsifiability, or refutability"[13] stands as the unique obligation of modern science. It is not sufficient to accumulate an incessant stream of confirmations, of observations that can be seen as verifying a theory: it must make risky predictions that can be objectively tested for falsity. Adopting this criterion greatly simplifies the question of what counts as a science.

CHAPTER 1

Evil Alchemists and Doctor Faustus

> We always met with failure in the end.
> Yet though we never reached the wished conclusion
> We still went raving on in our illusion,
> Sitting together, arguing on and on,
> And every one as wise as Solomon
>
> —*Chaucer*

Remote as they may seem from twentieth-century atomic physicists or industrial chemists in white lab coats, surrounded by equipment costing more than their life earnings, the medieval alchemists were the predecessors of modern scientists. Not only were they in their time at the cutting edge of experimental research into the mysteries of nature, but they endowed the profession with an aura of mystery, secrecy, suspicion, and, at times, heresy, from which it has never completely detached itself. Alchemy has provided a potent source of mythmaking for the critique of modern science, and its reputation continues to haunt science as depicted by writers and filmmakers to such a degree that, in order to understand the development of the scientist as a character in fiction and film, we need to begin with the alchemists and the particular historical events surrounding their appearance in Europe.

The Origins of Alchemy

Like alchemy itself, the character and preoccupations of the alchemist in medieval Europe were a concoction of many elements accumulated over centuries from diverse cultures. The pursuit of alchemy was not originally focused on desire for gold and wealth. Rather, the earliest alchemists were inspired by a vision of man made perfect and immortal, man freed from mental and physical illness and reflecting the One Divine Mind in its Perfection, Beauty, and Harmony.

References to this theoretical aspect of alchemy are found first in the myths and legends of ancient China dating from before the second century BCE, when two separate branches, *Waidan*, or external alchemy, and *Neidan*, or internal alchemy, evolved, associated with two aspects of Taoism.[1] The former involved the heating of substances such as gold, mercury, lead, arsenic, and jade in a crucible to produce elixirs and artificial "gold." These were ingested to achieve longevity and immortality, since gold (*jin*) represented constancy and immutability and *Jindan* was the Way of the Golden Elixir.[2] Neidan, on the other hand, was a form of the Taoist quest for purity, tranquility, and immortality within the person of the alchemist through breathing exercises, meditation, and observation. Both Waidan and Neidan were said to produce transcendence, immortality, longevity, healing, and protection from evil spirits.[3] The Chinese alchemist Go-Hung was believed to have perfected a recipe for a pill of immortality, and the search for this "elixir of life" became another central aspect of alchemy, leading to important contributions to medicine.[4]

It was not until the fourth century CE, however, that the first systematic treatises, integrating a number of different traditions, appeared in Alexandria and Hellenistic Egypt. Greek settlers in Egypt identified their god Hermes, the messenger god (and hence the one who crossed boundaries between divine and human worlds and taught divine wisdom) with the Egyptian deity Thoth, lord of knowledge, science, and magic. Hermes Trismegistus (thrice great Hermes), a conflation of the two, influenced both Christianity and Islam in that these religious and magical associations gave rise to the popular belief that alchemists had extraordinary power over nature.

Aristotle's thesis of the unity of matter and the interchangeable qualities of the four elements—earth, air, fire, and water—also contributed to the theoretical basis of Western alchemy. Aristotle had taught that everything in nature strives toward perfection, and since gold was considered the perfect and most noble state of matter, it followed that all base metals must strive to become gold. The alchemist's task therefore was merely to help nature achieve its innate desire. In Alexandria, this theory of transmutation became linked with the astrological charts of the Babylonians, with Chaldean traditions of magic, and with the practices of the Egyptian metalworkers, who, following the secret recipes of the priests of Isis, were adept at extending a given quantity of gold by producing alloys of gold with silver, copper, tin, and zinc. Thus, from the beginning, practical alchemy was closely associated with the "production" of gold, in both China and Egypt, and it was doubtless this factor that ensured both its popularity and its prolonged survival in the West.

During the eighth century, alchemy in the sense of metalworking was acquired by the Arabs, who gave it the name *alkimia* (*al*, the; *Khemia*, land of black earth, a name for Egypt), from which the word *alchemy* is derived.[5] It was the Arabs who integrated these diverse alchemical ideas from China, Egypt, and Greece with their religion. One of the great Islamic alchemists, Jabir ibn Hayyan, later known in Europe as Geber, affirmed that one could discover the secrets of perfection and the absolute only if one accepted the religious belief of the one God, Allah, the source of all.[6] For the Arabs, then, not only was there no conflict between alchemy and religion, but the Muslim faith provided the basis for a theoretical science. The Quran taught that "the scholar's ink is more sacred than the blood of martyrs" and the Prophet had promoted medical research by teaching that "for every disease Allah has given a cure."[7]

European Alchemists

For centuries alchemy remained under Islamic influence, but following the expulsion of the Moors from Europe and the return of the Islamic schools to Christian direction, the rare manuscripts they held, including those dealing with alchemy, were translated from Arabic into Latin, thereby providing sourcebooks for medieval alchemists.[8] It was at this point that alchemy became associated in European thinking with the so-called black arts, with heresy and magic. Alchemists were regarded as being at best sinister and most likely in league with the devil, an impression that was reinforced by the medieval suspicion of knowledge per se. Under the hegemony of the Catholic Church the Garden of Eden story was invoked to discourage independent thought about the causes and origins of phenomena, lest such knowledge constitute a rival authority. Soon the practice of secrecy, originally evolved to guard the formulas of the initiates, became necessary for sheer survival. Many alchemists lived in isolation, using a cryptic or cabalistic language to escape persecution similar to that accorded to witches, and for parallel reasons. Yet despite this reputation, alchemy exerted a fascination because of its fabulous promises, which came to include not only transmutation of base metals into gold but also other absolutes—eternal youth, perpetual motion, and the in vitro creation of human life in the form of a homunculus. Unlike the search for gold, these other aims appear at first to be of a philosophical rather than a materialistic nature, but they all involve the prospect of power—over people, over death, over natural laws.

A few alchemists were able to combine their alchemy with a career in the church, the most notable being the Dominicans Albertus Magnus and Thomas Aquinas and the Franciscan Roger Bacon. Not only did these practitioners add a dash of Christian symbolism to the existing brew of traditions in an attempt to make alchemy acceptable to the church; what is more important, they introduced the idea of testing the postulates of alchemy by reason and experiment. It was this questioning and insistence on demonstration, especially on the part of Roger Bacon, that was to form the basis of modern scientific method and to distinguish it from the reactionary alchemy still extant in the seventeenth century.

There was, however, increasing unease among the leaders of the religious orders about friars' involvement in alchemy. A series of acts was passed forbidding it,[9] and in 1380 Charles V of France outlawed alchemy entirely. Thus although alchemy continued to be practiced in defiance of the church, it was forced to become increasingly secretive as a defense against the spies of church and state.

Despite the uniform view of alchemists promulgated by the church, there were actually three quite distinct groups: the charlatans, who deliberately deluded others about their ability to make gold but were not themselves deceived; the "puffers," or laboratory assistants involved in the practical problems of "making gold" but not yet successful in the art (see fig. 1); and the scholarly alchemists, "philosophers by fire,"[10] who understood the secret language and who really believed that they knew, or were about to discover, the secret of transmutation. Representatives of all three groups feature in Chaucer's *Canon's Yeoman's Tale*.

By the sixteenth and seventeenth centuries, alchemy had risen in the social scale to acquire royal patronage. The court of Rudolph II at Prague was one of the centers where eminent intellectuals gathered from all over Europe to discuss and demonstrate various branches of the occult and their proposed relation to cosmology, medicine, and of course the production of gold by transmutation. Kepler relied on his astrological predictions at Rudolph's court to supplement his stipend as imperial mathematician and finance his astronomical research. Queen Elizabeth I of England was also very receptive to alchemy, employing Cornelius de Lannoy as her personal alchemist and encouraging a circle of practitioners about her court. Sir Walter Raleigh

Figure 1. The Alchemist, by Jacques Louis Perée (1769–1832), after David Teniers II (1610–90). Harvard Art Museums/Fogg Museum, Gift of Belinda L. Randall from the collection of John Witt Randall, R9316. Photo: Imaging Department © President and Fellows of Harvard College.

and Sir Philip Sidney were both enthusiastic students of alchemy, although there is no record of their experimental results. Dr. John Dee, a mathematician, astrologer, and alchemist, and Edward Kelley, an apothecary turned alchemist, were widely believed to conduct transmutations of base metals into gold, and the possibility of a commercial enterprise along these lines was suggested to the queen as a less risky operation than the plundering of Spanish galleons.

Even by the time of the Renaissance, however, the pursuit of gold had ceased to be the central obsession of alchemy. Theophrastus Bombastus von Hohenheim, better known by his self-bestowed name, Paracelsus, was more concerned with proving his theory of a universal order in the "latent forces of Nature" and with the practice of holistic medicine. He was the first to use the word *chemistry* to express the formulas and techniques of his treatments, which involved chemical drugs rather than the herbal medicines hitherto in favor. This preoccupation of alchemists with medicinal remedies, whether chemical or herbal, was the feature most used to distinguish them in nineteenth-century literature. Nathaniel Hawthorne's particular fascination with such alchemist characters will be discussed in chapter 6.

The other dream of the alchemists was the production of a homunculus, or minute human being (always masculine), by artificial means. Compared with the other aims, this project might seem unattractive, even bizarre, but it constituted an even greater threat to the social fabric and to the church's doctrines of the divine basis of life and the creation of the soul at the moment of conception. Such extreme hubris in attempting to bypass both the Creator and the divinely ordained method of reproduction evoked much the same furor as in vitro fertilization and genetic engineering were to produce in the twentieth century. The idea of a homunculus was first promulgated in the *Homilies* of Clement of Rome (ca. 250 CE), which described the alleged conjuring up of a such a being by the sorcerer Simon Magus. By the early sixteenth century Paracelsus claimed to have a precise recipe for the physical generation of a homunculus from a mixture of semen and blood without need of a female uterus,[11] and the concept has reemerged periodically. Goethe's *Faust* part 2 features a homunculus created through alchemy, and Frankenstein's monster can be seen as an outsized parody of a tiny person. This fascination with imitating the mystery of creation was also related to the legend of the golem, culminating in the story of the Golem of Prague.[12] Later developments of mechanical men, or automata, which would mimic or surpass human reasoning, from Roger Bacon's alleged talking head and Ramon Lull's automatons to the twentieth-century fascination with androids, derived from the same desire to create some semblance of human life. It is not surprising, therefore, that the essential features of the golem—the material origin of his body, derived from clay or dust; his role as servant to his human makers; and the implanted divine spark, which endowed the body with life—were retained, at least in the intention of those who sought to make homunculi or automata, until well into the twentieth century.

Although it enjoyed a minor revival in occult Rosicrucian doctrines, alchemy as conceived in the medieval period had begun to decay in the seventeenth century, superseded by the experimental procedures and mechanical philosophy of René Descartes and Thomas Hobbes. The immense attraction of this *philosophia mechanica* derived from its apparent explanatory power, its confidence that all events could be readily understood in terms of matter in motion and individual units of matter impacting on each other like billiard balls on a table. Compared with the obscurities and complexities of alchemy, this apparent simplicity and universality were intellectually alluring. From a post-Enlightenment perspective, it appears that there was a complete break between alchemy and modern chemistry and that Robert Boyle's publication, *The Sceptical Chymist* (1661), had effectively demolished the Aristotelian system of the four elements, one of the cornerstones of medieval alchemy, and introduced the modern idea of an element, thereby paving the way for chemistry as we know it. However, as we shall see in chapter 4, there was really no such hiatus. Some natural philosophers, including Robert Boyle and Isaac Newton, attempted to integrate the older system with the new mechanical philosophy, as exemplified in Boyle's *Origine of Formes and Qualities (According to the Corpuscular Philosophy)* (1666), and most of the early members of the Royal Society, including the most prestigious, were still eagerly experimenting with the ideas and practices of alchemy. That these activities have been largely suppressed in subsequent accounts of their work is due partly to their failure to establish the truth of these ideas but also, I believe, to the adverse associations of alchemy and in particular the degraded image of alchemists. The biographers of Boyle and Newton were concerned to omit any reference to alchemy. It was only in the twentieth century that this material was rediscovered and made widely available.[13]

The diverse traditions of alchemy determined many characteristics of both the art itself and its practitioners. Because of the intimate relation between alchemy and the alleged production of gold, the presence of charlatans trading on the greed of the populace was almost inevitable. The Hermetic element of secrecy, deriving from the priestly origins of alchemy, was also there from the beginning, giving rise to the popular belief that alchemists wielded extraordinary power over nature and transcended the limits of what it was considered proper for man to know. Further, the isolation enforced by both the alchemists' desire to protect their secrets and, later, the threat of persecution enabled them for a long time to retain a considerable degree of autonomy; no one asked questions, because it was too dangerous to know the answers.

Alchemists in Literature

Despite the large volume of early alchemical writings, very few alchemists are depicted in these texts.[14] The first literary portrait of alchemists occurs in Chaucer's *Canon's Yeoman's Tale* (1387), one of the *Canterbury Tales*. The yeoman embarks on a lengthy description of his master, a canon, as a failed alchemist, who has lost all his money in the pursuit of his art.

> We're always blundering, spilling things in the fire,
> But for all that we fail in our desire
> For our experiments reach no conclusion. (476–77)[15]

After seven years with the canon, the yeoman has lost "Al that I hadde" to "that slippery science" (478) and become ill and disfigured from poisonous lead fumes. From his disgruntled account of life in the laboratory, with its accidental breakages and the loss of precious concoctions, it is clear that the canon has never delivered the promised gold, either to himself or to his clients. The yeoman ruefully admits:

> We always met with failure in the end.
> Yet, though we never reached the wished conclusion
> We still went raving on in our illusion,
> Sitting together, arguing on and on,
> And every one as wise as Solomon. (*The Canon's Yeoman's Tale*, part 1, 478)

Even though they realize the futility of their pursuit of alchemy, the yeoman and his canon are so addicted to its promises that they cannot extricate themselves from its allure. "No one can stop until there's nothing left" (482). In the second part of his tale the yeoman tells of another canon, clearly a charlatan, who makes a living from the delusions and greed of humanity, in this case a gullible priest. After slyly introducing a little silver into his crucible, so that it appears to the onlooker to have been produced from mercury, the canon graciously consents to sell for forty pounds the magic powder allegedly responsible for this transmutation, a preparation that the yeoman dismisses, as

> Possibly chalk, or glass would do as well,
> Or anything else indeed not worth a fly. (490)

Although the yeoman's account concentrates on the trickery of the charlatan canon, there is a clear implication that such fraud could not succeed were it not for the greed of the client, who, already suspiciously wealthy for a priest, desires to multiply his money for nothing. This was the aspect of alchemy on which Ben Jonson was to focus in his play *The Alchemist*. Surprisingly, at the end of the tale, the yeoman proceeds to defend "true" alchemy, which, far from being irreligious, is "unto Christ so beloved and dear" (498); yet this defense of the "true" alchemy is linked with a warning against disclosing how to discover the philosopher's stone.[16]

The best-known alchemist figure in literature, arguably a stereotype in his own right, is Doctor Faustus. Many of his characteristics, notably his obsession with attaining forbidden knowledge, are derived from the older alchemist tradition, but the Faust legend evolved somewhat differently, and the significance attached to it has varied markedly at different periods and under different influences.

The original Doctor Johann Georg Faust, whose title was almost certainly spurious, was born around 1480, perhaps in the German town of Knittlingen. He seems to have been something between the extremes of traveling magician, astrologer, and quack doctor, on the one hand, and alchemist, perhaps even serious student of natural science, on the other.[17] With the passage of time, the legends about this figure became increasingly exaggerated, with anecdotes of magic tricks and familiars predominating. The first written account, the anonymous Spieß edition of *Historia von D. Johann Fausten* published in 1587, presents the story in a highly moralistic light. Faust is a presumptuous man, desirous of overstepping the God-given limits of human knowledge. To achieve this, he makes a pact with the devil, acquiring, in return for his soul at death, a variety of magical powers, which he employs in playing trivial tricks. Predictably, he comes to a suitably gruesome end, while the reader is referred to the scriptures and warned against such intellectual arrogance. Yet, despite the overtly didactic intention of the author, this Faust retains a certain implicit nobility in his determination to acquire knowledge.

It was on the English translation of this chapbook that Christopher Marlowe's play *The Tragical History of Doctor Faustus* (1604) was based. Compared with the German original, the translation stressed the theme of intellectual curiosity, and Marlowe emphasized it even more. From the beginning he presents what has come to be seen as the archetypal dilemma of the Faust figure: his Renaissance-humanist longing to transcend the limitations of the human intellect is accompanied by a fatalistic medieval awareness that such longing is doomed to failure. Reviewing the branches of current knowledge, Faustus rejects each in turn because of its restricted scope. Like Frankenstein, he complains that medicine cannot bestow eternal life, and he longs instead for "the metaphysics of magicians," since a "sound magician is a demi-god."[18]

Marlowe thus preserves the medieval association between intellectual arrogance and Lucifer's revolt against God, for ultimately Faust seeks to know

> unlawful things
> Whose deepness doth entice such forward wits
> To practise more than heavenly power permits. (158)

Given this association of the alchemist with the overreacher, defying God-given limits, it is no coincidence that, as a prelude to his pact with Mephistopheles, Faustus derides both theology and religion and flagrantly dedicates his ceremony to Satan (126). However, having gambled his soul away, Faustus discovers that he has made a poor bargain, for, ironically, he learns nothing of any importance. Mephistopheles's answers to his questions are as evasive and ambiguous as those of a Greek oracle, and the much-vaunted magical powers amount to little more than cheap conjuring tricks, a device that recurs in later satirical treatments of scientists.[19]

These aspects of Faustus's character derive from the German pietist tradition, but Marlowe also introduces a new theme, essentially a Renaissance rather than a medieval one: the tragic wasted potential of a gifted man. He therefore gives us a very different conclusion from the grimly exultant tone at the end of the Spieß edition of *Faust*. As Faustus sinks further into despair, his moral honesty and human decency increase. As he insists that his fellow scholars leave him to his fate and faces the full horror of the consequences of his past deeds, he arouses compassion and respect rather than condemnation, becoming, effectively, a tragic hero.

The intrinsic ambiguity of the Faust character has inspired a variety of treatments and evoked an even broader spectrum of responses, all of which have, in turn, attached to scientists. At one extreme is the Faust of Lutheran piety, a kind of minor Satan, who rebels against God's restrictions and refuses to repent and whose inevitable punishment is presented with all the didactic stops pulled out. At the other extreme is the Faust of German Romantic literature who embodies the noblest desire of man to transcend the limitations of the human condition and to extend his powers, for good as much as for evil, a Promethean figure who asserts the rights of man over a tyrannical order that seeks to enslave him. The aspects of the Faust stereotype that predominate at any particular time or place vary with the relative status accorded to the intellect, compared with the value placed on obedience to the prevailing hegemony, whether of church or state. During the medieval period, the pact with Mephistopheles was central, and Faust was predictably dragged down to hell with the full approval of the audience. However, during the Renaissance and preeminently in the period of German Romanticism, the nobility of Faust's quest (whether for knowledge or experience or for some more mystical truth) was stressed, and the significance of the satanic connection was left ambiguous. Whichever light he was seen in, however, Faust, like the alchemists, was a figure of both fascination and dread, exemplifying the moral dangers of intellectual aspirations and hubris.

The intimate connection between alchemy and magic was to issue in a range of works featuring some kind of magus who possessed special powers, often implicitly evil, over nature. For the most part, such magicians are presented as evil or, at best, awesome figures. A rare exception is Shakespeare's Prospero, the seemingly benign magus of *The Tempest* (1611), who not only has natural forces under his control but possesses magic similar (though superior) to that of the powerful witch Sycorax. Yet Prospero is depicted with some ambiguity. The morality of causing a shipwreck that, although it produces no actual casualties, nevertheless elicits grief and suffering in the minds of the survivors, who assume their relatives drowned, remains problematic. Moreover, Prospero's final admission that he himself, through preoccupation with his books and the study of magic, was partly responsible for the mismanagement of his former kingdom sets a retrospective qualification on both his magical powers and the morality of his revenge. Magic emerges as a doubtful good, possibly benign in the right hands (Prospero is virtually unique in renouncing his powers when justice appears to have been done) but suggestive also of manipulation (such as Prospero exercises over the innocent spirit Ariel) and, in other hands, a potential for evil.

Very different in purpose is Ben Jonson's play *The Alchemist* (1610). Whereas Chaucer had warned against the threat posed by charlatan alchemists, Marlowe had depicted the tragic grandeur of man's attempt to transcend his limitations, and Shakespeare had suggested the ethical dangers implicit in magical control over nature and spirits, Jonson, the Renaissance man of reason, presents the quest for transcendent knowledge as a foolish illusion that flourishes only because of human greed. It is significant that in *The Alchemist* there *is* no alchemist, only the low-life rogues Subtle and Face, who pretend, through a facility with jargon, to have special knowledge. Jonson makes it plain that their temporary "success" is due not to their cleverness but to the stupidity, avarice, and vanity of their willingly deluded victims.

The Alchemist was part of Jonson's more general intention to ridicule those whose pretensions led them to seek, or claim, transcendence in any sphere, but his immediate concern arose from an actual situation. The prototypes for Subtle and Face were most probably the well-known contemporary practitioners of magic and alchemy Dr. John Dee, Edward Kelley, and Simon Forman,[20] all of whom had gained considerable influence over the superstition-prone Elizabethan court, even though alchemy was officially outlawed. In Jonson's play, these educated and elevated personages are stripped of any claim to greatness and reduced to the level of a cunning servant and an itinerant rogue. More importantly, unlike Chaucer and Marlowe, Jonson has a powerful "hero," Lovewit, the urbane, unpretentious man of true reason, who sees the charlatans for what they are and is therefore immune to their "power." He expels the syndicate of pseudoalchemists from his house and demotes Face to his proper position as butler. Jonson's major innovation in the depiction of alchemists was to strip them of their exotic aura and show them as no more than ordinary cheats. As we shall see in chapter 3, Jonson's weapon of satiric deflation was so widely adopted that it became almost the norm in the late-seventeenth- and eighteenth-century representation of scientists, even in depictions of the serious founding members of the Royal Society.

Endurance of the Alchemist Stereotype

Because of its complexity and the diverse responses it has elicited at different times, the alchemist stereotype has proved the most resilient of all. It has recurred in representations of the scientist with both good and evil implications, from medieval times to the present, sometimes with specific allusion to magic, as in Somerset Maugham's *Magician* (1908) or the Harry Potter novels of J. K. Rowling. Because this stereotype is associated with a mystique that sets the alchemist figure apart from ordinary people, it is often taken to suggest communion with superhuman, even supernatural, powers (see fig. 2). Such powers may be seen as ennobling, as in the case of the Curies in real life and countless fictional examples of the noble scientist sacrificing his life to science, but they are more often darkly satanic, as in the Faustian tradition, luring their devotees to seek knowledge harmful to humanity.

It is not difficult to see how much of the popular image of the scientist derives from the alchemist tradition. The aspiration to know more than others about the causes of natural processes is inevitably associated with the desire for power over nature and hence with the attempts to manipulate or modify some aspect of nature for one's own purposes. Like their alchemist predecessors who held out the lure of gold, modern scientists often feel

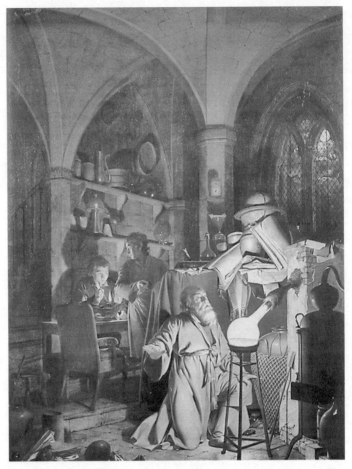

Figure 2. The Alchemist in Search of the Philosopher's Stone Discovers Phosphorus, by Joseph Wright (1771–95). Derby Museum and Art Gallery.

compelled to justify their research in terms that promise their funding bodies monetary gain or benefits to society.

There are numerous reasons for the persistence of the alchemist stereotype and for the imaginative power it has continued to exert.[21]

(1) The alchemist as illicit creator and the seeker after forbidden knowledge both have their roots in much older mythology that is deeply ingrained across cultures and possibly in the Jungian collective unconscious: the story of Eden, of Prometheus, of Daedalus and Icarus, of Pandora's box

and even Bluebeard's wife. Given this ancestry, their modern descendants resonate with cumulative effect. The scientist who discovers some power (be it a weapon, nuclear power, or the ability to create, clone, or modify life) that cannot be contained or controlled is Pandora trying vainly to return the escaping Troubles to the box. Nearly all alchemist narratives focus on a reversal of expectation and a corresponding nemesis: the glorious promises turn to ashes and destruction—sometimes because they are not achieved (as in Balzac's *La recherche de l'absolu*, 1834) but more often because they *are* achieved. As we shall see in chapter 7, Frankenstein's tragedy begins at the precise moment of his experimental success. And he is not alone in this.

(2) The medieval church was correct in regarding alchemy as a rival power. Francis Bacon's aphorism "knowledge is power" is nowhere so obvious as in the attractiveness of science. But power for those who possess the knowledge is simultaneously disempowering and excluding for those who do not.

(3) It would seem that our wish list has changed little since our medieval ancestors visited their local alchemist under cloak of darkness, fearful of being seen but greedy for results. For perpetual motion machines we have nuclear power; for transmutation of metals to gold, we have radioactivity and industrial processes with immense profits out of all proportion to outlay; for eternal youth we are offered antioxidants, Botox, and hormone therapy; for eternal life our strategies to cheat death include ever-new miracle drugs, organ transplants, and stem cell grafts; while on the homunculus front we have cloning, in vitro fertilization, artificial intelligence, donor eggs and sperm, surrogate parenting, and genetic engineering to produce the offspring of our choice. All have been greeted with a combination of fear and hope.

(4) The alchemist was a driven person, focused totally on his search to transcend the human condition. Frankenstein says: "Life and death appeared to me ideal bounds [that is, merely imaginary boundaries], which I should first break through, and pour a torrent of light into our dark world. A new species would bless me as its creator and source."[7] Like the alchemists, scientists are portrayed as having different allegiances from other people, being prepared to sacrifice people, animals, or even themselves for new knowledge. In our time this figure enacts the view that the pursuit of scientific knowledge justifies the means.

(5) A major element of the enduring appeal of the alchemist narrative is its ability to evoke perennially convincing patterns of horror, mystery, and evil. While most examples from past centuries (involving graveyards, corpses, ghosts, and monsters) have ceased to frighten us, the basic elements of the Frankenstein narrative seem perpetually relevant, and films have intensified this with special effects, reaching out to a far wider audience than the written word. Horror fiction and horror movies allow us to indulge our worst impulses and fears, to be vicariously complicit in what transcends the culturally sanctioned boundaries of "decency." In the case of the evil alchemist there is the additional lure of seeing the powerful individual dragged down (and he almost invariably is), and the natural order is restored.

(6) The perennial fascination of the master narrative of alchemy is that it tells a story of what we both desire and fear to know—the story of power beyond our dreams but also beyond our control. Caught between terror and desire, we are a captive audience for stories that embed our uncertain existence in the archetypal legend of the powerful mage, the sinister alchemist, the perplexed scientist.

(7) Then there is the appeal of a simplistic, universally understood image. The name "Frankenstein" has become journalistic shorthand for any experimentation popularly perceived as dangerous or likely to backfire. Whether in developing viruses for germ warfare, delivering genetically modified vegetables ("Frankenfoods"), cloning sheep, or growing new organs from embryonic stem cells, we can be sure that, in the media reports, Frankenstein will be implicated.

Our society still desires to do a range of secret deals with its alchemists, even while professing to treat them with suspicion: nuclear power plants and nuclear waste dumps, in vitro fertilization, cloning, surrogate parenting, the trade in organs and genes, antidiversity treaties with seed companies marketing the total package of genetically modified crops. DNA is the twenty-first-century philosopher's stone whereby we seek to transmute not metals but individuals and species.

It is also from the alchemists, with their arcane signs and symbols, that science has derived its reputation of involvement with processes that ordinary people cannot understand. The symbols and formulae of modern physics and chemistry, merely a convenient shorthand for their practitioners, represent an immense barrier to laypersons wishing to understand the science. The alchemist's dark cave may have morphed into the chemist's well-lit laboratory, but the view from the outside is much the same. The basic stereotypes and preoccupations associated with the alchemist—secrecy, illicit processes, forbidden knowledge associated with the creation or modification of life—continue to be recycled, particularly in films. *The Fly* (1986), *The Sixth Day* (2000), *Blueprint* (2003), *The Godsend* (2004) represent a few of the many examples that will be discussed in later chapters. They retain their popularity precisely because they continue to play on both the hopes for medical breakthroughs and the deeply engrained suspicion of powerful and exclusive knowledge that laypeople can neither share nor control.

CHAPTER 2

Bacon's New Scientists

> We have consultations, which of the inventions and experiences, which we have discovered, shall be published, and which not: . . . we do publish such new profitable inventions, as we think good.
> —*Francis Bacon*

Prior to the seventeenth century, the culture of Europe was still fundamentally hostile to the progress of science and, equally, to valuing scientists as respectable members of society. First, the Eden myth of the Old Testament associated desire for knowledge with disobedience to God, with pride and presumption, and, ultimately, with the Fall of Man. The pursuit of any branch of knowledge other than theology was therefore regarded with suspicion and fear as being likely to engender atheism, even diabolism, as portrayed in the various versions of the Faust legend. Further, the Renaissance revival of classical Greek and Roman culture reinforced the belief that humanity, so far from improving, was embarked on a decline: the golden age was located irretrievably in the distant past. There was thus a marked degree of pessimism about the future of mankind, which militated against that belief in progress essential for the development of science in the modern sense. In addition, there was no confidence that the individual had any ability to influence his own or society's future; that role was assigned solely to Providence, precluding any exploration of causality in the natural world.

Again, while the first flush of Renaissance humanism had temporarily promoted the physical world (as opposed to a spiritual dimension) to a position where it was considered worthy of study, by the last decades of the sixteenth century there was a growing sense of disenchantment with the high hopes and promises of the Renaissance and the Reformation: neither had cured social ills or ushered in the millennium. In England economic and religious troubles, as well as the political intrigues that had dogged the reign of the Tudor monarchs, fostered a widespread malaise and disillusionment with the power of reason to resolve the difficulties of the age. Despite the impetus given to scientific investigation by Copernicus's *On the Revolutions of the Heavenly Bodies* (1543) and Kepler's *New Astronomy* (1609), Sir Francis Bacon, Lord Verulam and sometime Lord Chancellor of England, feared a total collapse of learning and the onset of a second Dark Age. His own precarious personal situation after falling into disfavor with Queen Elizabeth may have contributed to his pessimism, but, for whatever reason, he wrote in 1603: "I see in the present time some kind of impending decline and fall of the knowledge and erudition now in use. Not that I apprehend any more barbarian invasions . . . but from the civil wars which may be expected I think . . . to spread through many countries, from the malignity of religious sects, and from those compendious artifices and devices which have crept into the place of solid erudition I have augured a storm not less fatal for literature and science."[1]

In his determination to stem this tide of ignorance and darkness and institute a complete reform of learning that he called *Instauratio Magna*, the "Great Instauration," Bacon's first step was to change the unfavorable image of scientific study and break the link with the Faust legend. This he did by what might seem a clever theological ruse, although there is no evidence to suggest disingenuousness on Bacon's part. By neatly relocating the basis of science in God's laws as embodied in nature, he at one stroke made the study of such laws not only ethically respectable but virtually mandatory for the believing Christian.[2] Indeed, Bacon went further, inverting the traditional interpretation of the story of the Fall to suggest the possibility of a glorious restoration "of man to the sovereignty and power (for whensoever he shall be able to call the creatures by their true names he shall again command them) which he had in the first state of creation."[3] Thus natural science, so far from being an instrument of the devil, is presented as a means of overcoming man's limitations resulting from Adam's sin.

Bacon's program involved not only a new conception of the goals and methodology of study but also a new image of the scholar who could be entrusted to carry out the great plan. This last premise was fundamental to the success of the whole undertaking, for Bacon was all too aware of that "sort of discredit . . . that groweth unto learning from learned men themselves" and "those errors and vanities which have intervened among the studies themselves of the learned."[4] "For men have entered into a desire of learning and knowledge, sometimes upon a natural curiosity and inquisitive appetite; sometimes to entertain their minds with variety and delight; sometimes for ornament and reputation; and sometimes to enable them to victory of wit and contradiction; and most times for lucre and profession; and seldom sincerely to give a true account of their gift of reason to the benefit and use of men."[5]

Bacon therefore spent considerable effort in recasting the image of his man of science. Apart from his many philosophical works, his fictional *New Atlantis*, published in 1626, the year of his death, was devised specifically to popularize his ideas about the central role of science and scientists in society. In this fable a party of European travelers, representing Bacon's contemporaries, comes by accident to the utopian island of Bensalem (the good Jerusalem?), where they investigate the college of natural philosophy, called Salomon's House, "dedicated to the study of the works and creatures of God . . . the noblest foundation (as we think) that ever was upon the earth, the lantern of this kingdom."[6] In Bensalem, scientists are respected and revered in the community as a new moral elite committed to research for the social good.[7]

In so characterizing his scientists, Bacon specifically rejected the traditional ideal of contemplation, adopted by both classical scholars and the medieval church, on the grounds that "it is reserved only for God and the Angels to be lookers on."[8] Instead he proposed an ideal of active service, *philanthropia*: the needs of the community were to determine what research should be pursued. Thus a primary characteristic of Bacon's new scientist is compassion for "the sorrows of mankind and the pilgrimage of this our life," and his chief desire is to alleviate these through the fruits of his learning. In his preface to the *Instauratio Magna* Bacon declares: "Lastly I would address one general admonition to all; that they consider what are the true ends of knowledge, and that they seek it not either for pleasure of the mind, or for contention, or for superiority to others, or for profit or fame, or power or any of these inferior things; *but for the benefit and use of life; and that they perfect and govern it in charity*."[9] It is clear from what follows that Bacon's premise is essentially utilitarian, interpreting human welfare in material terms of comfort and technological competence for the greatest number. This was in sharp contrast to the contemporary ecclesiastical view of compassion as secondary only to evangelism, and it looked forward to the modern welfare state based on humanist values.

In seeking to ameliorate the conditions of mankind in *this* world, Bacon's scientists are basically secular and future-oriented in their thinking. The golden age is no longer located in the classical past but waiting to be inaugurated through the efforts of noble and altruistic scientists. This leads to innovatory effects on both the methodology of the new science and the ethical behavior of the scientists. As distinct from the precedent-oriented system of the past, where some earlier authority, usually Aristotle, was invoked to determine the validity of all claims, Francis Bacon, like his earlier namesake, Friar Roger Bacon, stresses the importance of experimentation and observation of the particular as the necessary basis for inducing the general. Indeed, he explicitly accuses Aristotle of bending the facts to suit his hypotheses: "Nor let any weight be given to the fact that in his [Aristotle's] books on animals, and his problems, and other of his treatises, there is frequent dealing with experiments. For he had come to his conclusion before; he did not consult experience, as he should have done, in order [to proceed] to the framing of his decisions and axioms; but having first determined the question to his will, he then resorts to experience, and bending her unto conformity with his placets leads her about like a captive in a procession."[10]

Similarly, Bacon rejects the popular intellectual sport of logic chopping because "it commands assent therefore to the proposition, but does not take hold of the thing."[11] Instead, he claims that his own methodology is grounded firmly in observation, experimentation, and induction. Not content with examining natural phenomena under normal conditions, these scientists investigate the behavior of substances at immense altitude, depth, and temperature in order to generalize. This necessarily involves new equipment, and almost one-quarter of *New Atlantis* is devoted to a description of the "Preparations and Instruments" of Salomon's House.[12] This preoccupation with equipment was to characterize the development of British science from the early days of the Royal Society and played a significant role in forwarding the Industrial Revolution in England in advance of its European neighbors, who were still conducting science largely along theoretical lines. Bacon also introduced the notion of an international community of science, his "Merchants of Light," who, unlike that other contemporary hero figure, the patriot, transcend national boundaries and allegiances, traveling the world to collect and share their knowledge with all races.[13]

Since the accomplishment of these elaborate experiments and international travels was impossible for individuals or even for one generation of scientists, Bacon initiated the concept of a team effort, whereby scientists would labor not for their own personal goals but for a long-term communal project. They are therefore free to specialize in various areas and procedures, happy

Figure 3. New Atlantis. From *The New Atlantis*, by Sir Francis Bacon, 1627.

that their results will be absorbed into the growing corpus of knowledge (see fig. 3). In replacing the jealously guarded isolation of the alchemists with a group effort, Bacon was also demystifying scientific knowledge and making it common property. The four sets of institutional imperatives proposed by sociologist Robert K. Merton in 1942 as comprising the ethos of modern science, namely communalism, universalism, disinterestedness, and organized skepticism (abbreviated as CUDOS), are almost identical with Bacon's scientific ideals of 1623.[14]

This team concept presupposes, in turn, that scientists can dispense with any competitive scramble for individual success and honors. Bacon argues that the motivation of *philanthropia* can overcome the pride and self-interest apparent in even the most highly esteemed scholars of the past. Being part of a team also releases Bacon's scientists from the despair of completing such a momentous task in one lifetime. They are content to have contributed to the ongoing knowledge bank.[15]

Although he insisted on the need for an empirical basis for knowledge, Bacon did not underestimate the importance of theory. He realized that pure research, which he called *experimenta lucifera* (experiments bearing light), if based on correct observation, could in the long term be more productive of human good than could a narrow, technologically based science, *experimenta fructifera* (experiments bearing fruit), which sought only immediate and obvious gains.

Unlike the leaders of Sir Thomas More's utopia, the scientists of New Atlantis are treated with pomp, even reverence,[16] and dress with princely richness and finery. Bacon was promoting his scientists as a new elite class somewhere between the aristocracy and the priesthood, performing special religious ceremonies.[17]

Perhaps the most innovative aspect of *New Atlantis* is the self-imposed censorship that the scientists exert concerning which of their discoveries shall be made public on the basis of the common good: "We have consultations, which of the inventions and experiences, which we have discovered, shall be published, and which not: And take all an oath of secrecy, for the concealing of those which we think fit to keep secret."[18]

If Bacon was overly optimistic about the readiness with which society would embrace science, he seems to modern readers (who have reason to be hypersensitive to such matters) even more so in regard to the moral qualities of his scientists: "Only let man regain his right over Nature, which belongs to him by the gift of God; let there be given to him the power: right reason and sound religion will teach him how to apply it."[19] Despite his claims that the study of nature leads to a knowledge of the existence, character, and omnipotence of God, Bacon carefully avoided any reference to the argument from design so enthusiastically embraced by the next generation of natural philosophers, who sought the church's blessing. On the contrary, he actively insisted that "we do not presume by the contemplation of nature to attain to the mysteries of God" and "do not unwisely mingle or confound these learnings [i.e., natural knowledge and religion] together." In thus distinguishing between scientific and religious knowledge, Bacon paved the way for the future divergence of science from religion and the autonomy of the former.[20]

Despite his innovative program and methodology, Bacon himself was no scientist. Although he stressed the importance of experimental method, he ignored the importance of precise measurement and of mathematics; indeed, his view of science was almost wholly qualitative. He also had a naive faith in the use of inductive method, believing that if sufficient observational data were collected, a general scientific law would inevitably emerge. As we shall see in the next chapter, this was to have a profound effect on the early years of the Royal Society, encouraging indiscriminate collection of anything and everything as potentially useful. However, Bacon was extremely important as an entrepreneur of the new science, expounding, explaining, and, above all, making its practitioners respectable. He was the first to free scientists from the cloud cast upon them by the medieval church and to hold them up as the natural and acceptable leaders of a utopian state where material and ethical progress would develop in harmony.

New Atlantis in Fiction

The next narrative to incorporate Bacon's ideas was *The Man in the Moone; or, A Discourse of the Voyage thither by Domingo Gonsales* published anonymously in 1638 by Francis Godwin, bishop of Hereford. Gonsales, arguably the first fictional experimentalist, ignores Aristotelian dogma, building instead on the results of his controlled observations, as Bacon had advocated. Indeed, it seems likely that the basic idea for Gonsales's flying machine was suggested by Bacon's description of an experiment for flying in *Sylva Sylvarum*. Both make use of birds connected to a flying contraption by strings and pulleys and the initial strategy of taking off from a high cliff over water to take advantage of the updraft. Ignoring Aristotle's teaching that a body is always attracted to its ordained place in the universe, Godwin adopted Bacon's idea that the force of the earth's gravity decreases with distance.[21] Thus Gonsales, drawn upward by his twenty-five birds, discovers that after about an hour the birds need to strain less to bear his weight.

As a conscientious Baconian experimentalist, Gonsales also dismisses Aristotle's view that the upper atmosphere must be fiery hot and moist (since it was the home of the elements of fire and air) on the grounds that "my experience found [it] most untrue."[22] This manner of repudiating Aristotle by appeal to experience—"mine eyes having sufficiently informed me"—is peculiarly Baconian and constitutes the basis of his experimental method.

Gonsales's invention of his flying mechanism confirms Bacon's methodology, drawing on the chain effect of observation, hypothesis, experimentation, and induction. A chance observation of the characteristics of the native birds, or gansas, suggests a means of sending messages across the island; this is then linked with his knowledge of physics (pulleys) to suggest a mechanism for lifting materials. Gonsales then experiments with a sheep before trying it himself. His subsequent voyage to the moon, itself the result of this earlier scientific process, leads to observations that in turn overthrow old hypotheses and suggest new explanations of phenomena, exactly as Bacon had predicted in the House of Salomon.

Like his predecessor, Godwin never doubts that the pursuit of science will be a blessing, for the scientist is an instrument of God's purpose, and certainly God will prevent the discovery of whatever might prove harmful. When Gonsales, having arrived on the moon, asks the Lunars whether they have discovered how to make themselves invisible, they answer: "God would not suffer it to be revealed to us creatures subject to so many imperfections, being a thing so apt to be abused to ill purposes."[23] By this simple expedient Godwin, like Bacon, claims divine blessing on all scientific enterprise and vindicates whatever discoveries may come to light. It is a neat piece of theological argument, for to doubt the conclusion is to be caught doubting the implicit major premise of God's omnipotence.

The most influential of Bacon's contemporaries to adopt his agenda was Samuel Hartlib (1600–62), who was interested in reforming education and promoting the open dissemination of knowledge. Hartlib not only collected about him a circle of prominent intellectuals from various disciplines who were interested in integrating their ideas but set up a collection center for all kinds of information, for which he hoped to obtain government funds. In this he was unsuccessful, but at his death the role was taken over by Henry Oldenburg, who, as secretary of the Royal Society, first maintained it as a personal initiative and then, in 1665, formalized such dissemination in the *Philosophical Transactions of the Royal Society*. This venture exemplified Bacon's ideal of the widespread collation and open communication of scientific knowledge and contributed significantly to negating the associations of secrecy and mysticism that had characterized the image of the alchemist.

Bacon's Legacy: The Royal Society

Bacon's House of Salomon inspired not only the "Philosophical Colledge" of Abraham Cowley's imagination[24] but, more concretely, the institution in London of the College of Philosophy (1645), which after the restoration of the monarchy became the Royal Society of London for the Improving of Natural Knowledge (1662). The mathematician John Wallis and the chemist Robert Boyle, two founding members of the Royal Society, explicitly acknowledged Bacon's model, while Thomas Sprat wrote in his official *History of the Royal Society of London* (1667): "I shall onely mention one great Man who had the true Imagination of the whole extent of this Enterprize, as it is now set on foot, and that is Lord Bacon. In whose Books there are every where scattered the best arguments that can be produc'd for the defence of Experimental Philosophy: and the best directions that are needful to promote it. . . . If my desires could have prevail'd . . . there should have been no other Preface to the History of the Royal Society but some of his Writings."[25] The frontispiece of Sprat's *History* features an engraving of Charles II and Bacon in which Charles II is described as "Author et Patronus" and Bacon as "Artium Instaurator" (see fig. 4), while Abraham Cowley's ode "To the Royal Society," printed in the same volume, compares Bacon to Moses in rescuing mankind from the wilderness of ignorance.

The Royal Society's indebtedness to Bacon rested chiefly on its adoption of his interlocking propositions of experimentalism, open communication, and usefulness. Sprat, contrasting the new science with the former scholastic philosophy, stressed this aspect of usefulness, and from its inception, the Royal Society enjoined its members to share information. In his introduction to the first issue of the *Philosophical Transactions* (1666), Henry Oldenburg, the then secretary, pointed out that the new periodical was being published "to the end that such Productions being clearly and truly communicated, desires after solid and useful knowledge may be further entertained . . . and those addicted to . . . such matters, may be invited and encouraged to try and find out new things, impart their knowledge to one another, and contribute what they can to the Grand design of improving Natural knowledge."[26]

True to the ideals of their spiritual founder, Sir Francis Bacon, the members of the Royal Society took as their motto *Nullius in verba*, Take no one's word for it. The physicist Robert Hooke, one of the founding members, described the aims of the Royal Society in terms that read like a summary of the practices in New Atlantis: "This Society will not own any Hypothesis, System or

Figure 4. Frontispiece from Thomas Sprat's *History of the Royal Society of London*, 1667, showing, *from left to right*: William, Viscount Brouncker, president of the Royal Society; Charles II as "Author et Patronus"; and Sir Francis Bacon as "Artium Instaurator." Wikimedia Commons.

doctrine of the Principles of Naturall Philosophy, ... nor dogmatically define, nor fix axioms of scientific Things, but will question and canvass all till by mature debate and clear argumentism, chiefly such as are deduced from legitimate Experiments, the Truth of such Experiments be demonstrated."[27]

Theoretical science was viewed skeptically by the early members of the Royal Society, for it savored too much of the clean-hands approach of Aristotle and the ancients and was thus open to the abuses of the logicians. They voiced their reservations even toward Descartes, because he "did not perfectly tread in his [Bacon's] Steps, since he was for doing too great a part of his work in his Closet, concluding too soon, before he had made Experiments enough."[28]

Bacon's proposal for the responsible censorship of knowledge by the members of the House of Salomon was also adopted by some members of the Royal Society. Robert Boyle published in the *Philosophical Transactions* an account of an experiment whereby he had succeeded in combining a particularly "noble" form of mercury with gold to produce an exothermic reaction that he believed would be therapeutic. (Such an incalescent reaction had long been the dream of alchemists.) However, Boyle was uncertain about the possible "political inconveniences" that might result if such mercury should "fall into ill hands"; he was therefore using his article to consult with the learned world about the relative merits of keeping silent about such a substance. Isaac Newton had no doubt that Boyle should maintain "high silence" and informed Secretary Oldenburg of this.[29]

New Atlantis also inspired both the founders of other academies in Europe and the French encyclopedists in their resistance to attempts by the church to limit and control science. The encyclopedists adopted with enthusiasm Bacon's program of cataloging all knowledge about the natural world, as Diderot was to acknowledge in the "Prospectus" to his *Encyclopédie*: "If we have come to it successfully, we shall owe most to the Chancellor Bacon."[30]

When we recall the secrecy associated with the work of the alchemists and the way they protected the mystery of their activities in order to preserve and heighten their magical power over the uninitiated, it is immediately apparent how completely Bacon recast the scientist as an idealized figure that has been rivaled in only a handful of utopian proposals and that has continued to exert influence over modern science. It is therefore somewhat ironic that Bacon derived many of the tenets of his natural philosophy, including suggestions for transmuting other metals into gold, from alchemy, with which he was well acquainted.[31] This, however, is not apparent in *New Atlantis*, and it is significant that this text was selected as a model for the scientific utopia that not only appealed to Bacon's century but reemerged in almost the same terms in the early twentieth century until the Second World War.

Bacon's call for empirical procedures, the preeminence of experimental results over dogma and authority, his ideal of an international cooperative effort by scientists devoted to the ongoing pursuit of truth, the requirement that research be open and its results communicated to other scientists, and, perhaps more contentiously, his stipulation that science should serve the needs of the community rather than its own purposes have all become accepted principles of modern science, even when, as in wartime, for instance, it pays only lip service to them. Moreover, although Bacon's account of the methodology of science as empirical observation, hypothesis, experiment, and inductive generalization has been challenged by twentieth-century philosophers of science Karl Popper, Thomas Kuhn, and Paul Feyerabend, who deny that this is how science in fact operates, it still corresponds closely to the popular conception of how scientists work and has influenced how scientists are portrayed in fiction.

It is apparent that all these ideals hinge on the moral caliber of the scientists involved. Hence it is not surprising that the proposals of Bacon, like those of H. G. Wells, his twentieth-century successor in proposing a scientifically based utopia, depend heavily on the assumption that scientists have a highly developed sense of dedication and morality. This was perceived by their critics as the weakest point in the utopian argument, but in Bacon's case this reaction was longer in coming, for with the institution of the Royal Society, it did indeed seem that the "New Atlantis" would soon emerge in England, if not throughout Europe. Bacon's infectious belief in the glorious future of learning, if men would but sail beyond the Pillars of Hercules,[32] first cast the scientist in the role of adventurer and world savior, an idea that was to reemerge with more problematic implications in the twentieth century (see chapters 9 and 11).

Bacon and Nature

In elevating and idealizing scientists, however, Bacon simultaneously degraded nature to the status of a relatively impotent opponent to be subdued: "If a man endeavour to establish and extend the power and dominion of the human race over the universe, his ambition (if ambition it can be called) is without doubt a more wholesome thing and a more noble than the other two [those who seek individual or national prestige]."[33] Prior to Bacon's rallying call "knowledge itself is power," natural philosophers, whether alchemists, physicians, or herbalists, perceived themselves as allies of nature, but Bacon's ideas hinge on a revival of the prelapsarian concept of man as being set over nature, to dominate and control it. This assumption was to survive longer even than that of the noble scientist, and except for a brief period coinciding with the Romantic movement (discussed in chapter 6), it has continued to dominate both the methodology of science and the attitudes of Western society until the present day. However, the checkered estimate that characterized

Bacon's unfortunate political life has been reflected in the evaluation of his philosophical ideas. After enjoying in the early twentieth century the adulation of H. G. Wells, who effectively adopted the scheme of New Atlantis for his scientific utopias, Bacon is again under review. It is not accidental that Bacon should be best remembered for the aphorism "knowledge is power," for his scientists appear committed to subduing nature to the will of man. The father of the House of Salomon asserts: "The End of our Foundation is the knowledge of Causes and secret motions of things, and the enlarging of the bounds of Human Empire, to the effecting of all things possible."[34] It is for this reason that in recent years Bacon has been held responsible for the mechanistic and reductionist rationale of science, separating the (male) scientist observer from the object of study, (female) nature. In the view of many environmentalist philosophers of science, and especially of ecofeminists, Bacon bears a major part of the responsibility for the deep division between science and nature and for the perception of nature as a passive object available for exploitation, manipulation, and domination, precluding an ecological ethic in which humans are an integral part of nature.[35] He has therefore, by extension, been regarded as responsible for the environmental crisis of the late twentieth century. However, Kate Aughterson[36] argues that, on the contrary, *New Atlantis* questions the existing sexist assumptions of Western science in Bacon's time by creating Bensalem as a masculine entity (Bensalem also signifies "the perfect son") rather than the conventional feminized land, and by displacing the usual Eurocentric relationship of man to the land. By describing the marriage between science and nature as one of mutuality and equity and by insisting that "nature to be commanded must be obeyed,"[37] Bacon subverts "conventional binary oppositions, including that of a feminised, subjugated nature."[38] Indeed, in Aughterson's reading, "Bacon's new scientist may, in his or her role reversals, and equal respect for both the natural world and technology, remind us more of Haraway's cyborg than the masculinist ideologue described by many ecofeminists" and "offers a way of enabling us to re-articulate the legacy of Baconian science for the twenty-first century."[39]

CHAPTER 3

Foolish Virtuosi

> The smallest worm insults the sage's hand;
> All Gresham's vanquish'd by a grain of sand.
>
> —Henry Jones

The Royal Society

It now seems a strange irony that the first systematic satires of scientists should have been directed at the founding members of the Royal Society, that is, at the first people we may consider scientists in the modern sense. Yet they, along with their contemporaries, the amateur virtuosi,[1] were ridiculed on the Restoration stage as pursuing useless, irrelevant, and usually disgusting research into topics with which no gentleman should concern himself. Even Charles II, their first and most prestigious patron, was not above mocking their endeavors.[2] Later they were rapped on the knuckles by the leading moralists of the day for intellectual arrogance and, perhaps worse, for the "Dulness" of the factual and unadorned style of writing they adopted. The complex reasons underlying this irreverent reception of some of the great names in the history of science are still debated, but the following factors were significant in producing that climate of opinion.

In the latter half of the seventeenth century the ideas put forward in Bacon's *New Atlantis* suddenly gained momentum. Of particular importance was his proposal that the new learning should reject authority per se and substitute for it the inductive method based on observation and experimentation. This emphasis and a corresponding skepticism about any hypotheses not grounded in sense data implied a distrust of pure reason at a time when, through a combination of factors, the temper of the age was more strongly rationalistic than at any other time in British history.[3] Further, the

Royal Society's emphasis on mechanical rather than theoretical science, in accordance with Bacon's insistence that research should be socially useful, appeared to the urbane Restoration wits a continuation of the unlovely utilitarianism of the Puritans.

The association of science with Puritanism was indeed more than coincidental, although the exact nature of the connection is still disputed. Sociologist Robert K. Merton noted an affinity between seventeenth-century science and the Puritan values of asceticism, orderliness, hard work, faith in progress and reason, and a democratic acceptance of knowledge, irrespective of rank.[4] During the Interregnum, Puritan enthusiasts had tried to instigate revolutionary changes in the university curriculum, curtailing humanities subjects to make way for experimental science on the Baconian model, that is, vocational and practical knowledge of such matters as planting orchards and grinding lenses. Christ College, Oxford, was proposed as the first institution to be radically "reformed" along these lines. It is small wonder that when the danger passed, along with the Commonwealth, the threatened faculties retaliated with attacks on the Royal Society, seemingly the representative of these abhorrent views, and the bishop of Chichester repeatedly directed sermons against the Royal Society.[5]

Even more implacable enemies of Puritanism were the wits of the Restoration stage. The Puritans had closed the theaters on the grounds that they were the seedbeds of immorality, so if the new science was associated with Puritanism, it is not surprising that playwrights were in the vanguard of the scientist-bashing movement. In his *History of the Royal Society*, Thomas Sprat records a real fear of the attacks of the "Wits and Railleurs": "For they perhaps by making it [the New Philosophy] ridiculous, becaus it is *new*, and becaus they themselves are unwilling to take pains about it, may do it more injury than all the Arguments of our severe and frowning and dogmatical *Adversaries*."[6]

Another serious liability for the scientific cause was the charge of atheism arising from its perceived association with Descartes's mechanical philosophy. Descartes had developed his *philosophia mechanica*, purporting to explain all natural phenomena mathematically on the basis of matter and motion, rather than teleologically, as Aristotle had done. As a result of Harvey's work on the circulatory system of the blood, even animate nature came to be understood mechanically in terms of pumps, valves, and levers, differing from inanimate nature only in supposedly containing a spirit, an immaterial substance, as well as matter. Bacon's followers initially welcomed the

Cartesian dualism as being conducive to their aim of ultimately explaining all phenomena in terms of a few basic physical laws, and at first this alliance with a respectable rationalist philosopher worked to their advantage. However, when the atheistic English philosopher Thomas Hobbes ridiculed the immaterial aspect of the dualistic system, it began to appear that the Cartesian model was fundamentally irreligious and that the scientists were intent on disposing of God, leaving only a mechanical creation. In a climate of atheist hunting, this made the scientists a ready target, and had the Royal Society not numbered so many bishops among its founding members, it might have fared much worse.[7] Hobbes's enthusiastic support became a liability, and in their zeal to deny any such alliance and show that they gave due weight to spiritual, "immaterial substance," some members of the Royal Society, notably Joseph Glanvill, Robert Boyle, Ralph Cudworth, and John Ray, attempted to gather data in support of ghosts and witches.[8] The most enduring vindication of the pursuit of science in relation to religious faith was John Ray's *Wisdom of God Manifested in the Works of the Creation* (1691), in which he, like Boyle, argued that the scientist was the priest of natural theology. Westfall has suggested that the amount of time spent by these eminent naturalists in vindicating religion was related to their own uncertainty: "More than answering hypothetical atheists, they were trying to satisfy their own doubts."[9]

As suggested earlier, it would be a mistake to assume that the Royal Society members had entirely renounced their interest in alchemy. Many, including some of the most eminent names, such as Boyle and Newton, were still deeply involved in the riddles of alchemy, although they attempted to resolve them in a different way. Thus much of the censure applied to alchemists still attached to them, at least in the early years of the society, when excursions into ghost hunting and the occult furnished another stick with which the rationalists could beat Royal Society members.

The Virtuosi

In terms of their representation in literature, however, the Society's worst misfortune was its perceived association with the virtuosi. An interesting feature of Restoration science was the strong bond of collaboration between recognized scientists (equivalent to today's trained, professional scientists) and the much larger body of "enthusiasts" (corresponding to today's amateurs), who had less educational background in science but unbounded enthusiasm for miscellaneous scientific projects, believing that they were thereby contributing important data to a cumulative natural history. These wealthy amateurs, frequently called virtuosi to distinguish them from the scientific elite or natural philosophers, were a doubtful asset to the Royal Society, for although some of them amassed collections of scientific value, most were indiscriminate hoarders of trivia. Their private museums, or "cabinets," were carefully cataloged but absurdly overvalued, yet many scientists of the time were reluctant to discard them, in case they should prove important in Bacon's long-term scheme (see fig. 5).[10] The fashion for assembling collections, in part a spin-off from gentlemanly travel, took root in English culture during the first half of the seventeenth century, expressing a desire for some meaningful correspondence between the individual and the world around him, as well as advertising the extent of his travels.[11] With their huge, indiscriminate collections, the virtuosi effectively parodied one of Bacon's fundamental ideas, the amassing of as many separate facts as possible in order eventually to understand the whole. The virtuosi also fulfilled other, more useful functions. Many contributed considerable funds to the Royal Society,[12] served as officers, and eagerly undertook the expensive and time-consuming collection of vast amounts of data for the "scientists" to examine and interpret. Moreover, without the enlarged market provided by the virtuosi, it is unlikely that scientific apparatus and instruments would have developed as rapidly, or become as affordable, as they did.[13]

However, many of the virtuosi were merely dilettantes, jumping on any fashionable bandwagon. Avid purchasers of anything strange and rare, they were a gift to the satirist, and it is not surprising that a significant number of Restoration plays feature foolish would-be scientists collecting useless or disgusting trivia, oblivious to what is going on around them. These characters were the predecessors of the absentminded professors and the hapless inventors of twentieth-century comic films like *The Absent-Minded Professor* (1961), *The Nutty Professor* (1964, 1996), *Honey, I Shrunk the Kids* (1989), and the *Back to the Future* trilogy (1985), which parody the would-be powerful scientist. The scientist characters in these films are "mad" but not intentionally evil. Nevertheless they are not harmless, and their inventions are temporarily disastrous. Wayne Szalinski's electromagnetic shrink ray reduces his and neighboring children to half a centimeter in height, and Dr. Emmett "Doc" Brown of *Back to the Future*, inventor of a time machine, finally admits to regretting his invention, which has altered history and "caused nothing but disaster."

Another feature of contemporary science to evoke ridicule was the de-

Figure 5. Engraving from *Dell'Historia Naturale*, by Ferrante Imperato, Naples, 1599. Wikimedia Commons.

velopment of the telescope. Attempts to improve on Kepler's refracting telescope by using weaker convex objective lenses led to the production of longer and longer instruments. Johannes Hevelius produced his *Selenographia* (1647), the first atlas of the moon, using telescopes from 8 to 10 feet long, and his 150-foot instrument erected in Danzig proved so difficult to point, and so unstable in even the slightest breeze, that its usefulness was severely limited.[14] It became a gift for visual caricature. Stage telescopes became so long and cumbersome as to require several bearers, and astronomers were represented as either gullible and prey to wild speculations or devious, insofar as they attempted to deceive others. Since astronomy had played an important role in the researches of the Royal Society from its inception in Wren's Invisible Colledge, such satirical representations necessarily rebounded upon its members.

It is against this background that we must look at the representations of scientists in literature from the Restoration to the mid-eighteenth century.

Scientists in Restoration Literature

Satiric representations of scientists in this period show them as wholly comic, almost invariably stupid, and out of touch with the real world. They imagine that they have great power over nature or are about to acquire it through some new discovery, but, not having the sense to realize their own stupidity, they are easily outwitted or manipulated. One of the first such satires was *The Description of a New World called the Blazing World* (1666), by Margaret Cavendish, duchess of Newcastle, an early feminist who prided herself on her scientific interests and knowledge. To the consternation of the Royal Society members, she insisted on honoring this exclusively male preserve with a visit, an event that elicited several comic ballads as well as a caustic description by the diarist Samuel Pepys of the lady's eccentric style of dress: "Her dress so antick, and her deportment so ordinary that I do not like her at all; nor did I hear her say anything that was worth hearing, but that she was full of admiration, all admiration."[15]

However, far from being the frivolous and fatuous person that Pepys presents, Cavendish was intent on exposing the pettiness and self-seeking character of contemporary science and its hegemonic claims to interpret nature.[16] *The Blazing World*, appended to Cavendish's *Observations upon Experimental Philosophy* (1668), is essentially a moral beast fable set in a new world where the scientists combine human qualities with the characteristics of various animal species.[17] The Bear-men are experimental philosophers, the Bird-men are astronomers, the Fly-worm- and Fish-men are natural philosophers, the Ape-men are chemists, and the Spider- and Lice-men are mathematicians. The empress of this fabled world, the first fictional female scientist, undertakes multiple responsibilities: promoting science research, creating her own royal society of virtuosi, examining existing knowledge, and speculating on natural philosophy. The "Blazing World" of the title is an exploration of a contemporary controversy—the existence of parallel worlds. The empress soon finds that, instead of accumulating knowledge, her subjects are embroiled in disputes and quarrels over contemporary scientific issues—the cause of the sun's light, why the sun and moon change their shape and size, and why there are spots on the sun. When the empress requests her experimental philosophers to resolve the disputes by observing the appropriate bodies through their telescopes, she finds that they cannot agree either about what they see or about what it means: "The Empress began to grow angry at their Telescopes, that they could give no better Intelligence; for, said she, now I do plainly perceive, that your Glasses are false Informers, and instead of discovering the Truth, delude your senses; . . . for you may observe the progressive motions of Celestial bodies with your natural eyes better than through Artificial Glasses."[18] This passage involves scarcely veiled criticism of Francis Bacon, who had advocated the need for continual invention of new tools, since sense and reason unaided were incapable of discovering the truth. However, the response Cavendish attributes to the astronomers is even more damning: they beg to be allowed to keep their telescopes *in order* to prolong their disputes, "For, said they, we take more delight in Artificial delusions, then in natural truths."[19] This early satire introduces many of the criticisms that were later leveled against the virtuosi, condemnation of their delight in disputation as an end in itself, their concern with "delusions" rather than with the truth, and their preoccupation with instruments for their own sake rather than for any usefulness to society.

It seems unlikely that Margaret Cavendish intended this condemnation to be applied specifically to the members of the Royal Society, but certainly the anonymous "Ballad of Gresham Colledge" (so called; its actual title was more explanatory: "In Praise of that choice Company of Witts and Philosophers who meet on Wednesdays weekly att Gresham Colledge") was explicitly directed against this august body. The poem satirizes the scientists' preoccupation with quantitative measurement to the exclusion of qualitative considerations and their determination to reduce natural complexity to simple demonstration. Today, when most of these scientific procedures are taken for granted, we may well miss the intended irony concerning these men who "take nothing upon trust" (a clear reference to the motto of the Royal Society), but "knowe all things by Demonstration":

> To the Danish Agent late was showne
> That where noe Ayre is, there's noe breath.
> A glasse this secret did make knowne
> Where[in] a Catt was put to death.
> Out of the glasse the Ayre being screwed,
> Pusse dyed and ne're so much as mewed.[20]

The ballad describes other investigations being carried out by Royal Society members to improve the material quality of life.[21] Because, with hindsight, we now know that the projects mentioned were ultimately successful, it is at first difficult to realize that the poem was intended to be farcical, but the author clearly implies that these investigations are either so obvious as to be apparent to common sense (a cat dying when deprived of air) or else ridiculously far-fetched, like measuring longitude or descending deep into the sea. To appreciate the point, we need to recall that a century was to pass before longitude could be measured accurately, since it depended on the perfecting of the chronometer, and two hundred years before the diving bell was a practicality.

The satirist Samuel Butler also mocked the virtuosi as stupid and self-deceived, but in addition he stressed their moral deficiencies. Incapable of perceiving the beauty and wonder of ordinary nature, they were intent only on novelty and, in their insatiable desire for fame, were prepared to sacrifice mere truth. For Butler, astronomy was only a more fashionable form of astrology, and in his astrologer character Sidrophel (star lover) of *Hudibras* (1664) he ridiculed indiscriminately all the scientific preoccupations of his day—Descartes's vortices, the atomic theory in physiology, blood transfusion, the stentorophonic tube, pendulum watches, chemistry, and bottled air. In *The Elephant in the Moon* (1676) Butler's satirical net was cast more widely still, and he appeared bent on deriding virtually the whole membership of the Royal Society, since characters corresponding to Sir Paul Neal, Sir John Evelyn, Robert Hooke, Antonie van Leeuwenhoek, and even Robert Boyle have been identified.[22] Butler's charge against the new philosophers was the same as he had leveled at the virtuosi: that they sought marvels rather than the truth. In the poem, which has clear echoes of Margaret Cavendish's satirical fantasy, a group of astronomers gathers to observe the moon, constructing elaborate theories of lunar politics and sociology on the basis of no observable evidence. Indeed, they are determined in advance to observe a "wonder." Soon they are overjoyed to find, as they think, a battle in process on the moon and, in the midst of the fray, an elephant. Delighted at this visual "evidence," they enthusiastically plan an article for the next issue of *Philosophical Transactions*, the journal of the Royal Society, until it is suggested by a servant, the representative of common sense, that the "elephant" is a mouse that has been trapped in the telescope (see fig. 6). The telescope symbolizes the distorted vision of the scientists, who are interested only in a hypothetical elephant, not an actual mouse. Even when apprised of their mistake, Butler's astronomers are anxious to suppress the truth in order to further their scientific reputations, but eventually they open the telescope and find, along with the "elephant" mouse, a swarm of flies and gnats, the "combatants" in the lunar battle. Butler's moral is heavily underlined.

Figure 6. The Astronomer, by Thomas Rowlandson (1756–1827), published by Rudolph Ackermann, 1815. The caption reads: "Why I was looking at the Bear. / But what strange Planet see I there!"

> That learned men, who greedily pursue
> Things that are rather wonderful than true,
> And, in their nicest speculations, choose
> To make their own discoveries strange news;
> ....
> In vain endeavour Nature to suborn,
> And, for their pains, are justly paid with scorn.[23]

Butler judges the scientists as not merely foolish but hypocritical; they profess to seek truth but actively suppress it if it interferes with their pursuit of fame. Instruments designed to improve man's vision—the telescope and the microscope—are of no use against moral blindness and the refusal to see.

Not satisfied with this tour de force, Butler apparently contemplated a more extended satire on contemporary scientists. The incomplete fragment "A Satire on the Royal Society" reduces the scientific debates of the day to ridiculous irrelevancies:

> And all their constant occupations:
> To measure wind, and weigh the air,
> And turn a circle to a square;
> ....

To find the North-west passage out,
Although the farthest way about;
....
To stew th' Elixir in a bath
Of hope, credulity, and faith.[24]

The belief that the virtuosi were interested only in the unnatural and the monstrous was still current some seventy years later. The novelist Henry Fielding, describing the would-be serious young men of his day, includes among their amusements "natural philosophy, or rather unnatural, which deals in the wonderful and knows nothing of Nature, except her monsters and imperfections."[25]

The first developed satirical portrait of the new scientist, and certainly the most influential in its time, was Sir Nicholas Gimcrack in Thomas Shadwell's box office success *The Virtuoso* (1676). Because many of Gimcrack's experiments and theories were only slightly changed from reports in the *Transactions of the Royal Society* or in Robert Hooke's *Micrographia*, it has been widely assumed that Shadwell was poking fun at the Royal Society members, but this is a simplified judgement both of the play's main character and of its overall place in the tradition of Restoration satire.[26] Not only is it explicitly stated at the beginning of the play that Sir Nicholas has been refused membership in the college[27] but, so far from mocking the practices of the Royal Society, Shadwell actually uses them as the yardstick with which to censure Gimcrack, the dilettante virtuoso who flits from wonder to wonder without method or mental discipline. Nevertheless, in the character of Gimcrack, Shadwell effectively deconstructs the pretensions of contemporary scientists, their cultivated marks of difference, and their arrogance toward their fellows. There is thus an implicit moral beneath the overt satire.

Like all virtuosi, Gimcrack is an indiscriminate collector of oddities, an easy prey for charlatans. He has been duped into paying ten shillings each for eggs alleged to have been laid with hairs in them from a tradesman who had inserted the hairs through a fine hole, and has outlaid £2,000 on telescopes, air pumps, and microscopes. He has spent twenty years studying the nature of "lice, spiders and insects" (1.2, p. 17), a reference to Robert Hooke's *Micrographia* (1665) recording his microscopic observations of flies, fleas, and lice. Perhaps the most damning indictment of the virtuoso, and one that was to be echoed by nearly all the eighteenth-century satirists and moralists, is that voiced by Gimcrack's niece Miranda. She describes him as "one who has broken his brains about the nature of maggots, who has studied these twenty years to find out the several sorts of spiders, and never cares for understanding mankind" (1.2, p. 10). Indeed, he has traveled in Italy without taking any interest in its culture: "'Tis below a virtuoso to trouble himself with men and manners. I study insects" (3.3, p. 43). This aspect of the virtuoso was an ongoing source of humor well into the eighteenth century, and Gimcrack was as well known as any contemporary celebrity.

Gimcrack's many useless experiments provide much of the stage comedy. Lying awkwardly, facedown on a table, holding between his teeth a string attached to a frog in a bowl of water, he claims to be learning to swim by imitating the movements of the frog—not because he ever intends to swim, for he hates water, but for the theory. "I content myself with the speculative part of swimming; I care not for the practic. I seldom bring anything to use; 'tis not my way. Knowledge is my ultimate end" (2.2, p. 97). This dedication to the cult of uselessness is elaborated throughout the play, as Gimcrack frolics from one experiment to the next, without any interest in their implications, his motto being "so it be knowledge, 'tis no matter of what" (3.3, p. 42).

Unlike the members of the Royal Society, Gimcrack never follows through on the results of his experiments. Robert Boyle measured the weight of air at different altitudes in order to determine the relation between mass, pressure, and volume; Gimcrack collects bottles of air from different areas merely for the novelty, "to let it fly in my chamber" (4.1, p. 66). Nevertheless, for Shadwell's audience the connection with Robert Boyle would have been clear, and was so to Boyle himself when he saw the play. He wrote in his diary: "Dammed dogs. Vindica me deus (God grant me revenge). People almost pointed."[28] Similarly, Gimcrack's blood transfusions between men and sheep are ludicrous, because they are done merely out of curiosity and are wholly destructive; however, Shadwell is careful to have two other characters point out that the practice of transfusion per se (a topical issue in the Royal Society)[29] is by no means ridiculous.

Sir Nicholas is also typical of the virtuoso rather than of the serious scientist in that he makes excessive claims for his experiments merely in order to make them appear more marvelous and unnatural, for, like Butler's astronomers, he has no interest in the normal. He must claim not only to swim but also to fly: "I can already out fly that ponderous animal called a bustard" (2.2., p. 27). Not content to describe the habits of spiders, he insists that he has a tame spider called Nick that follows him around—"the best natur'd, best condition'd spider that ever was met with" (3, p. 44). He does not merely profess to have accomplished a successful blood transfusion between a man and a sheep; he must assert that the man thereupon assumed an ovine character and grew a tail and wool. He claims to have improved the stentorophonic tube, so that he can hear things being said eight miles away (5.2., p. 71), yet he does not hear his wife and nieces plotting against him in the next room. These exaggerations result partly from his pretentious language, a comment on the inflated and obscure language of the Royal Society's *Transactions*,[30] and his class consciousness, but also from his inability to distinguish between cause and effect, between science and superstition. We discover that, like many of the early members of the Royal Society, he is also a Rosicrucian, absorbed in magic and spiritualism.

Although most of the comedy arises from the ridiculousness of his pseudoscientific antics, Gimcrack stands condemned on moral grounds as well, notably for his hypocrisy, which is seen in his closet lechery and his belief that mankind is beneath his notice. These exaggerated vices contribute to the successful comedy, but Sir Nicholas *is* the play to such an extent that the usual intrigues of Restoration comedy are subsidiary to his experiments and his ludicrous theories. Shadwell's play was so popular that Gimcrack became the literary progenitor of a whole tribe of virtuosi of both sexes, all having traits modeled on his. Like him, Dr. Boliardo, of Aphra Behn's *The Emperor of the Moon* (1687), who in line with the current fashion for astronomy has studied the moon through his telescope, is ridiculed for his pretensions, which include wearing all manner of mathematical instruments hanging from his belt and having a servant bear after him a thirty-foot-long telescope. Like Gimcrack, he is easily manipulated by the more astute gallants courting his pretty wards, thereby demonstrating that intellectual and moral pretensions are merely a façade for the universal failings of the human condition.

Given the gender imbalance of scientists, far more extreme then than now, it is interesting that not all virtuosi characters were male. For the first time in literary history, there was a rash of female "scientists" as, in the wake of Margaret Cavendish, scientific ladies made their stage debut. Usually they were first ridiculed and then, if young and attractive, reformed, ending up as sensible, biddable wives in the socially acceptable mold. Shadwell's *The Sullen Lovers* (1688) presents Lady Vaine Knowall, a would-be virtuosa, a pedant with little if any accurate knowledge and, like Gimcrack, totally ignorant of the world around her. In Thomas Wright's *The Female Virtuoso's* (1693), an adaptation of Molière's *Les femmes savantes*, the female wits prepare to form a learned society, the Academy of Beaux Esprits, but Lady Meanwell, the leading virtuosa, is full of useless projects, all totally impractical, and (clearly a far worse charge) is an unsatisfactory wife.

In *The Basset Table* (1706) Susannah Centlivre created a somewhat more appealing virtuosa, Valeria, "a Daughter run mad after Philosophy," whose passion is collecting insects and examining them microscopically. She enters pursuing "a huge Flesh Fly," which she has just received from her suitor, Mr. Lovely, for dissection. Indeed, Valeria goes further than Gimcrack in her devotion to science. Unhampered by any feminine tenderness, she is an ardent vivisectionist, eagerly dissecting her pet dove to test "whether it is true that doves lack gall," opening up a dog to find a tapeworm, and examining a fish under her microscope. "Can Animals, Insects or Reptiles be put to a nobler Use than to improve our Knowledge?" she asks rhetorically.[31] However, she is eventually brought to see the error of her ways and marries Mr. Lovely. It is clearly implied that her aspiration to scientific learning is unnatural and she must be reeducated to fit the accepted female role.

A more vigorous attack on the female virtuosa is the character of Lady Science in James Miller's anonymously published comedy *The Humours of Oxford* (1726). Lady Science has none of Valeria's pert charm; she is "a Female Bookworm," "a Pretender to Learning and Philosophy"[32] who suffers from Sir Nicholas Gimcrack's fault of being out of touch with life. A precursor of Oscar Wilde's Lady Bracknell, she interrogates her prospective son-in-law not about his morals or religion but about his cosmology—whether he holds to the Ptolemaic or the Copernican hypothesis and whether he believes it "ever possible to find out the Longitude" (61). The main condemnation of Lady Science, however, hinges on her daring to aspire to a province ordained for men. "The Dressing-Room, not the Study, is the Lady's Province—and a Woman makes as ridiculous a Figure poring over Globes, or thro' a Telescope, as a Man would with a Pair of *Preservers* mending Lace" (79). It is not scientists per se who are being satirized here, but only women who have dared to surrender their femininity to science.

Scientist bashing was good theater, especially when supplemented with preposterous visual props and farcical situations, and most of the ridiculous scientist characters in Restoration drama derived from no more serious purpose than novelty and a popular backlash against the Puritans who had closed the theatres, thereby reflecting the accumulated resentments of the previous century. It was only when the investigations of the Royal Society were recognized as socially and economically important that the satire became not just comic but, as we shall see in chapter 5, serious, vicious and more sternly moral. Ironically, the person who was preeminent in elevating the status of science in England, and who became a figure of national eulogy, was also to provoke the most extreme reaction against the premises of scientific materialism and the reductionist procedures associated with it. That person was, of course, Isaac Newton, one of the few actual scientists to feature as a literary character in propria persona.[33]

CHAPTER 4

Newton

A Scientist for God

> Nature and Nature's laws lay hid in night;
> God said, "Let Newton be!" and all was light.
> —*Alexander Pope*

It was primarily through the influence of one person, Isaac Newton (1642–1727), that the popular image of the scientist changed from being either a sinister or a stupid character to a highly respected man of genius, embodying the ultimate attainments of reason (see fig. 7). At the time of his death, and in the years immediately following, Newton received the highest literary accolades ever accorded to a scientist, and this adulation was the more unusual in being associated with an actual scientist rather than a fictional character, although, as we shall see, the figure of Newton as represented in contemporary verse became increasingly mythologized. Newton was regarded as Britain's national treasure, and to praise him became an act of patriotism, even of piety.

When we think of Newton, we think of his three laws of motion and his law of universal gravitation, but these were not what Newton himself regarded as his main research interest, nor the reason why he was so acclaimed by his contemporaries. Without in any way detracting from Newton's real and enduring achievements, it is important to realize the extent to which Newtonian iconography, bordering on hagiography, was constructed through a rigorous selection of the physical and mathematical elements of his work and a corresponding suppression of other aspects of his thinking, notably his preoccupation with alchemy, to produce the image that was intellectually, politically, and psychologically congenial to the climate of opinion in England

Figure 7. Portrait of Sir Isaac Newton, ca. 1715–20, attributed to the English School. Wikimedia Commons.

around the time of his death. Whether conscious or not, this exercise in censorship and propaganda was so effective that his own persistent attempts to link the physics of the *Principia* to his alchemical research and thereby evolve a vast unifying theory linking all matter and reactions throughout the universe remained virtually unknown until the mid-twentieth century. The first indication of disparity between the person and the legend was John Maynard Keynes's paper "Newton, the Man," produced for the Royal Society's tercentenary celebrations of 1946. Keynes concluded from his study of some 57 of Newton's alchemical manuscripts, which he had acquired from 121 auctioned lots, that "Newton was not the first of the age of reason. He was the last of the magicians, the last of the Babylonians and Sumerians, the last great mind which looked out on the visible and intellectual world with the same eyes as those who began to build our intellectual inheritance rather less than 10,000 years ago."[1] Betty Jo Teeter Dobbs, having explored Newton's overriding concern with deriving a unified truth linking his alchemy, his Arian theology, and his writings on the causal relations of matter and forces, concludes that "in a sense the whole of his career after 1675 may be seen as one long attempt to integrate alchemy and the mechanical philosophy," to find evidence for a "vegetative principle" operating in the natural world, equivalent to the secret, animating principle of the alchemists.[2] We know from some of the manuscripts auctioned in 1936 that Newton was greatly interested in procuring or developing the philosopher's stone and was strongly influenced by the Rosicrucians, who claimed to have possession of both the philosopher's stone and the elixir of life.[3] Regarding the centrality of alchemy to his thought, Karin Figala concludes that Newton's "alchemy cannot be seen solely in connection with his chemical experiments but was also a link between his religious beliefs and his scientific aims."[4]

Throughout his life Newton had sought the structure of the world in alchemy—a system of the small world to match with his system of the greater. This longing is clearly expressed in the preface to the *Principia*: "I wish that we could derive the rest of the phenomena of Nature by the same kind of reasoning from mechanical principles."[5] In the *Principia* he even suggested that vapors from the sun, stars, and comets might be condensed into water and humid spirits and then, by continued fermentation, into all the more dense substances.

Despite the height of his subsequent fame, recognition of Newton's work was slow in coming. The young Isaac showed no inclination for the farming career proposed by his widowed mother and in 1661 had the good fortune to be sent to Trinity College, Cambridge, by a maternal uncle who had studied there. At Cambridge Newton encountered the intellectual ferment elicited by Descartes's *philosophia mechanica*, by Kepler's optics and his laws of planetary motion, by Galileo's mechanics, and by the new mathematics of his own teacher, Isaac Barrow.[6] In 1664 the university was closed because of plague, and the newly graduated Newton was forced to return to his mother's house in Lincolnshire. During this year of "exile" from Cambridge, he devised his three laws of motion and his theory of gravitation, although he did not publish these results for another twenty-three years. His earliest scientific paper on light and color, read on his behalf at the Royal Society in 1672, was accorded a far from favorable reception,[7] and the ensuing controversy caused him to eschew publication of his work. It was only at the urging of Edmund Halley, later Astronomer Royal, that he consented to write down the elaborate proofs (which he had mislaid!) for his theory of planetary motion, including the infinitesimal calculus he had had to derive for this purpose. This work on planetary motion, together with the treatise *On the System of the World*, was at Halley's instigation published as *Philosophiae Naturalis Principia Mathematica* (universally known as the *Principia*) in 1687. The work was almost immediately recognized as vindicating a new method of analysis, one that quickly came to be seen as the necessary and sufficient instrument for scientific progress. The first book of the *Principia* deals with definitions of space, mass, and time and motion in a nonresistant medium. The second book considers motion in a resisting medium and forms the basis of hydrostatics and hydrodynamics. But it was the third book, entitled *De Systemate Mundi*, or *On the System of the World*, that seized the popular imagination. Its introduction confidently affirms, "Superstat ut ex iisdem principiis doceamus constitutionem systematis mundani" (It remains that from the same principles we demonstrate the form of the system of the world), and it did indeed provide the complete framework for the celestial mechanics of the next two centuries. "The true system of the world has been perceived, developed, and perfected," wrote the French historian D'Alembert, and even Laplace was moved to exclaim that "it [the *Principia*] has all the certainty which can result from the immense number and variety of phenomena, which it rigorously explains, and from the simplicity of the principle which serves to explain them."[8] Certainly the *Principia*, ending with the "General Scholium," in which Newton asserted, "Hypotheses non fingo" (I do not create hypotheses), was taken as the ultimate justification of Bacon's experimental science against the theoretical constructions of Descartes. The first words of Newton's *Opticks* (1704) were later to reaffirm this in English: "My Design in this Book is not to explain the Properties of Light by Hypotheses, but to propose and prove them by Reason and Experiments."[9]

The Attraction of the *Principia*

Although the *Principia* was eagerly bought, it was less widely understood. Newton, in order "to avoid being baited by little smatterers in mathematics," had designed its obscure presentation "to be understood only by able mathematicians," and it has been claimed that only a handful of Newton's contemporaries, including one woman, the Marquise de Châtelet, were able to understand its complex geometrical arguments.[10] It was not long, however, before simplified versions were current in England. On the continent, acceptance was slower, partly because of French allegiance to the Cartesian system, but Voltaire's *Eléments de la philosophie de Newton* eventually ensured the popularization of the work, even in France. Indeed, referring to the increasing influence of scientific discourse on the language of literature, Voltaire remarked in 1735: "Verses are hardly fashionable any longer in Paris. Everyone begins to play the mathematician and the physicist. Everyone wants to reason. Sentiment, imagination and charm are banished."[11]

Given the difficulty of the *Principia*, we might well ask why a mathematical text should have elicited such widespread public veneration, even mystique. There were at least two major reasons and a number of minor ones. Most important was its image of nature that typified order, simplicity, and harmony and appeared accessible, reasonable, and predictable. Taken together, these qualities engendered in Newton's contemporaries a sense of empowerment and a belief that man, far from being insignificant in the expanding universe revealed by the new telescopes, had attained a position of superiority, since he alone was capable of understanding the workings of the whole celestial system. The anonymous *A Philosophic Ode on the Sun and the Universe* (1750) praises Newton specifically for revealing the order of the universe:

> Newton, immortal Newton rose;
> This mighty frame, its orders, laws,
> His piercing eyes beheld:
> That Sun of Science pour'd his streams,
> All darkness fled before his beams,
> And Nature stood reveal'd.[12]

In formulating his basic rules of scientific procedure, Newton had asserted: "We are to admit no more causes of natural things than such as are both true and sufficient to explain their appearance," since "Nature is pleased with simplicity." He insisted that "to the same natural effects we must, as far as possible assign the same causes"[13] so that the pull of gravity or the reflection of light on earth and on the planets would be attributed to the same causes. It was the apparent simplicity of Newton's system that took hold of the popular mind, cutting like Occam's razor through Cartesian vortices and other complex theories adduced to account for observed phenomena. There is a definite sense of relief in James Thomson's lines on Newton: "The heavens are all his own; from the wild Rule / Of whirling VORTICES, and circling SPHERES / To their first great simplicity restored."[14] The simplicity of the Newtonian scheme was extolled by many other contemporary poets, as in this stanza by William Tasker:

> Th'eccentric comet's course he knew,
> From principles sublimely few,
> Explain'd all Nature's laws.
> Th'Attractive and repulsive force,
> He taught to solve the planets' course
> Encircling thee, O Sun![15]

Newton's unified model of the cosmos was based on his three relatively simple laws of motion, which, together with the inverse square law, explained not only terrestrial motion but also the large-scale movements of the planets and of comets. Celestial events such as comets, hitherto regarded as so capricious as to be attributable only to divine retribution, and the precession of the equinoxes, which had not formerly been explicable mathematically, were now entirely predictable. The fact that the so-called laws of science are actually no more than the descriptions that best fit the observed phenomena was submerged in the transferred sense of power and authority conveyed by the notion of the lawmaker, in this case Newton, who was credited (as in the poems quoted above) with having imposed them on a hitherto unruly universe.

The second major attraction of the *Principia* for Newton's contemporaries was its vindication of quantitative science and the experimental method expounded by Bacon as powerful tools applicable to all spheres of operation.[16] Even Newton's work on light, although less obviously related to the mechanical scheme, was characterized by measurement and calculation rather than by generalities and thus seemed to carry an inherent guarantee of truth. In defining mass, inertia, and force and their relation to velocity and acceleration and the hitherto mysterious force of gravity, the *Principia* dispensed with mystical causes for phenomena and refashioned the seemingly random and hostile universe as a well-oiled and entirely reliable clockwork model.

Descartes had believed that mathematization of the natural world and even of human society would reveal the order underlying apparently unpredictable events. Newton's success in applying the mathematical principles to explain universal gravitation and the movements of the planets in the solar system was perceived not only as establishing a universal law that could encompass the cosmic order but as suggesting that all branches of knowledge, not just physical science, might be susceptible to the same universal laws. In 1728, the year after Newton's death, J. T. Desaguliers produced a ponderous application of Newtonian physics to politics: *The Newtonian System of the World: The Best Model of Government, an Allegorical Plan*. In a period when the religious and political turmoil of the previous century were still fresh in living memory, it was alluring to believe that scientific principles could resolve these deep divisions once and for all. R. G. Olson writes: "The scientific-mathematical-empirical method seemed to receive almost pontifical authority by its association with Newton's accomplishments in natural philosophy; and while it may be an oversimplification, I think it is not really a major distortion to say that the enlightenment of the eighteenth century was in large measure an attempt to extend the domain of application of scientific method to include not only inanimate nature, but also the laws of human thought, morality, society, and religion."[17] In brief, it can be said that before Newton's work it was considered necessary to invoke ad hoc principles and occult causes based on human and divine analogies for all but the simplest terrestrial phenomena, while after the assimilation of the *Principia* it was assumed that all terrestrial and celestial movements were explicable in precise and numerical terms, by calculations based on a few general laws.

Although Newton's ideas about gravity were related to alchemical concepts and were at first branded by his contemporaries as occult, since they involved forces acting at a distance, this was conveniently ignored by the image-makers of the Enlightenment.[18] The intellectual climate of eighteenth century rationalism, ironically referable in considerable part to the Newtonian heritage, could not have reconciled such an idea with the image of the great mathematical physicist.

The ultimate success of the *Principia* at both a practical and a theoretical level was also instrumental in propelling Western culture toward the scientism that has largely characterized it ever since. Newton's mechanistic and rationalistic procedures destroyed, at least temporarily, the climate of suspicion that had attended the Faust figure, and the quest for knowledge was no longer associated with sin but regarded as the highest pursuit of mankind. Samuel Johnson, who in many ways epitomized the educated Englishman of the Enlightenment, asserted categorically, "Sir, a desire for knowledge is the natural feeling of mankind; and every human being, whose mind is not debauched, will be willing to give all that he has to get knowledge."[19]

The manifest practical and economic spin-offs from Newtonian mechanics also rapidly quelled the accusation, alluded to in the preceding chapter, that science and scientific instruments were useless. Apart from the widely applicable laws of motion, Newton's lunar tables, from which, given an accurate chronometer, longitude could be calculated,[20] and his lunar theory of tides were important improvements in naval science and cartography. Among his other activities, Newton also devised the reflecting telescope, which overcame both the problem of chromatic aberration and the inconvenient length required for adequate magnification in the older refracting models. The new, shorter telescopes were much easier to use and cheaper to construct and mount. As they increased in popularity, more people were able to observe for themselves the greatly expanded universe they revealed. In the context of such spaces the individual seemed no longer important enough to draw the fire of devils and angels, even if they were assumed to exist. Magic, sin, and spiritual aspirations seemed increasingly irrelevant in this orderly, mechanical cosmos.

E. N. da C. Andrade has summed up the impact of Newton on his contemporaries in the following terms:

> If we are to try to represent Newton's achievements by some modern analogy, to construct some imaginary figure who should be to our times what Newton was to his, we must credit this synthetic representative with, I think, the whole of relativity up to, and somewhat further than, the stage at present reached—we must suppose our modern Newton to have satisfactorily completed a unitary field theory. In light we must credit him both with having established the existence of spectral regularities and with their explanation in terms of the quantum theory. Possibly, too, we must give him the Rutherford atom model and its theoretical development, a simple astronomy in little to correspond to the solar system. Let us, then, think of one man who, starting in 1900, say, had done the fundamental work of Einstein, Planck, Bohr and Schrödinger, and much of that of Rutherford, Alfred Fowler and Paschen, say, by 1930, and had then become, say, Governor of the Bank of England, besides writing two books of Hibbert lectures and spending much of his time on psychical research, to correspond with Newton's theological and mystical interests. Let such a man represent our modern Newton and think how we should regard him. Only so, I think, can we see Newton as he appeared to his contemporaries at the end of his life.[21]

Impressive as it is, this list does not convey perhaps the most important attribute that Newton embodied for his contemporaries, namely, a sense of confidence in the harmony and order of the universe. Whereas in 1611 Donne had written, "And new Philosophy calls all in doubt," Newton had seemingly

Figure 8. A Philosopher Giving a Lecture on the Orrery, by Joseph Wright of Derby, 1764–66. Derby Museum and Art Gallery. Wikimedia Commons.

restored order and predictability through the application of a very few basic principles that appeared to explain all natural phenomena. Thus the response was not merely, or even primarily, to Newton the person so much as to the picture that he spread before them of a nature magnificent and expansive yet governed by simplicity and uniformity, a system that domesticated the expanding universe that geographers and astronomers were revealing. One indication of this contemporary fascination with order in nature was the popularity of the orrery, an intricate clockwork model of the solar system named after Charles Boyle, Fourth Earl of Orrery. Although its original purpose was to explain the principles of astronomy, it became a fashionable acquisition, suggesting an aesthetic as well as a pedagogic attraction. Joseph Wright's painting *A Philosopher Giving a Lecture on the Orrery* (ca. 1764–66) vividly conveys both a sense of devout contemplation on the part of the onlookers, ranging from young children to middle-aged men and women, and the power of the scientist figure who explains this complex system (see fig. 8).[22]

Moreover, despite the mechanistic basis of Newtonian physics, a large part of Newton's appeal to his contemporaries resulted from the religious implications they saw in his work, for although the new science in effect excluded purpose (divine or otherwise) as a legitimate consideration, it was not at first seen as antireligious. On the contrary, Newton affirmed the need for the continuing involvement of God in the universe "to provide a cause for gravitational attraction and to adjust certain imperfections in the system."[23] As occult and supernatural explanations of the world were replaced by mechanics, so the former image of the godless scientist was replaced by that of a wise religious teacher who interpreted the ways of God to man. The respected physician Sir Thomas Browne reaffirmed Bacon's assertion that the study of God's works was almost as pious as the study of God's word: "There are two Books from whence I collect my Divinity; besides that written one of GOD, another of His servant Nature, that universal and publick Manuscript, that lies expans'd unto the Eyes of all."[24] This natural theology was to become the recourse of many nineteenth-century Christians who had come to doubt biblical revelation.

In the first edition of the *Principia* (1687), Newton had not explicitly mentioned God; nor did the first edition of the *Opticks* (1704) refer to God as the creator of the world. However, Roger Cotes, Plumian Professor of Astronomy at Cambridge and editor of the second edition of the *Principia* (1713), urged Newton to "add something by which your Book may be cleared from some prejudices which have been industriously laid against it."[25] He was referring to a letter from Leibniz to Harsöker criticizing the omission of God from a work that presumed to present the system of the world. In response Newton added the "General Scholium" at the end of book 3 as a tacit acceptance of the limitations of knowledge, beyond which faith is the only appropriate attitude.[26] Similarly, he later added to the *Opticks* two more queries, which treat of God as creator and perceiver of the physical world.[27]

In Query 31, Newton made an explicit connection between natural and moral philosophy, so important to his contemporaries: "And if natural Philosophy in all its Parts, by pursuing this Method, shall at length be perfected, the Bounds of Moral Philosophy will be also enlarged.... Our Duty towards him, as well as that towards one another, will appear to us by the Light of Nature." Wisdom is here associated with humility and reverence for the works of the Creator, encouraging the image of the high priestly scientist, coworker with God in the revelation of his wisdom.

Eulogizing Newton

In this regard, Newton's work can be seen as the high point of many efforts by natural theologians to establish the presence of God in the world. Where

John Ray discovered *The Wisdom of God Manifested in the Works of the Creation* [1691], Richard Lower in the design of the heart, and Nehemiah Grew in botanical structure, Newton affirmed divinity in the whole system of the universe and gave to his contemporaries a new meaning for the words of the psalmist, "The Heavens declare the glory of God." However, neither Newton nor his contemporaries recognized that a God derived from nature, especially a nature viewed from the perspective of mechanics, could be only a reflection of the values of physics, a Great Mechanic. The simplicity and order of the solar system are evidence of divine intelligence only if we assume that these values, rather than complexity and maximum degrees of freedom, represent the divine purpose. The poet-mystic William Blake was the first to recognize this fundamental flaw in Newton's natural theology, but even if Newton's contemporaries had perceived the circularity of his arguments, they would still have applauded them. In the fight against atheism a slight bending of scientific principles troubled no one's conscience, certainly not Newton's. He enthusiastically helped Richard Bentley prepare the first Boyle lectures,[28] entitled *A Confutation of Atheism*, which invoked Newton's cosmology as evidence for the theological argument from design.

So widespread was the mythology associated with Newton that after his death his soul was popularly supposed to have winged its way straight to the throne of God, and in many poems of the period he was depicted as a space traveler embarking on a cosmic voyage to learn from God those few truths that might have eluded him on earth. Alan Ramsay's fulsome "Ode to the Memory of Sir Isaac Newton: Inscribed to the Royal Society" includes such stanzas as the following:

> The god-like man now mounts the sky,
> Exploring all yon radiant spheres;
> And in one view can more descry
> Than here below in eighty years.
> Now with full Joy he can survey
> These Worlds, and ev'ry shining Blaze,
> That countless in the *Milky Way*
> Only through Glasses show their Rays.[29]

Thus, after his death Newton became an important role model for scientists in the climate of ambivalence that developed during the eighteenth century. On the one hand, there was acclaim for the discoveries of science, particularly for the demonstration of harmony in the universe, which could be interpreted in religious terms as evidence of the greatness of the Creator; on the other hand, there was a wariness about ascribing too much praise to any mortal and thereby encouraging a godless pride in the human intellect. Newton's death conveniently resolved this dilemma, allowing him to be universally endorsed as a national treasure. Indeed, in James Thomson's poem "Sacred to the Memory of Sir Isaac Newton," through a complex system of analogy, Newton is cast as the new Adam, realigning man to the Creation. The poem emphasizes Newton's work on light, astronomy, and the tides as the counterparts of the first elements of the Creation described in Genesis and contrasts fallen, "erring" man with the "all-piercing Sage."[30] Henry Grove expressed much the same popular myth of the great soul illuminating the darkness of ignorance and the religious revelation implicit in scientific discovery. To him Newton appeared "like one of another Species," so great was his understanding.[31]

From Newton's fame it was a short step to Britain's fame, and a great part of the reverence accorded Newton by his compatriots, especially in the spate of eulogies occasioned by his death, was nationalistic in inspiration. The British Newton had surpassed the French Descartes, and no European could stand against him. Even Huygens and Leibniz, who claimed priority over Newton in inventing the calculus, were finally unable to refute Newton's conclusions. David Mallet's poem *The Excursion* (1728), written a year after Newton's death, is an example of such nationalistic, self-congratulatory verse: "To thee, great Newton! Britain's justest pride, / The boast of human race."[32]

Although Newton was the prototype of the great scientist, other lesser lights were also included in the general enthusiasm for British science. Dryden's poem "To my Honour'd Friend, Dr. Charleton" (1663) paid tribute to the recipient as well as to Bacon, Gilbert, Boyle, and Harvey in tones of restrained applause. Sprat's *History of the Royal Society* (1667) was prefaced by Abraham Cowley's ode "To the Royal Society," in which the members of that body are compared in rapid succession to Moses's Israelites led by Bacon to the promised land, to Gideon's picked band, and to the infant Hercules.[33] Thomson linked Bacon, Boyle, and Locke with Newton as exemplars of man's deliverance from ignorance by the revelation of God in nature.[34] In 1705 Edmund Halley, the second Astronomer Royal, had captured the public imagination with his synopsis of known comets[35] and after his death in 1729 was rewarded with the promise of a cosmic tour of discovery similar to the one assigned to Newton. Richard Savage confidently asserted that the heavenly tourist would thereby have his faith confirmed:

Hence Halley's soul etherial flight essays;
Instructive there from orb to orb she strays;
Sees, round new countless suns, new systems roll!
Sees God in all! and magnifies the whole.[36]

This liaison between astronomy and religious faith was, however, always fragile if subjected to scrutiny, and it is not surprising that it failed to survive for long. Inevitably, after such excesses of adulation, a reaction set in, and scientists fell under attack from literary moralists and satirists, this time on the charges of arrogance, delusion, and heresy.

CHAPTER 5

Arrogant and Godless

Scientists in Eighteenth-Century Satire

> We nobly take the high Priori Road,
> And reason downwards till we doubt of God
> Thrust some Mechanic Cause into his place:
> Or bind in Matter, or diffuse in space. —*Pope*

Although it was slow by today's standards, when fame automatically elicits a desire to expose the deficiencies of public figures, there was inevitably a "sciencegate"—a reaction against the excessive adulation of Newton. Not all his contemporaries were as easily persuaded of the religious motivation of scientists as the writers quoted in the preceding chapter. After the witty Restoration satires, which emphasized the comic foolishness of the virtuosi, the mood of the age became more serious, and the charges leveled against scientists changed also. In particular, those educated in the humanities and steeped in the Greek and Roman classics had little sympathy with a worldview that denigrated the past and looked instead toward a glorious future to be ushered in through the power of science. They accused scientists of ignoring the moral dimension and of failing to see that, in Pope's words, "The proper study of mankind is man."

The eminent classical scholar Meric Casaubon was one of the first to deny explicitly that a knowledge of natural philosophy could civilize a man or teach him ethical responsibility,[1] but major literary figures, such as Alexander Pope, Jonathan Swift, and Samuel Johnson, soon became increasingly critical of the moral failings that they believed followed on the success of the new science. Resurrecting the Faustus image, they asserted that scientists were attempting to discover more than it was proper for humanity to know. They also shared a barely suppressed anger at what they saw as the arrogance of scientists, especially those who believed in Bacon's premise that eventually man would fully understand and exploit the mysteries of the universe.

The third and essentially new component in the eighteenth-century criticism of scientists was the fear that science might actually succeed in deriving a self-sufficient, purely mechanistic system, with no moral dimension and no need of God. Newtonian physics, by explaining the movement of all bodies, terrestrial and celestial, great and small, fostered the expectation that all scientific explanations could be couched in terms of particles and atoms moving in space—the so-called billiard-ball universe of Pierre-Simon Laplace, who claimed that, given the original position and momentum of every particle in the universe, he could predict every subsequent event without recourse to the God "hypothesis."[2] God could be accommodated in the cosmic drama only in the walk-on, walk-off part of First Cause. Although, as we saw in the preceding chapter, Newton himself was far from subscribing to this view, after his death the Newtonian system was increasingly interpreted almost exclusively in mechanistic terms. Thus, if primary properties constituted the only reality, the nonmeasurable qualities of beauty, truth, and goodness were, by implication, unreal, nonexistent, and irrelevant. It was no wonder that eighteenth-century moralists feared the advances of science on moral as well as theological grounds and attacked the enemy with the strongest ammunition that an age steeped in satire could contrive.

Alexander Pope's social satire *The Dunciad* (1728–42) includes an attack on the intellectual arrogance of contemporary scientists who, Pope believed, were replacing God with scientific laws. One of the scientists in this poem declaims:

> All-seeing in thy mists, we want no guide,
> Mother of Arrogance, and Source of Pride!
> We nobly take the high Priori Road,
> And reason downwards till we doubt of God:
> Thrust some Mechanic Cause into his place:
> Or bind in Matter, or diffuse in space.[3]

Pope's later *Essay on Man* (1733), while conceding the religious value of science in illustrating the wisdom of God in the Creation, points out the limitations of natural philosophers, who cannot explain the causes of the wonders they observe, and warns against the pride of those who profess to know the secrets of God. Because of its immense scale and because of the double connotation of the phrase "the heavens," astronomy in particular came to epitomize for

Pope the arrogance and pretensions of science. "Is the great chain, that draws all to agree, / And drawn, supports, upheld by God, or thee?"[4]

Whereas Pope, in rebuking Newton's contemporary scientists, had exonerated Newton himself,[5] Jonathan Swift made no such concessions. He wrote, "It is hard to assign one Art or Science which has not annexed to itself some Fanatic Branch: such are, The Philosophers' Stone; The Grand Elixir; The Planetary Worlds; The Squaring of the Circle." He ridiculed the pretentious and absurd claims of those who, like Bacon, set out to take all knowledge for their province: "The whole school of the Greshamites [the Royal Society] are too wild in their claims; the whole realm of human knowledge is too broad for human nature to conquer."[6]

It was already clear in his early work *The Battle of the Books* (1697) that Swift's sympathies were with the values of the ancients rather than the moderns, represented chiefly by those sympathetic to the new science, but by the time he published *Gulliver's Travels* (1726), humor had changed to bitter condemnation. In part this reflects the changed social status of scientists, who by the eighteenth century were no longer isolated eccentrics but economically and politically influential figures in the state.[7] Yet Swift saw that, whatever their intellectual attainments, they were prey to the same moral failings as other men, and thus the very success and power they enjoyed made them potentially more dangerous and sinister. These powerful scientists Swift called "Projectors," a term loaded with contemporary significance. Projectors were speculators whose fantastic schemes threatened innocent people with financial ruin, and the more frequent use of the word pertained to precarious commercial speculations such as the notorious "South Sea Bubble." By applying the term to scientific projects, Swift emphasized the widespread destruction that could ensue from the ill-considered schemes of irresponsible scientists.

Like Pope's *Essay on Man*, *Gulliver's Travels* is concerned with man's pride and reaffirms, in effect, that the proper study of mankind is not nature but man, regarded in moral terms. In book 3, which describes the ivory-tower abstractions of Laputa and the ludicrous excesses of Balnibarbi, Swift deals with the pretensions to reason and intellectual pride of the natural philosophers. As examples he focuses on astronomers, who typify for him a perverted preoccupation with the most remote areas of the physical world to the exclusion of their own moral and spiritual condition.

What does Swift have against astronomy? In the first place, for Swift, a clergyman, it exemplified the sin of pride, insofar as it represented a desire to transcend human limitations and aspire to the wisdom of God by trying to understand the universe. The well-documented parallels between the experiments of the Academy of the Projectors in Lagado and those reported in the *Philosophical Transactions* of the Royal Society left no doubt about the target of Swift's satire.[8] This theme is developed with increasing venom throughout book 3 as Gulliver encounters the flying island of Laputa, whose carefully recorded dimensions are proportional both to those of the earth as calculated by Newton and to the terrella[9] of William Gilbert's experiments in magnetism.[10] Laputa thus symbolizes the astronomers' determination to force nature into a rigorous mathematical framework of their own devising and their formulation of universal laws that were only one step from usurping the role of God. It is significant, therefore, that the Laputans are convinced that they can hear the music of the spheres (traditionally heard only by angels) and that the music they themselves produce in response to this is a cacophony, indicating their delusion.

Related to this aspiration to acquire the knowledge of absolutes was the scientists' emphasis on abstract thinking, their obsession with other planets rather than their own, and their consequent contempt for everyday experience. The Laputans no longer even touch the ground but live on a flying island. Their minds "are so taken up with intense speculations that they can neither speak nor attend to the discourses of others, without being rouzed by some external taction upon the organs of speech and hearing,"[11] a parody of John Locke's idea that knowledge is acquired only through stimuli impacting on the senses. Such is their preoccupation that the Laputans, having no instinct of preservation, have to be forcibly restrained from falling over a cliff. It is not accidental that their chief interests are mathematics and music, the main occupations of the members of the Royal Society. As the chapter title "A Phenomenon Solved by Modern Philosophy and Astronomy" forewarns, the Laputans' abstract speculation about flight and its allegedly practical application in the complicated movement of their flying island "in an oblique Direction" is an intended parody of Newton's "resolution of any one direct force ... into two oblique forces"[12] in the plotting of planetary orbits and the paths of comets.

In Swift's satire, this preoccupation with the abstract is exacerbated by the Laputans' obsession with elaborate scientific instruments, another gibe at the Royal Society. Gulliver's Laputan tailor constructs an ill-fitting suit because he insists on taking his client's measurements with "scientific" instruments—a quadrant and compass—and on surveying everything, terrestrial and lunar, in the process. Again, because of their contempt for practical geometry, which they despise as "vulgar and mechanic," the Laputans' houses are as ramshackle as their clothes, for, like many of Newton's calculations, their complex mathematics cannot be followed by the builders. Some of the effects of their willful abstraction are far from harmless. The flying island proceeds at the whim of the king, who is either oblivious of the damage done to his subjects down below in Balnibarbi when the island comes between them and the sun or else uses this as a threat to suppress them. In Balnibarbi itself the Projectors, fresh from training in Laputa, look with one eye upward to the stars and with the other into their own minds, and in their mania to usher in a utopian future of their own design they take no account of nature. They have devastated the once fertile land with projects designed to change the naturally useful into its useless opposite—producing naked sheep, soft marble, and tangible air; sowing the fields with chaff; reducing excrement to its original food; and petrifying the hooves of a living horse. This too is an attack on the experiments conducted by the Royal Society, a more vitriolic version of Shadwell's gibes at the virtuosi.

Clearly Swift, so far from adopting the optimism of eighteenth-century science, believed pessimistically that scientific discoveries would be, at best, a temporary benefit: even if they led to increased wealth, this would only hasten luxury, corruption, and social decay. He also attacked the notion that knowledge, especially of astronomy, would bring peace and happiness. The Laputans, who have cataloged ten thousand stars, discovered two moons of Mars, and charted the path of ninety-three comets, live in a state of perpetual anxiety, "never enjoying a minute's peace of mind. . . . [They fear] that the earth by the continual approaches of the sun towards it, must in course of time be absorbed or swallowed up. . . . And that the next [comet], which they have calculated for one and thirty years hence, will probably destroy us. . . . [So] that they can neither sleep quietly in their beds, nor have any relish for the common pleasures or amusements of life" (175–76). The passage includes allusions to the predicted return of Halley's comet; to the contemporary fascination with sunspots, seen here as a pox on the face of the sun; and to the obsession with cataloging the increasing number of stars observable with telescopes of ever-bigger aperture.

Although only book 3 of *Gulliver's Travels* explicitly focuses on scientists as the representatives of intellectual pride, it is important also to see it in the context of book 4, where the rationalist, horselike Houyhnhnms represent a more generalized image of disembodied intellect. The Houyhnhnms have many admirable qualities, particularly when contrasted with the barbaric Yahoos, but Swift ridicules the notion that we can escape our human limitations and become Houyhnhnm-like. The more Gulliver tries to do so, the more ridiculous he becomes. There is, moreover, a sinister suggestion of the cruelty and inhumanity contingent upon the passionless state of the Houyhnhnms, motivated, as they are, by pure reason and impervious to emotional considerations, a view that was to be developed extensively by the Romantic writers of the next century. Thus, ultimately, Swift's moral purpose, like Pope's, is to redirect mankind from excessive and optimistic reliance on rationalism and the pride of the intellect, and to reaffirm the limitations of the human condition, in the face of which the only proper response is humility. That he should have chosen contemporary science to typify intellectual arrogance is an indication of both the increasing influence of scientists and the extent to which natural philosophy had come to represent the cutting edge of contemporary knowledge.

The next stage after the acute anxiety experienced by the Laputans over the projected arrival of comets is madness, and eighteenth-century scientists did not escape this charge. In the final plate of his series of engravings *The Rake's Progress* (1735) William Hogarth depicted a scene from Bedlam (Bethlehem Hospital for the reception of lunatics), where two of the inmates have been rendered insane by their pursuit of astronomy or its application (see fig. 9). One astronomer is shown peering at the ceiling through a roll of paper that he imagines to be a telescope, while another inmate, who has drawn on the wall a ship, the earth, the moon, and various geometric figures, is evidently attempting to square the circle (calculus) and to use astronomy to calculate longitude—another pressing preoccupation of the period.

A similar delusion about his power over the heavens besets the nameless astronomer in Samuel Johnson's moral essay *Rasselas* (1759). This learned man "has spent forty years in unwearied attention to the motions and appearances of the celestial bodies, and has drawn out his soul in endless calculations."[13] His long years of isolation and contemplation have persuaded him that the weather and the seasons are dependent on him, and the strain of such a responsibility has proved too much for his mind. Johnson diagnoses his case as resulting from too much solitude, with consequent overindulgence in the imagination. (This charge is particularly interesting in the light of the Romantics' accusations, less than half a century later, that science destroyed the imaginative faculty.) Since astronomy demands many hours of uninterrupted solitude, during which such fantasies can take root, it would seem almost impossible, according to Johnson, to combine astronomy and sanity.

Figure 9. Bedlam: A Rake's Progress, Plate VIII, by William Hogarth, 1735. Hogarth's engraving of the Bethlehem Hospital (Bedlam) for mental patients shows two victims of science. The man peering at the ceiling through a roll of paper imagines he is an astronomer. Behind him, a man who has drawn a ship, the earth, the moon, and various geometric patterns, has become mad attempting to work out a means of calculating longitude. Sir John Soane's Museum, London. Wikimedia Commons.

Severe as this eighteenth-century moral censure was, it paled beside the fundamentally new and much more radical attack on the Newtonian system mounted by William Blake, who accepted none of the assumptions of the scientific method. For him, Newton, revered not fifty years before as the champion of religion, shared with Bacon and Locke the doubtful distinction of constituting an infernal trinity, and his dismissal of scientists in general as deficient human beings was to be reiterated and elaborated upon by Romantic writers for much of the nineteenth century.

CHAPTER 6

Inhuman Scientists
The Romantic Perception

> May God us keep
> From single vision and Newton's sleep!
>
> —Blake

The Culture of Mechanism

The eighteenth-century satirists had attacked scientists on the grounds of arrogance, for believing that they could transcend human limitations. But their criticism carried little weight in view of the lucrative spin-offs from science that drove the Industrial Revolution. In theoretical terms as well, the unprecedented success of Newtonian physics in explaining the behavior of physical entities, from billiard balls to planets, seemed set to realize Descartes's dream that reason, exemplified in mathematics, would ultimately resolve all problems facing mankind.

In this climate, it was inevitable that a mechanistic model of the animate world would evolve. Descartes had proposed that the bodies of animals could be described and understood as complex machines, and although he was careful not to say so explicitly, it is clear that he believed human beings were no exception, for his description of the human body closely parallels accounts of the automata of his time.[1] Another French philosopher, Julien Offray de La Mettrie, had no such reservations. Using an analogy with complex clocks, he declared in his provocative *L'Homme machine* (1747) that all animals, including human beings, are only complex animal-machines.[2] In England the philosopher John Locke, Fellow of the Royal Society and an admirer of Boyle and Newton, had already concluded that all thought and knowledge were derived from external sensations registered by the five senses and subsequently built up into a system of knowledge by an essentially mechanistic process. Scientific procedures became the new orthodoxy, infiltrating every branch of knowledge from economics to political thought, providing a new vocabulary for literature as well as science and influencing even the conventions of eighteenth-century European art.[3]

The attack on this entrenched supremacy of science by the Romantic writers was more radical than that of their predecessors. Their case against Enlightenment science hinged on its reductionist philosophy, for they saw that limiting the universe to the sum of separate, measurable entities limited man as well, since it denied the validity of emotions, nonrational experiences, spiritual longings, and individuality. Consequently, they portrayed scientists not as arrogant overreachers but as deficient in human qualities. The villains of Romanticism were not alchemists or shaman figures but those who reduced the world to a mechanism and isolated themselves from human relationships and the healing power of nature. These remain central characteristics of the mad scientist.[4]

As their alternative to the mechanistic Newtonian universe the Romantics proposed a universal life force, with which man could communicate. In fact, there was much in contemporary science to support this vitalist approach. The cutting edge of science was now chemistry and the little-understood phenomena of electricity and magnetism, areas where the mysterious concept of force at a distance was observed but not yet accounted for. The work of Henry Cavendish and Joseph Priestley in England on electricity and the composition of gases, of Antoine Lavoisier in chemistry in France, and Benjamin Franklin's experiments with electricity in America and Galvani's in Italy seemed, to the popular mind, not far removed from alchemy and magic. Electrolysis appeared to fulfill the alchemists' dream of depositing pure gold, while Galvani's experiments with static electricity applied to a dead frog's leg, causing it to twitch, were suggestive of restoring, if not creating, life and held promise of breaking down the barriers between living and nonliving. In chemistry the theory of "elective affinities," evolved to account for the bonding of chemical elements to form stable compounds only with specific others, had clear analogies with human behavior, suggesting a unitary principle applicable to both the animate and the inanimate world.

The German Romantics

Despite the similarity of scientific knowledge throughout Western Europe, the attitudes toward science and scientists adopted by Romantic writers show national variations. In Germany the philosopher Friedrich Schelling developed a *Naturphilosophie*, based on the premise that nature was one huge living organism, and hence the true goal of science was to discover the *Weltseele*, or world-soul, of this organism. He suggested that this "soul," or *Geist*, was characterized by polarity between opposing forces that produced a "cosmic heartbeat." Magnetic and electric polarity and chemical "affinities" were all interpreted as manifestations of this universal *Urpolarität*, which could be traced throughout nature from crystal structures to the spiritual states of man. Goethe famously developed this idea in his novel *Die Wahlverwandtschaften* (*Elective Affinities*) (1809), which explores the notion that human relationships are governed by the laws of chemical affinity. By contrast, the term *mechanisch*, associated with the classical mechanics of the Newtonian system, became, for Schelling and the circle of Romantic writers who collected around him in Jena, *the* polemic adjective of abuse, the antithesis of their organic imagery.[5]

However, many of the German Romantics also had a background in science. Novalis had studied mineralogy, physics, chemistry, mathematics, and physiology; Friedrich Schlegel had studied physics; Goethe had specialized in botany and read widely in all the contemporary sciences; and J. W. Ritter was a pharmacist, physiologist, chemist, and physicist. In their diverse but comparable ways, these writers believed not only that science yielded material for poetry but equally that their Romantic perspective would lead to a breakthrough in science. Goethe's theory of light and color perception, based on his own observations that shadows had color and the phenomenon of the "after image," supported his belief that the observer played an active part in the recognition of colors and was not merely a passive recipient of light particles falling upon the eye, as Newton and Locke had proposed. So passionately did Goethe believe this that he conducted a lifelong battle, at considerable personal expense, against the Newtonian theory of light.[6] Ritter's interest in electrophysiology led him to study the response of mimosa leaves to electric stimuli, a careful and highly regarded piece of experimental research triggered by his Romantic belief that the whole of nature partook of sensitivity. Similarly, Ritter's discovery of ultraviolet rays in the spectrum of the sun, was a direct result of his theory of polarity, which for aesthetic reasons required wavelengths to balance the infrared rays at the other end of the spectrum.

Given this close involvement with contemporary science and the interest of the German Romantics in portraying psychological and philosophical truths about scientists, it is not surprising that the archetype of the isolated alchemist, epitomized in the Faust character pursuing truth and power, was revived and reinterpreted as a noble and heroic figure.[7] In these Romantic reenactments there is no suggestion that knowledge is evil, only that it is insufficient to satisfy the aspirations of the heroic genius. Friedrich von Klinger, whose play *Sturm und Drang* (1776) gave its name to the artistic period, also published *Fausts Leben, Taten und Höllenfahrt* (1791). His protagonist embarks on an intellectual enquiry into cosmic principles and purpose, but by the end of book 1 he has discarded this program in favor of an investigation of evil and injustice. Like the traditional Faust he is a rebel, but less against the restrictions of the human condition than against the moral limitations of the universe. His hubris resides in his assumption that he himself can right injustices. But so far from being condemned for this, he is portrayed as a moral idealist, albeit misguided and doomed to failure, a Promethean figure denouncing the immorality of the Creator of such a flawed universe.[8] Similarly, in Friedrich (Maler) Müller's play *Fausts Leben dramatisiert* (1778) the craving for knowledge is only one of many goals, and again the protagonist is presented as a heroic character, whose desire to transcend his humanity is endorsed.

The most famous Romantic Faust is of course Goethe's (part 1, 1805; part 2, 1832). An eminent scholar who has mastered all the separate branches of learning, this Faust shares the alchemists' obsession with discovering the unifying principle of nature: "grant me a vison of Nature's forces / That bind the world, all its seeds and sources / And innermost life."[9] In his opening monologue on the vanity of learning Faust explains that he has turned to magic because he believes it will disclose more of the mystical secrets of nature than orthodox science does. Knowledge qua rational explanation of the theoretical laws of the universe is of little interest to him, because from the beginning of the play he already knows everything science has to teach, and in his second conversation with Mephisto he describes himself as "purged the lust for knowledge from my soul" (p. 54, l. 1768). Rather, he seeks an immediate, intuitive understanding of nature and ultimate experience of reality through the senses. Thus Goethe transforms the traditional Faust's desire for knowledge into a quest for a spiritual unity with nature, a kind of pansophism, and aligns this quest with sensuous experience. He thereby manages to accommodate two seemingly contradictory traits of the medieval Fausts—their desire for limitless knowledge and their apparently debased actions. Faust himself articulates this dualism in his lament, "In me there are two souls, alas, and their / Division tears my life in two" (p. 35, ll. 1112–13), and the uniting of these contraries in the drama becomes, itself, symbolic of the wider Romantic quest for unity out of diversity.

Goethe thus dispensed with the earlier trappings of the Faust story, presenting instead a scientist who seeks confirmation of his theoretical ideals not in the laboratory but in the experiences of life itself. As J. W. Smeed points out, "What distinguishes the *Sturm und Drang* Fausts from Lessing's Faust and, to some extent, from the Faust of the original chapbook is that they do not so much want to *know* more than other men as to *be* more than other men."[10]

German Romanticism produced more than a dozen works dealing with the Faust character, but Goethe was the last to enunciate the ideal of a genuinely universal worldview, one embracing the intellect, the emotions, and transcendent experience. After Goethe, the rapid growth of information in all scientific disciplines led to specialization in science that is reflected in the literature. Experimental scientists and natural philosophers are increasingly depicted with divergent interests, and there is no longer any expectation that a unifying *Weltanschauung* might be possible—or even desirable.

The English Romantics

Unlike their German counterparts, the English Romantic poets, few of whom had any training in science,[11] showed little faith in scientists' ability to engage with an organic, feeling model of the world. Hence none of these writers created a character who could reasonably be regarded as a scientist.[12] Rather, they emphasized the gulf between intuition and emotion on the one hand and scientific rationalism and mechanism on the other.

These national differences reflect the divergent development of science in England and Germany. In England, technology and, by extension, science were closely associated with the Industrial Revolution, which, besides producing immense national wealth, had also brought in its train destruction to large tracts of countryside and appalling conditions for factory workers. It is therefore not surprising that the English Romantic writers identified science with technology, with the ugliness of the factories desecrating nature, and with the reification of the individual as an extension of the machine at which he worked. In Germany and France, on the other hand, the Industrial Revolution came much later, and for their Romantic writers, science and its practitioners were more closely associated with philosophical theories.

The attack of the English Romantic writers was directed against both the content and the methodology of science. In place of scientific materialism and reductionism they affirmed a transcendent, metaphysical reality and an organic view of nature in which the whole could not be reduced to the sum of its parts, nor the animate explained in terms of the inanimate.[13] Related to their protest against mechanism was their rejection of rationalism as the only valid means of knowing. On the contrary, they privileged nonrational experiences antithetical to the procedures of science—visions, intuition, the subconscious as accessed in dreams, "divine madness," *the furor poeticus*,[14] and, above all, the imagination.

Virtually all these aspects of English Romanticism can be seen as reactions against the figure of Newton, who necessarily represented for the British the prototypical scientist. With his compact, internally coherent model of the universe as a machine, explicable in mathematical terms, he epitomized the materialistic, rationalist view of the world. Although, as we saw in chapter 4, Newton was by no means a thoroughgoing mechanist and was even, in his own way, a mystic and a visionary,[15] his public image was synonymous, for later generations, with analytical and reductionist procedures, which were anathema to the Romantics.

William Blake, the earliest and the most extreme of the English Romantic poets, was also Newton's most uncompromising critic. In the eighteenth century Newton had been welcomed by the deists as demonstrating the existence of God—who else, they argued, could have invented a system so mathematically perfect and artistically satisfying as that encompassed in Newton's laws?[16] But to Blake, Newton, with his three-dimensional, mechanistic model of the universe, was a dangerous heretic, blinded by materialism from seeing the complexity of truth:

> Now I a fourfold vision see,
> And a fourfold vision is given to me.
> ....
> ... May God us keep
> From single vision and Newton's sleep![17]

Blake bracketed Newton the archmechanist with Bacon, the advocate of experimentalism, and Locke, representing the philosophy of the five senses, as an infernal trinity of intellectual arrogance:

But the Spectre, like a hoar-frost and a mildew, rose over Albion,
Saying: "I am God, O Sons of Men! I am your Rational Power!
Am I not Bacon and Newton and Locke, who teach Humility to Man,
Who teach Doubt and Experiment? and my two wings, Voltaire, Rousseau?"[18]

In the *Opticks*, Newton had explained light as a stream of particles, which Blake equated with the atomic theory of Democritus. Against this reductionism he asserts the primacy of a transcendent, spiritual reality:

You throw the sand against the wind,
And the wind blows it back again.

And every sand becomes a Gem
Reflected in the beams divine;
Blown back they blind the mocking Eye,
But still in Israel's paths they shine.

The Atoms of Democritus
And Newton's Particles of Light
Are sands upon the Red Sea shore,
Where Israel's tents do shine so bright.[19]

Yet Blake also credited Newton with an extraordinary feat of the imagination. His engraving *Newton* (1795) reflects the complexity of his own response (see fig. 10). It shows a muscular youth measuring with dividers the base of an equilateral triangle drawn on a scroll. The curve of the youth's bent back parallels the arc inscribed in the triangle; the muscles of his back are a row of rhomboids ending in two triangles at the hip; and the triangular forms of his right leg, left wrist, and stretched fingers of both hands mimic the angle of the dividers and of the drawn triangle, implying that his humanity has been reduced to a mathematical parody of itself.[20] However, this youthful Newton expresses great intensity and energy as he draws his diagram, not on a stone tablet or in a book, but on a scroll, which in Blake's iconography invariably signifies imaginative creation.[21] In the final Apocalypse represented in his poem *Jerusalem* (1804), Bacon, Newton, and Locke appear in the heavens as the foremost representatives of Science, counterbalancing Milton, Shakespeare, and Chaucer as the champions of Art,[22] for Blake regarded true science as eternal and essential; what he rejected was a science divorced from humanity.[23]

Figure 10. Newton, by William Blake, 1795. For Blake, Newton symbolized the dangers of rationalism. Obsessively absorbed in his diagram, which his body has come to emulate, Newton thinks the whole of life can be measured mathematically. Tate Gallery, London. Wikimedia Commons.

William Wordsworth showed a similar ambivalence toward scientists. On the one hand he elevated Newton, or at least the Newton represented in Roubiliac's statue in Trinity College, Cambridge, as an imaginative discoverer of far-off worlds, making him, in effect, an honorary Romantic hero, if not poet:

Newton with his prism and silent face,
The marble index of a mind for ever,
Voyaging through strange seas of Thought alone.[24]

But he regarded science as an essentially lonely pursuit in contrast to the social involvement of poetry: "If the labours of men of Science should ever create any material revolution, direct or indirect, in our condition and in the impressions which we habitually receive; the Poet . . . will be ready to follow the steps of the Man of Science . . . carrying sensation into the midst of the objects of Science itself."[25] Indeed Wordsworth is more often remembered for condemning the man who would "peep and botanise on his mother's grave" and the scientific dissociation of subject and object: "Our meddling intellect / Misshapes the beauteous forms of things— / We murder to dissect."[26]

The English painter Joseph Wright of Derby vividly captured this spectrum of contemporary responses to science in his painting *An Experiment on a Bird in the Air Pump* (see fig. 11). The man who presides over the demonstration, designed to show that air is really necessary to support life, has the obsessive expression and anachronistic garb suggestive of an alchemist totally preoccupied with his experiment. His young assistant on the right stands poised ready to reintroduce air into the jar so that the bird may revive, but in order to dramatize the effect, he delays this until the critical moment when the bird is on the point of death. The man in the foreground, who carefully records the time taken for the bird to lose consciousness, is concerned only with the mathematical results. By contrast, the children, whose values are always, for the Romantics, a better guide to truth, are concerned chiefly for the suffering of the bird.[27]

Figure 11. *An Experiment on a Bird in the Air Pump*, by Joseph Wright of Derby, ca. 1767–68. National Gallery, London. Wikimedia Commons.

Percy Bysshe Shelley, although sympathetic to many aspects of science and particularly of chemistry, also rejected the imperative to dominate nature, on the grounds that, in so doing, man would diminish himself: "The cultivation of those sciences which have enlarged the limits of the empire of man over the external world has, for want of the poetic faculty, proportionally circumscribed those of the internal world; and man, having enslaved the elements, remains himself a slave."[28] For similar reasons the American poet Walt Whitman, a late Romantic, expressed contempt for the astronomer, who, imprisoned physically in his lecture room and mentally by charts and figures, presumes to explain the immensity of the stars:

> When I heard the learn'd astronomer;
> When the proofs, the figures, were ranged in columns before me;
> When I was shown the charts and the diagrams, to add, divide, and measure them;
> When I, sitting, heard the astronomer, where he lectured with much applause in the lecture-room,
> How soon, unaccountably, I became tired and sick;
> Till rising and gliding out, I wandered off by myself,
> In the mystical moist night-air, and from time to time,
> Look'd up in perfect silence at the stars.[29]

The Irish poet W. B. Yeats, who regarded himself as one of the "last Romantics," expresses a similar view:

> Seek, then,
> No learning from the starry men,
> Who follow with the optic glass
> The whirling ways of stars that pass—
> Seek, then, for this is also sooth
> No word of theirs—the cold star-bane
> Has cloven and rent their hearts in twain,
> And dead is all their human truth.[30]

The Prose Writers

The deep rift between science and Romanticism is reflected not only in poetry but in nineteenth-century prose. Dickens and Carlyle, who saw spiritual values being increasingly eroded by the march of industrialism and the economically profitable alliance between science and technology, depicted the scientist as lacking in feeling and avoiding human relationships. "Men are grown mechanical in head and in heart, as well as in hand," lamented Carlyle,[31] and Dickens makes a similar point in his satire on the British Association for the Advancement of Science, which met for the first time in 1831. The "great scientific stars, the brilliant and extraordinary luminaries" gathered there, are characterized by their complete insensitivity to the suffering of a pet dog, which they proceed to vivisect. The "scientific reporter" records: "You cannot imagine the feverish state of irritation we are in, lest the interests of science should be sacrificed to the prejudices of a brute creature, who is not endowed with sufficient sense to foresee the incalculable benefits which the whole human race may derive from so very slight a concession on his part."[32]

Sometimes, although rarely, the unfeeling scientist in literature is pitied and permitted to reform. Browning's long dramatic poem *Paracelsus* (1835), based on the life of the Swiss alchemist, depicts a scientist who, in his proud and single-minded pursuit of knowledge, sacrifices everything else in life only to realize at the moment of death that he has omitted the one important factor, love. Similarly, in Dickens's story "The Haunted Man" (1848) the introspective chemist Redlaw, who is full of regrets for the past and emotionally unfulfilled in the present, is reclaimed by the simple, loving Milly Swidger, a Dickensian "angel in the house."

But more often there is no reprieve for the scientist, whether alchemist or chemist, who has devoted his life to knowledge, to the exclusion of the emotions. In such accounts, mathematicians feature prominently as exemplars of the dehumanizing process. This is comically expressed in the poem "The Mathematician in Love" (1874) by W. J. M. Rankine,[33] himself an engineer and mathematician. The mathematician is mocked for his inability to relate emotionally to a young lady, and his obsession with formulas is duly punished in the living world, where emotions rather than abstractions are the accepted currency.

More usually the treatment is tragic rather than comic. Two stories of Fitz-James O'Brien explore the psychology of an obsessed scientist. In "The Diamond Lens" (1858) Linley, a microscopist, is haunted by the desire to construct a perfect microscope, one that will provide hitherto unimagined magnification. To acquire the diamond necessary for such a lens, he kills a fellow lodger and then labors in solitude for months to produce the microscope. With the resulting lens he discovers in a drop of water an infinitesimal world inhabited by a beautiful girl, whom he calls Animula and with whom he falls in love. This love is hopeless, since the lovers are separated by that very barrier of magnitude the scientist had sought to transcend. He is released from his obsession only by the extinction of the girl when the water droplet evaporates, whereupon he is declared insane. The story combines two traditions: the introverted, isolated scientist who stops at nothing to pursue his research and the Faustian overreacher, whose nemesis arises from his very success in constructing the microscope. In "The Golden Ingot" (1858) O'Brien depicts another scientist, Blacklock, who is obsessed with the alchemists' goal of transforming base metals into gold. When one day he finds a gold ingot in his crucible, he believes he has at last succeeded, until he discovers that his devoted daughter, intending to bring him happiness, has pawned everything to purchase the ingot and has secreted it there, whereupon he dies of a stroke.

Many of the best-known fictional scientists of this type destroy not only themselves but those close to them as well. Balthazar Claës of Balzac's *La recherche de l'absolu* (1834) is one of the most complex of these studies. Although Balzac's major interest is the psychological, almost clinical, study of a genius and the effect on his family, the underlying moral expresses the Romantic view that preoccupation with science atrophies the emotions that sustain personal relations and social responsibilities. Claës's devoted wife Josephine pleads for an emotional response when she tells him, "Science has eaten away your heart."[34] His answer, a piece of unwitting self-parody, is to redefine feelings in the current chemical term *affinities*. Claës is portrayed as a latter-day alchemist (the local townspeople mockingly call him "Claës the Alchemist"), with all the stereotypical trappings—his physical appearance, his alchemist's laboratory, and an assistant who is popularly supposed to be akin to the devil. Balzac also includes a Mephistopheles figure in the Polish stranger, through whose persuasive agency Claës effectively sells his soul to the devil as he becomes obsessed with finding a method to produce diamonds by transmutation, ruining his family in the process. Balzac's analysis is more complex, however, than this indicates. The word "absolu" in Balzac's title suggests a transcendent reality beyond the analytical procedures and chemical terms of the particular experiment. Thus although Claës lacks feelings toward his family, an aura of grandeur attaches to him, not only in his acknowledged genius but in his devotion to an ideal, his self-sacrifice, and the insults he endures. Like Browning's Paracelsus, Claës discovers the secret of the absolute only at the moment of death, for it is a metaphysical rather than a physical truth.

Two of the stories in E. T. A. Hoffmann's *Nachtstücke* (1817), "Ignaz Denner" and "Der Sandmann," explore the destruction of innocent, gullible victims by evil scientists.[35] In "Der Sandmann" the title character, Dr. Coppelius, overtly a lawyer, is a closet alchemist who exerts extraordinary influence over the protagonist's father. As the child Nathanael watches from his hiding place with fascinated horror, his father and Coppelius, in a setting heavily suggestive of an alchemist's laboratory, engage in an experiment intended to produce an automaton (representative of mechanistic science). On discovering the child, Coppelius threatens to extract his eyes, and the boy faints with terror. A year later, Coppelius is implicated in the death of Nathanael's father from an explosion in their laboratory. Thus, from the beginning he is associated with the sinister, the secretive, and the death of innocent people. His experiments are also mysteriously linked with the removal of natural eyes and the substitution of artificial vision. This latter theme is amplified when Coppelius subsequently returns, disguised as Coppola, an Italian hawker of scientific glasses—barometers and telescopes. The latter distort the vision of those who look through them, bewitching them with an illusory image, so that, symbolically, they have indeed lost their natural sight. Gazing through such a telescope Nathanael sees the beautiful automaton Olimpia, technically flawless but lacking emotions and spontaneity. Infatuated with Olimpia's apparent perfection, he dances with her at a ball but is perturbed to discover how stiffly she holds herself and how mechanically she dances. She plays and sings like a clockwork model, and when Nathanael bends to kiss her, her lips are ice cold.[36] Her mechanical nature is finally exposed when he hurls her to the floor, causing her dismemberment. But Nathanael is also destroyed. Under the spell of Coppola's telescope, he flings himself to his death from a tower, itself a symbol of isolation from both society and nature. Hoffmann thereby reinforces his moral that the products of scientific obsession, whether in the form of an automaton or the fabricated glass that distorts nature's truth, finally destroy life itself.

The American novelist Nathaniel Hawthorne also distrusted rationality, regarding the emotions as a truer guide to morality and spiritual and mental health. In this he may have been reflecting the frontier values of nineteenth-century America, where settlers experienced a world that was not orderly and rational but violent and unpredictable. In Hawthorne's novels and stories excessive cultivation of the intellect is always destructive, not only of the one so obsessed but also of those whose lives intersect with his. In a *Notebook* entry of 1844 Hawthorne wrote: "The Unpardonable Sin might consist in a want of love and reverence for the Human Soul; in consequence of which, the investigator pried into its dark depths, not with a hope or purpose of making it better, but from a cold philosophical curiosity—content that it should be wicked in whatever kind or degree, and only desiring to study it out. Would not this, in other words, be the separation of the intellect from the heart?"[37]

Hawthorne was interested in psychology and in the many contemporary fringe sciences, from phrenology to mesmerism, and his scientists are symbolic rather than realistic, most being cast in the alchemist mold. Aylmer, protagonist of "The Birthmark" (1845), is a chemist whose affection for his beautiful and loving wife is marred by his obsession with her birthmark, which he increasingly sees as abhorrent and determines to erase it in order to make her perfect. Her subsequent death after taking the potion he expects will remove the mark is both literal and symbolic. This sacrifice of life and emotions for an abstract ideal is seen as the end result of his arrogant desire to explore what "seemed to open paths into the region of *miracle*" (my italics).[38] Science has become a religion for Aylmer, who has "faith in man's ultimate control over Nature" (12). His inability to accept the "symbol of imperfection" (15), the birthmark, with its implications of original sin and the human condition, represents his determination to perfect nature through science. Here Hawthorne makes a pertinent comment on the meliorist claims of nineteenth-century scientists, who assumed they could eliminate sin along with disease and usher in a utopia cleansed of physical and moral blemishes.

In "Rappaccini's Daughter" (1844) Hawthorne creates another cold and inhuman alchemist figure. Rappaccini is "as true a man of science as ever distilled his own heart in an alembic."[39] Like Aylmer, his physical attributes have atrophied as a result of his intellectual obsession, and again this condition is associated with heartlessness. He is "a tall, emaciated, sallow, and sickly-looking man, dressed in a scholar's garb of black. He was beyond the middle term of life, with gray hair, a thin gray beard, and a face singularly marked with intellect and cultivation, but which could never, even in his more youthful days, have expressed much warmth of heart" (180). Rappaccini has cultivated a garden of exquisite but poisonous flowers, not to enjoy their beauty, but in order to study their fatal effect on other forms of life. His colleague Baglioni remarks: "I know that look of his: it is the same that coldly illuminates his face as he bends over a bird, a mouse, or a butterfly, which, in pursuance of some experiment, he has killed by the perfume of a flower" (192). By a process of gradual assimilation, his lovely daughter Beatrice has become immune to the fatal perfume emanating from the flowers, but as a result she too exudes the toxin, so that whatever she breathes upon dies. As an extension of his experiment with insects, Rappaccini encourages a student, Giovanni, to woo his daughter. On finding that he, too, has become poisonous by association with Beatrice, Giovanni brings an antidote to cure them both so they can flee from her father, but, rather than curing her, the potion kills her. The characterization of Rappaccini clearly expresses Hawthorne's mistrust of the scientific mind and his conviction that its exclusive cultivation leads to death.

Characteristically, in Hawthorne's work as in much Romantic writing, the rationalist head is contrasted with the emotional heart. This is rendered almost diagrammatically in Roger Chillingworth, the learned physician cum alchemist of Hawthorne's best-known novel, *The Scarlet Letter* (1850). Like Faust and Frankenstein, Chillingworth has studied at a German university but also relies strongly on herbs and cures learned from the Indians and concocted in his alchemist's laboratory.[40] In the medieval tradition of witchcraft, Chillingworth, cold and unfeeling as his name suggests, becomes a "deformed old figure . . . stooping away along the earth" (193) as an index of his spiritual corruption. During his long absence at sea, when he is presumed dead, Chillingworth's young wife, Hester Prynne, has an adulterous affair with the much-admired minister Arthur Dimmesdale and conceives a child. This causes a scandal in the Puritan community, leading to her imprisonment and later ostracism. On his return Chillingworth is obsessed with discovering the identity of his wife's lover and taking his revenge. Hawthorne focuses on Chillingworth's behavior as he studies and assumes control over his victim in the manner of the scientist observing a passive specimen. For Hawthorne Chillingworth's great sin lies in coldly studying Dimmesdale and exploiting his psychosomatic dependence upon him in order to destroy him. Dimmesdale speaks for Hawthorne when he says, "That old man's revenge has been blacker than my sin. He has violated, in cold blood, the sanctity of a human heart" (212).

Hawthorne's scientists clearly owe their basic conception to the Faust and Frankenstein narratives, and like their prototypes, most began as idealists. Aylmer's life had formerly been devoted to lofty scientific ideals, and Chillingworth was once a dedicated physician intent on helping others. Like their predecessors, they all succumb to hubris, desiring to transcend the limitations of human knowledge. But Hawthorne imparts a specifically Romantic cast to the Faust myth by including the psychological implication of the effect on the human personality when the head dominates the heart.

The Romantic writers' reaction against rationalism and reductionism and their development of an alternative approach to the natural world spanned nearly a century; therefore inevitably their emphasis varied at different times and places. But there are certain constant factors that can be taken as categorizing the Romantic attitude toward natural science and scientists. These include a deep suspicion of abstraction, a rejection of materialism, an assertion of the importance of secondary characteristics and values, and rejection of mechanism in favor of organism, which in the last decades of the nineteenth century became formalized as the doctrine of vitalism.[41] In much Romantic writing scientists are characterized by their failure to embrace these principles, a failure that is depicted as so closely affecting their relationships with people and nature that it functions as an index of moral and spiritual ill health.

The Romantic image of the scientist as cold, inhuman, and unable to relate to others became one of the strongest influences on twentieth-century stereotypes of the scientist in both literature and film, as will be seen in chapters 13 and 14, but a particular version of this Romantic image is the archetypal character of Frankenstein.

CHAPTER 7

Frankenstein and the Creature

> Life and death appeared to me ideal bounds which I should first break through, and pour a torrent of light into our dark world. A new species would bless me as its creator and source; many happy and excellent natures would owe their being to me.... I thought that if I could bestow animation upon lifeless matter, I might, in process of time ... renew life where death had apparently devoted the body to corruption.
>
> —Mary Shelley

The Composition of *Frankenstein*

Although *Frankenstein* (begun in 1816, when its author, Mary Shelley, was just eighteen, and published in 1818) is clearly a Romantic work and indeed predates many of the works considered in the preceding chapter, it merits separate discussion because of the extraordinary influence it has continued to exert on subsequent presentations of the scientist. Frankenstein has become an archetype in his own right, universally referred to and providing the dominant image of the scientist in twentieth-century fiction and film and the media. Not only has his name become synonymous with any experiment out of control,[1] but his relationship with the monster he created has become, in the popular mind at least, complete identification: Frankenstein *is* the monster. The archetypal power of the Frankenstein story arises from the fact that in its essentials it was, according to its author, a product of the subconscious rather than the conscious mind and thus, in Jungian terms, draws on the collective unconscious.

The circumstances of the composition of *Frankenstein* are almost as well known as the story itself and have themselves inspired other fictional accounts, including a film, an opera, and a novel.[2] Yet it is worth stressing here that, in Shelley's own account, the story was produced by the concurrence of two specific factors: the need to produce a horror story and the account of an alleged scientific experiment. Mary and Percy Shelley, their baby son, and Mary's stepsister Claire Clairmont were spending the summer of 1816 near Geneva, where they were neighbors of the poet Lord Byron and his personal physician Polidori. As they were kept indoors by bad weather, Byron suggested that, to pass the time, they should each write a ghost story as entertainment. Mary records that she found great difficulty in thinking of a suitable plot until one evening when the others had been discussing the latest experiments allegedly conducted by Erasmus Darwin. The latter was said to have "preserved a piece of vermicelli in a glass case till by some extraordinary means it began to move with voluntary motion. Not thus, after all, would life be given. Perhaps a corpse would be reanimated; galvanism had given token of such things. Perhaps the component parts of a creature might be manufactured, brought together, and endued with vital warmth." After hearing this discussion, Mary's imagination produced a series of vivid images, which were to form the central scene of her novel. The esteemed Doctor Darwin was translated into "the pale student of unhallowed arts, kneeling beside the thing he had put together." This suggests that the very attempt to create life was already associated, at least in Mary's subconscious mind, with the demonic and the horrific. The problem of finding a subject for her story was instantly solved: "What terrified me will terrify others; and I need only describe the spectre which had haunted my midnight pillow ... making a transcript of the grim terrors of my waking dream."[3]

It is not difficult to supply reasons why the account of Darwin's alleged experiments should have had such a disturbing effect on Mary Shelley. As the youngest and least assured person present, and clearly intellectually overawed by the discussion (she tells us that she was "a devout but nearly silent listener"), Mary, who had only the preceding year lost her first child, a daughter born prematurely, and who had recently undergone a second, difficult confinement, would have felt emotionally distressed, even violated, by a discussion that not only abolished the role of the female in the creation of life but trivialized the process by reducing it to "a piece of vermicelli in a glass case." Unable to argue with the intellectual giants Byron and Shelley, she doubtless suppressed her disquiet, which erupted violently in her subsequent dream. What is more interesting for the purpose of this exploration of images is her immediate identification of the highly visual nightmare image of the attempt to create life with her earlier aim "to think of a story . . . which would speak to the mysterious fears of our nature" (8).

The power and immediacy of *Frankenstein* owe much to its highly visual sequences (hence the power of the film versions that exploit these), but its more subtle and pervasive implications derive from the older myths relating to the desire for knowledge and creativity on which it draws—Pandora's box, both the Prometheus myths, the Genesis stories of the Creation and the Fall, and the Faust legend, particularly Goethe's treatment of it, as well as *St Leon: A Tale of the Sixteenth Century*, a novel by Mary's father, William Godwin, to whom she dedicated *Frankenstein*.[4] However, Shelley's treatment also involves a more complex assessment of the issues presented by Goethe or Godwin, insofar as Frankenstein embraces both the scientific rationalism and reductionism condemned by the Romantics and the ultimate Romantic quest for knowledge of the absolutes of life and death. She thereby suggests that these apparently contrary positions are, in the final analysis, merely variations of the same basic type—the overreacher whose aspirations lead inevitably to the destruction of himself and others, yet who is admired as the heroic genius.

Frankenstein and Contemporary Science

Although the scientific background of the novel, with its Gothic trappings, may now seem fantastic, Mary Shelley was careful to make it consistent with current scientific theories, including ideas about electricity. Percy Shelley had long been interested in electricity and galvanism. He had personally experimented with making a large-scale battery and had repeated Benjamin Franklin's experiment with a kite in an electrical storm,[5] so that the methods whereby Frankenstein brings his monster to life—an electric discharge—would have been regarded by many contemporaries as at least feasible. In his preface to the novel, Percy Shelley explicitly insisted that "the event on which this fiction is founded has been supposed, by Dr. Darwin and some of the physiological writers of Germany, as not of impossible occurrence."[6] This adherence to what was scientifically credible at the time was essential to Mary's purpose of exploring both the premises of scientific materialism and the values of the Enlightenment.

Frankenstein is not only the Romantic overreacher determined to transcend human limitations; he is also the heir of Baconian optimism and Enlightenment confidence that everything can ultimately be known and that such knowledge will inevitably be for the good: "I doubted not that I should ultimately succeed" (53). Frankenstein also accepts uncritically the reductionist premise of the eighteenth-century mechanists, that an organism is no more than the sum of its parts and apparently has no sense of the extraordinary irony involved when he sets out to create a "being like myself" from dead and inanimate components. Even in retrospect he sees no anomaly in this, for he tells Walton, not without pride: "In my education my father had taken the greatest precautions that my mind should be impressed with no supernatural horrors. I do not ever remember to have trembled at a tale of superstition, or to have feared the apparition of a spirit" (51).

But the being he creates is not merely a mechanism, the sum of its inanimate parts; it is indeed a being like himself with free will and a soul, which are not subject to Frankenstein's control. As such, it is a re-creation of Frankenstein's own unconscious desires, both good and evil, which have been suppressed by the discipline of his research program and by cultural censorship.[7] The monster functions as both an alter ego and a substitute for the natural child to whom Frankenstein has denied existence by deferring his marriage with Elizabeth. This doppelgänger relationship symbolizes the essential duality of man, the complex of rational and emotional selves (Hawthorne's notion of the head versus the heart), mutually alienated but finally inseparable.[8] In the image of the larger-than-human monster, Shelley reaffirms the Romantic position that the unconscious is an intrinsic and more powerful part of the human experience than the rational mind and, if suppressed, will ultimately emerge to destroy the latter, an idea that Stevenson was to exploit in *Dr. Jekyll and Mr. Hyde*.

Shelley further aligns herself with the Romantic protest against scientific rationalism in emphasizing the fact that the obsessive desire to acquire scientific knowledge is achieved at the cost of physical and emotional well-being. In order to pursue his experiments (which depend on raiding graves and charnel houses, all associated with death) Frankenstein denies the affections of the living, isolating himself from his loving family, fiancée, and friends who would both impede his studies and save him from his fatal obsession. He becomes insensitive to natural beauty, and his health and emotional state deteriorate. He tells Walton: "My cheek had grown pale with study, and my person had become emaciated with confinement.[9] . . . It was a most beautiful season . . . but my eyes were insensible to the charms of nature. And the same feelings which made me neglect the scenes about me, caused me also to forget those friends who were so many miles absent" (54–55). These aspects of

Frankenstein's character were to form the basis for many later representations of the emotionally crippled scientist isolating himself from life, nature, and human affections. Some of these have already been considered in the preceding chapter, but they became more numerous in twentieth-century fiction and films (see chapter 13 below).

Shelley's critique of science and its assertion of the right to pursue knowledge wherever it might lead is both more complex and more radical than the Romantic condemnation of intellectualism. Frankenstein is simultaneously both the scientific rationalist and the passionate idealist. Dissatisfied with the mere accumulation of facts, he desires to know the unknowable, to press beyond the boundaries of human knowledge, "to penetrate the secrets of Nature." No secret is more powerful to the imagination than the creation of life, the equivalent of the alchemists' quest for the homunculus, the diminutive human, which Frankenstein's oversize monster parodies. His research program using the materials of death to create life also references the characteristically Romantic attempt to reconcile opposites.

Again, although Frankenstein is explicitly exonerated from any desire for mere wealth and justifies his research on the grounds that it will benefit the whole human race, Shelley makes it clear that his search is by no means disinterested: "Wealth was an inferior object; but what glory would attend the discovery, if I could banish disease from the human race and render man invulnerable to any but a violent death" (40). It becomes apparent that "glory" is closely akin to power, and in relating the story of his life to Walton, Frankenstein unwittingly reveals how strongly he is attracted to such power, regardless of the devastation it leaves in its wake. He tells how he stood enraptured by the violence of an electrical storm, undismayed by the consequent destruction of the living oak tree. And it is no coincidence that his inspiration at the University of Ingolstadt is not the empiricist M. Krempe but the charismatic M. Waldman, who, in language resonating with a combination of biblical and Baconian rhetoric, claims for modern chemistry supremacy over other branches of knowledge because of the power it offers: "[The chemists] have indeed performed miracles. They penetrate into the recesses of nature and show how she works in her hiding places. They ascend into the heavens; they have discovered how the blood circulates and the nature of the air we breathe. They have acquired new and almost unlimited powers; they can command the thunders of heaven, mimic the earthquake, and even mock the invisible world with its own shadows" (47–48).

Thus, what captivates Frankenstein is less the lure of knowledge for its own sake than the promise of the power and distinction it will confer. The reference to Faust's bargain with Mephistopheles is clear. Frankenstein says, "I felt as if my soul were grappling with a palpable enemy," and he formulates his obsessive purpose in terms of a spiritual conquest. "More, far more will I achieve. . . . I will pioneer a new way, explore unknown powers, and unfold to the world the deepest mysteries of creation" (48). "Life and death appeared to me ideal bounds which I should first break through, and pour a torrent of light into our dark world" (54). That is, he sees himself as reenacting the role of the Creator and, in accordance with the Romantic quest, wrenching opposites into unity at his will.

Frankenstein and Romanticism

Another interesting facet of Shelley's portrayal is that Frankenstein's idealism is presented as one of his most dangerous qualities. In this she was almost certainly referring to the contemporary political situation in France, where the original ideals of the French Revolution (warmly welcomed at first by the Romantic writers) had plunged the country into the Reign of Terror and led, in the long term, to the Napoleonic Wars. Indeed, there are striking similarities between the descriptions of the monster and the iconography associated with Napoleon Bonaparte, whose defeat at Waterloo had occurred only the year before the novel was begun.[10]

It was largely as a result of her ambivalent characterization of Frankenstein as both scientist and Romantic idealist that Mary Shelley was able to suggest new and complex facets of the motivation of the scientist and his relation to his work. In so doing, she preempted many of the philosophical and psychological considerations that have subsequently been recognized as inextricably linked to scientific research. First, since almost everything that pertains to Frankenstein is equally relevant to the career of the artist, Shelley indicates the close nexus between scientific research and the creative imagination and suggests that they entail both opportunity and risk. Like art, science represents the highest creative endeavor of humanity and has the power to produce either benefits or disasters. Second, her chief purpose, as outlined in her introduction, is to explore the ethical consequences of the *success* of Frankenstein's experiment. In scientific terms, the creation of the monster is a brilliant achievement; yet Frankenstein's horror begins at the precise moment when the creature opens its eye, the moment when for the first time Frankenstein himself is no longer in control of his experiment. His creation is now autonomous and cannot be uncreated any more than the results of scientific research can be unlearned, or the contents of Pandora's box recaptured. Hence, paradoxically, it is at the moment of his anticipated triumph that Frankenstein qua scientist first realizes his inadequacy (see fig. 12).

As soon as he perceives the power and horror of the living monster, Frankenstein rushes from the room, and when the creature seeks him out, he flees and hides from him. Subsequently he falls into a "nervous fever" for several months and later blames his unconscious state for his failure to prevent the monster's escape. He thus effectively puns on the idea that he was

Figure 12. Frontispiece to *Frankenstein*, by T. Holst (1831 Standard Novels edition). This frontispiece depicts the critical moment when the monster comes to life and Frankenstein flees in horror. Wikimedia Commons.

unconscious (i.e., unaware) of the consequences of his work. Yet the hideousness of the monster and the potential for power and destruction contingent on his size were apparent throughout the construction program. Shelley thus suggests that claims of ignorance on the part of scientists for their failure to foresee the consequences of their work are too glib to be credible and cannot be accepted. By his voluntary isolation, Frankenstein had deliberately excluded those, such as his father and Clerval, who might have given him the ethical advice he did not wish to hear.

Frankenstein and Social Guilt

Shelley also explores the nexus between Frankenstein's pursuit of scientific success, his failure as a human being, and his social guilt. The inevitable neglect of human ties involved in the scientist's dedication to his research results not only in his own isolation and loneliness but in a moral and emotional loss to society. Whereas many other Romantic treatments of the scientist's isolation assumed that this was a voluntary state that could be reversed at will (Paracelsus, Redlaw, Aylmer, and Claës were indicted for their refusal to reform), Shelley suggests that there is an inevitable loneliness and guilt contingent on scientific research. Frankenstein begins by frequenting remote and lonely places because his isolation is dictated by the requirements of his research, but subsequently his separation from society becomes a necessity imposed by the progress of his illegal experiment. In relating his tale to Walton, another scientist pursuing an obsession in contravention of the natural ties of affection, Frankenstein digresses to moralize explicitly: "If the study to which you apply yourself has a tendency to weaken your affections, and to destroy your taste for those simple pleasures in which no alloy can possibly mix, then that study is certainly unlawful, that is to say, not befitting the human mind" (56).

Similarly, the monster, deprived of any affection from his creator and driven away from the company of those he wishes to befriend, becomes through this enforced isolation from society a ruthless murderer. Both, therefore, through isolation from society lose their sense of social morality. Significantly, Mary dedicated *Frankenstein* to her father, William Godwin, who, like Victor Frankenstein's father, rejected religious sanctions for behavior and sought to ground morality wholly on a social and psychological basis.[11]

Shelley also probes the intellectual pride of the scientist, masquerading as a desire to serve mankind. Like many researchers, Frankenstein justifies his experimental program as a moral endeavor designed to "pour a torrent of light into our dark world. A new species would bless me as its creator and source; many happy and excellent natures would owe their being to me.... I might, in process of time, ... renew life where death had apparently devoted the body to corruption" (54). She thus points to the opportunities for self-delusion that accompany the pursuit of scientific fame.

Philosophically, Shelley also examines the paradox involved in the pursuit of freedom through knowledge. The intrinsic dilemma of *Frankenstein* resides in the irony that the mind, which can conceive of freedom from limitations, is also the source of man's most acute agony, because it perceives both its own restrictions (the more Frankenstein learns, the more aware he is of his own ignorance) and the bondage that it has itself constructed. This recognition, symbolized in Frankenstein's involuntary subjection to the very creature that represents his triumph over the former boundaries of knowledge, was to be further explored by H. G. Wells in "The Time Machine," *The Island of Doctor Moreau*, and *The Invisible Man*.

There is a related irony attaching to Frankenstein's desire to act the part of God. In usurping this role, Frankenstein acquires all the burdens of the creator and none of the glory. Like the God of the Genesis story, he plans a paradise from which evil and death will be absent, only to have his creature turn against him in defiance and introduce sin into the world. Indeed, like Milton's Adam, the monster blames his creator for this sin.[12] But unlike the Christian God, Frankenstein fails to engineer a redemptive scheme for his creature, even though the monster begs him: "Make me happy and I shall again be virtuous" (100). In a parody of the God of the deists, Frankenstein isolates himself from his creation; unable to bring himself to accept it in its sinful state, he leaves it to fend for itself.

The presence in the novel of Walton, the polar explorer, another type of scientist, is important in suggesting the universality of the statements concerning Frankenstein. There are many significant parallels between the two men, both in their situations and in their attitudes. Both have left the security and affection of family in the pursuit of a goal that will inevitably cost the lives of innocent people. Both are motivated by the desire for glory and personal achievement but attempt to justify their search in terms of the benefits it will confer on a grateful posterity. Both seek to yoke together natural opposites. Frankenstein seeks to eliminate the boundaries between life and death; Walton is bent on discovering the tropical paradise in the polar wastes that one contemporary theory proposed. The poles, being the furthest points from the inhabited world, also represented boundaries and possessed an almost mystic significance symbolized in their supposed link with the tropical lands of the equator. To reach them symbolized completion, the final definition of the world—hence the nineteenth- and early-twentieth-century fascination with polar exploration and Walton's personal obsession with achieving this ultimate transcendence of geographical limitations. The charismatic effect on Frankenstein of Waldman's words about the acquisition of "new and almost unlimited powers" is mirrored in Walton's admiration for Frankenstein and the compulsive eagerness with which he listens to his story.

Ironically, Mary Shelley also provided an image for the concept of women in relation to science, an image that was to continue well into the twentieth century, if indeed it has ever been significantly revised. Her own subconscious protest against the idea that women were unnecessary in the creation of life has already been discussed, but it was also to issue in the critically neglected figure of Frankenstein's fiancée, Elizabeth. In the terms of his research proposal, she too is rendered redundant to the life process, as Frankenstein, while professing to be in love with her, keeps deferring his marriage and hence the creation of a child by natural means. Officially this delay is only until he has perfected the unnatural creation of his monstrous "child"; however, the monster's subsequent murder of Elizabeth enacts Frankenstein's own repressed desire to rid himself both of his social responsibilities and of the much simpler, natural method of procreation.[13] It is significant that, immediately after the animation of his monster, Frankenstein dreams of Elizabeth transformed by his embrace into the corpse of his dead mother.

This aspect of the story was to provide a new and pertinent twist to the powerful Gothic image depicted in Henry Fuseli's painting *The Nightmare* (1782): the vulnerable female stretched passively on a bed, her head hanging down, menaced by the surrealist intruders (see fig. 13). In *Frankenstein* the menacing intruder is no supernatural visitor that can be banished by the rational, waking mind, but the material creation of science. Frankenstein awakens to find his nightmare actualized. The sexual connotations of this neo-Gothic image, which filmmakers were quick to exploit, also dominated the pulp science fiction of the twentieth century and were clearly regarded as an important aspect of their attraction, since variations of this icon formed a staple part of both the film posters of *Frankenstein* and the lurid cover designs of the pulp magazines.[14] By extension, the suggested rape and subsequent death of the beautiful and vivacious Elizabeth by the monster created by science can also be seen as a figurative enactment of Bacon's perception of science as penetrating the secrets of a symbolically female nature, laid inert on the rack.

By bringing together in Frankenstein the apparently opposite qualities of the scientist and the Romantic visionary, Mary Shelley not only enriched immeasurably her depiction of the scientist but extended the basic Romantic protest against materialism and rationalism. She showed Frankenstein, apparently so rational, so desirous of secularizing the world and removing its mysteries, to be, at crucial points, highly irrational, suppressing those considerations that might conflict with his obsession. George Levine points out that *Frankenstein* "as a modern metaphor implies the conception of the divided self, the creator and his world at odds. The civilised man or woman contains within the self a monstrous, destructive, and self-destructive energy."[15] The novel thus becomes a modern, scientific formulation of the archetypal myths of *psychomachia*, or the conflict within the soul. In this wholly secular version, science and technology are a concretization of inner desires,

Figure 13. Nightmare, by John Henry Fuseli, 1781. Detroit Institute of Arts, Detroit. Wikimedia Commons.

masquerading as rational but, like the monster, equally capable of springing from the dark, unacknowledged depths of their creator's subconscious.

This perception suggests some important qualifications of the Enlightenment belief that the pursuit of knowledge is, by definition, rational and good and should not be restricted by any sociomoral considerations. Few nineteenth-century readers, however, were able to follow these implications. It has taken such twentieth-century monsters as nuclear power, in vitro fertilization, and genetic engineering, bursting upon an ethically unprepared world with their dual potential for good and evil, to illuminate fully the depths of meaning in *Frankenstein*. Not surprisingly, playwrights and filmmakers have returned with great frequency to the story, modifying it to suit the prevailing tastes, values, and scientific debates of their time.

Later Versions of Frankenstein

The first dramatic presentation of *Frankenstein* was H. M. Milner's play of 1826, *Frankenstein; or, the Man and the Monster*, which became the subject of one of the earliest films, the Edison Company's *Frankenstein* (1910). This film eliminated most of the repulsive physical situations and concentrated instead on the psychological aspects of the story, emphasizing that the creation of the monster was possible only because Frankenstein allowed his normally healthy mind to be overcome by evil and unnatural thoughts. Edison's ending, in particular, was far more positive and romantic, echoing contemporary optimism about science: the monster finally fades away, leaving only his reflection in a mirror, and even this is subsequently dissolved into Frankenstein's own image by the power of Elizabeth's love. Frankenstein has been restored to mental health; thus the monster can no longer exist.

Carlos Clerens, historian of horror films, rates the 1931 Universal film classic *Frankenstein*, which introduced Boris Karloff as the monster, as "the most famous horror movie of all time";[16] yet, compared with Shelley's novel, the film is hardly horrific at all. The heavily underlined moral, stated at the beginning, "It is the story of Frankenstein, a man of science who sought to create a man after his own image without reckoning upon God," restores a religious dimension of supernatural order and justice to Mary Shelley's entirely secular and unredeemed situation. In this version, Henry Frankenstein (who, following Peggy Webling's 1930 play, on which the film is based, has changed given names with Clerval) is presented as the innocent victim of a mistake whereby his careless assistant has brought him the brain of a murderer instead of one from a noble person for insertion into his creature. The implication is that the creation of the monster per se posed no real or abiding problem and that with due precautions, a better result could be obtained the next time. This treatment of the story, including the otherwise anomalous introductory moral, was consistent with the adulation of scientists in the United States during the 1930s. The evil character of the monster is reduced to an experimental error rather than being associated with Frankenstein's own id and the result of his hubris.

In contrast to Mary Shelley's novel, where Frankenstein's laboratory is not described but his own mental processes are detailed at length, the fifty-six or more film versions[17] have focused on the laboratory (in the 1931 version Frankenstein works in an old watchtower strongly resembling the one used in the 1926 silent horror film *The Magician*, in which an alchemist seeks a virgin's blood to create life), the frightening appearance of the monster, and his killing rampage.

Although the 1931 film ended with the monster being burned to death and the celebration of Frankenstein's wedding to Elizabeth, the box office success indicated a sequel. The final scenes of the film were cut from all prints in circulation, and *Bride of Frankenstein* (1935) opened with Mary Shelley telling Shelley and Byron the sequel to "her" novel. In this film, Frankenstein becomes the pawn of another scientist, the mad and evil Dr. Pretorius, who, having constructed various homunculi, now wishes to produce something larger. He forces Frankenstein to create the mate for which the monster of the novel had begged. The female monster (in an extension of the doppelgänger effect in the novel, she is played by the same actress who plays Mary Shelley, Elsa Lanchester) is striking but not hideous, and she immediately rejects the monster, who in despair electrocutes her, Dr. Pretorius, and himself. In this film Frankenstein is entirely absolved of guilt, and the function of the evil scientist bent on creating life has passed to the alchemist figure Pretorius.

Bride of Frankenstein was followed by a long succession of Frankenstein derivatives whose titles are sufficiently indicative of their content and of the way in which Frankenstein has been integrated into Western culture as an ever-contemporary byword, almost as a real person. At different periods the emphasis falls variously on horror, space travel, sexuality, stem cells, or comedy associated with the figure of the scientist. One of the most interesting films in terms of the application of the Frankenstein story to a contemporary scientific debate is *Frankenstein 1970* (1958), in which Boris Karloff returned to the screen as the disfigured Victor Frankenstein, victim of Nazi torture. By means of an atomic reactor, he revives the monster from his ancestor's experiment, but they both die a horrible death from radioactivity when the reactor blows up. Only then is the monster's face revealed. It is the face of a youthful Victor Frankenstein, symbolizing in startling visual imagery the identification of creator and creature, in this case the atomic scientist and his dangerous and faulty creation, atomic power.

The allure of the Frankenstein story shows no sign of diminishing. Theodore Roszac's novel *Memoirs of Elizabeth Frankenstein* (1995) retells the narrative from the point of view of Victor's adopted sister and designated bride, who dies at the hand of his creation. Recent film versions have imported details from Shelley's life as a frame for the main plot. Ken Russell's film *Gothic* (1986) incorporates Mary Shelley's miscarried child, Frankenstein's abortion of the female monster, and the image of the dead baby floating in Lake Geneva at the close of the film, and Kenneth Branagh's *Mary Shelley's Frankenstein* (1994) attempts to recreate the scene of the composition of *Frankenstein*. Other versions transpose the story into a modern scenario. In David Wickes's television film *Frankenstein: The Real Story* (1992) Frankenstein clones himself to become the monster, with whom he shares a psychic bond, while in *Frankenstein* (2007) Dr. Victoria Frankenstein, a geneticist, is trying to culture organs from stem cells to save her son William. However, a creature, referred to as UX, begins to grow rapidly in the experimental tank and escapes. Liam Scarlett's ballet *Frankenstein* (2016) plays on the irresponsibility of Victor toward his creature. These recent versions of the story stress the enduring relevance of the novel to modern research. In a kind of paratextual parody these ever-multiplying versions of *Frankenstein* reenact Shelley's narrative: her novel has itself become a monster that has escaped and continues to produce monstrous offspring.

Shelley published another science fiction novel, *The Last Man* (1826), an apocalyptic end-of-the-human-race story in which a technologically sophisticated society is unable to survive the ravages of a plague attack that eventually kills the human race. The novel is less about science than the many intrigues of those struggling to survive, but it does include one scientist character, Merrival, an astronomer who, like Frankenstein, is so absorbed in speculating about the condition of the earth in six thousand years that he is oblivious to the plague raging around him and the poverty his obsession enforces on his wife and children. In this he is a precursor of Balzac's more complex study of an obsessive scientist, Balthazar Claës in *La recherche de l'absolu* (1834), discussed in the previous chapter.

CHAPTER 8

Victorian Scientists

Doubt and Struggle

"The stars," she whispers, "blindly run;
A web is wov'n across the sky;
From out waste places comes a cry,
And murmurs from the dying sun;

"And all the phantom, Nature, stands
With all the music in her tone
A hollow echo of my own,--
A hollow form with empty hands."   —*Tennyson*

At no other time in history has a scientific theory evoked so much public controversy and private anguish as occurred in Britain during the latter half of the nineteenth century. Although this was associated directly with the publication in 1859 of Charles Darwin's book *On the Origin of Species by Means of Natural Selection*, the unease had been building for at least a decade, fueled at one level by a fear of social change and possible political revolution (such as occurred in many European countries during the 1840s). At another level there was insecurity resulting from the eroding of traditional religious faith, partly by the new Higher Criticism that approached the Bible as literary history rather than as inerrant revelation, but also by a realization derived from popular accounts of astronomy and geology that humanity occupied only a tiny speck in an alien universe that was inconceivably vast in time and space.[1] Thus in the second half of the nineteenth century the main anxiety concerned not the moral failings of scientists but the fragmentation of knowledge,[2] the relation between science and religion, and the reputation of science relative to that of the humanities. Effectively this was a battle for the hearts and minds of readers about what values make and sustain a society's culture and who were the proper guardians of that culture.

The extent of the popular controversy over Darwinism and the degree to which loss of religious faith was attributed to science are clear indicators of a radical change in the public awareness of science and of scientists during Queen Victoria's reign. Whereas in the 1830s the only scientific research was that carried out by wealthy amateurs or doctors whose practice allowed them sufficient time for medical research, by the end of the century there was increasing eagerness for scientific education, and science as a paid profession carried out in institutions was becoming a possibility. This change did not emanate from the universities. For most of the nineteenth century the curriculum at Oxford and Cambridge remained essentially that of the grammar school, with Oxford emphasizing the classics and Cambridge pure mathematics. Experimental science was considered unworthy of study and inappropriate for the awarding of degrees. The first change in this reactionary attitude occurred in Cambridge in the 1870s when Michael Foster introduced a course in general biology at Trinity College and James Clerk Maxwell was appointed Cavendish Professor of Experimental Physics and charged with setting up the Cavendish Laboratory.[3] However, the proliferation of mechanics' institutes offering scientific studies and the popularity of the Society for the Diffusion of Useful Knowledge indicate the determination by members of the working class to acquire the kind of education they themselves considered relevant. In contrast to the specialist scientific societies, which, like the Royal Society, had been originally established for the benefit of wealthy amateurs, the British Association for the Advancement of Science (BAAS) was set up in 1831 as a democratic organization specifically to counter this kind of elitism[4] by organizing popular public lectures on science in towns all over England and encouraging noted scientists to accept responsibility for communicating science to the general public. In 1825 Michael Faraday had founded the popular Christmas lectures at the Royal Institution for the general public, including children (see fig. 14).

In addition, Victorian Britain produced a formidable array of eminent scientists, including Sir Frederick William Herschel, Sir Humphrey Davy, John Dalton, Michael Faraday, Robert Chambers, Charles Babbage, Sir Charles Lyell, James Joule, Charles Darwin, Alfred Russel Wallace, Thomas Henry Huxley, John Tyndall, and James Clerk Maxwell. These men, many skilled communicators of science, captured the public imagination and enhanced the status of science in an increasingly literate society.

Figure 14. Detail of a lithograph, *Michael Faraday Delivering a Christmas Lecture in 1856*, by Alexander Blaikley. Wikimedia Commons.

However, all was not harmonious within the scientific edifice itself. The hitherto close liaison between scientists and philosophers represented in the term *natural philosophers* was fracturing. Disputes between empirical scientists and theoreticians became increasingly acrimonious as science became more specialized, more occupied with specific problems, less concerned with questions of truth or universal values and more committed to materialism. This trend was reflected in the Victorian novel as characters came to be described as geologists, astronomers, biologists, or mathematicians rather than merely as scientists or naturalists. Causality, too, in the form of social and psychological determinism, became an integral part not only of plot but of characterization, as the new science of psychology began to arouse wide public interest and, later in the century, social Darwinists attempted to account for socioeconomic disparities in biological terms.[5]

Physics and chemistry, the prominent sciences in literature before the 1850s, were replaced in the latter half of the century by astronomy, geology, and biology, which provoked widespread debate and anxiety. In 1831 Thomas Henderson had announced that the nearest star to the sun was twenty-four trillion miles distant, threatening the order and security of the compact Newtonian model of the solar system with unimaginable spaces. The fear engendered by the vastness of space was exacerbated by the hitherto unthinkable time scales revealed by geology. In the seventeenth century Bishop Ussher, using Old Testament genealogies and events, had calculated the date of the Creation as nightfall on 22 October 4004 BCE; but after the publication of Sir Charles Lyell's three-volume *Principles of Geology* (1830–33), where the word *evolution* was first used in its modern sense, and his later *Geological Evidences of the Antiquity of Man* (1863), both proposing a far longer history of the planet, it became increasingly difficult to reconcile the idea of such a vast universe, in which man was effectively a mere speck of matter marooned on a tiny planet, with the Christian belief in a personal, caring God and the timescale of the Old Testament.

In this social context it would be hard to overestimate the impact of Darwin's theory when *On the Origin of Species* exploded into print in 1859, the first edition selling out in one day. While religious orthodoxy had been able to assimilate Newton by declaring his cosmic scheme a triumph for God, who emerged as a first-class Watchmaker, it was much less able to integrate Darwin. It now seemed that not only the inanimate world but the realm of biology and even, after Darwin's yet more iconoclastic *Descent of Man* (1871), man himself could be wholly accounted for in terms of cause and effect. The German chemist Jacob Moleschott voiced the extreme materialist case when he asserted that the secret of life was chemistry and that people were no more than the sum of their heredity, food, and environment.[6]

Even more shattering for Darwin's contemporaries than a mechanistic biology was the absence of purpose in the Darwinian scheme. The ideas of evolution put forward in earlier works, such as Erasmus Darwin's long poem *Zoonomia* (1794–96) and Robert Chambers's *Vestiges of the Natural History of Creation* (1844), had posited a benevolent purpose on the part of the Creator in producing life forms exquisitely adapted to their circumstances. But with the Darwinian scheme the concept of teleology, of purpose or design on the part of Creator, was swept away. The bases of the new theory were chance, waste, and suffering, all of which appeared irreconcilable with the Christian belief in a loving God. To many it now seemed that concepts such as free will, sin, redemption, and the other foundation stones of Christianity were at best irrelevant, at worst a long-standing deception. The mental and emotional turmoil that this caused among thinking people of the late nineteenth century was the theme of countless novels, poems, and plays, of which only a small number are now read, and while the angst was not exclusively associated with science,[7] it was frequently examined in terms of the dilemma facing the religious scientist or the doctor engaged in medical research. Several of the writers whose work is discussed in this chapter had some personal acquaintance with actual scientists or with doctors who engaged in medical research, and Charles Kingsley, Elizabeth Gaskell, George Eliot, and Thomas Hardy also took the trouble to acquire a working knowledge of the science they wrote about. This in itself is an indication both of the seriousness with which science was being approached in literature and of the interest in scientific questions on the part of the reading public. Because of the increased differentiation of sciences and the complexity of the issues involved, we shall consider Victorian scientist characters under the following groupings, which indicate the major concerns raised in relation to science during this period: idealized natural scientists, professional scientists, scientists and religion, immensities of space and time, and the exploiters.

Idealized Natural Scientists

Unlike the unflattering eighteenth-century caricatures, portrayals of nineteenth-century medical researchers are, almost without exception, complimentary to the point of eulogy. Largely as a result of the widespread public benefits deriving from Edward Jenner's introduction of vaccination against smallpox and Joseph Lister's pioneering work in antisepsis, doctors in Victorian literature are depicted as heroic figures of unfailing integrity, battling the diseases of the poor for little financial reward, and early portrayals of natural scientists are endowed with a similar catalog of virtues, moral and physical as well as intellectual.

One of the first such portraits is Tom Thurnall, doctor-scientist of Charles Kingsley's novel *Two Years Ago* (1857). Kingsley, an enthusiastic amateur zoologist as well as a novelist and clergyman, believed passionately in the ideal of *mens sana in corpore sano* (a healthy mind in a healthy body). Tom, modeled on the author's brother George,[8] has impeccable qualifications, having studied in the laboratories and hospitals of Glasgow and Paris and having been a ship's surgeon and army doctor. Ever ready with his microscope and collecting equipment, he is no mere theoretician but a genuine, muscular scientist, extolling the virtues of experience as the basis of scientific method: "We doctors, you see, get into the way of looking at things as men of science; and the ground of science is experience."[9]

Like his author, Tom is a zealous exponent of sanitary reform.[10] When an outbreak of cholera threatens his village, it is, for him, a holy war against disease. As he tells the curate: "You hate sin, you know. Well, I hate disease. Moral evil is your devil, and physical evil is mine" (212). Kingsley introduces other contemporary medical issues as well. Scrupulously honest, Tom goes against the normal practice in refusing to prescribe medicines he thinks unnecessary and invariably charging the minimum for those he does prescribe. Tom's interest in collecting zoological specimens along the Devonshire coast, which makes him the first medical research scientist in literature,[11] arises not only from his fascination with the specimens per se but also from their value in medical practice and their social usefulness. "This little zoophyte lives by the same laws as you and I. . . . He and the sea-weeds, and so forth, teach us doctors certain little rules concerning life and death" (150). In Tom Thurnall, the scientist-doctor, Kingsley contrived to combine the qualities he most admired in the scientist, the soldier, and the social reformer by identifying the doctor's war against disease and unsanitary conditions with a sociomoral war against greed and stupidity.

Idealized as pursuing scientific research for the benefit of a suffering humanity, doctors acquired at this time the elevated social and moral status that they were to retain for more than a century.[12] It was more difficult to justify the pure scientist, about whom there still lingered the suspicion of intellectual and perhaps spiritual pride, the taint of Faustus. Considerable public relations work had to be done before a pure scientist could weather the realistic novel with his moral reputation intact.

Again, one of the first in the field was Charles Kingsley. His prolific correspondence with such scientists as Huxley, Darwin, and Philip Gosse was mutually fruitful, and he was honored by being elected a fellow of both the Linnean Society (1857) and the Geological Society (1863). Despite the stereotyped view of the Victorian age as a battleground between science and religion, there is no suggestion of this in Kingsley's writings. He, like the seventeenth- and eighteenth-century exponents of natural theology, believed that a better knowledge of the marvels of nature would engender a deeper love of the Creator. "I am sure that science and the creeds will shake hands at last . . . and that by God's grace I may help them to do so."[13] His fairy tale *The Water Babies* (1863) was written in order to introduce children to the concept of evolution in a Christian context.[14] It is therefore not surprising that the several naturalists who appear in his works, far from being isolated intellectuals, are all of the vigorous, "muscular Christian" stamp favored by Kingsley. In *Glaucus; or, The Wonders of the Shore* (1855) Kingsley extols the natural scientist as a kind of latter-day crusader, embodying bravery, patience, modesty, reverence, and chivalry: "It is these qualities, . . . which make our scientific men, as a class, the wholesomest and pleasantest of companions abroad."[15] Such praise from a well-known and respected clergyman did not pass unheeded. Despite the impending clash with the religious establishment, an aura of selfless devotion to truth began to flicker around the scientific cranium.

An unusually late example of a scientist idealized without qualification is to be found in *Melampus* (1883), a long and now little-read poem by George Meredith, who alone among the Victorian poets attempted to integrate the Darwinian picture of nature with Romantic values. His scientist-poet Melampus, an early environmentalist, observes and learns from nature the secrets that will allow man to overcome his alienation from the natural world. Essentially he is an intermediary through whom the healing power of nature can flow to others, curing the psychosomatic diseases contingent upon their alienation from nature. In *Melampus*, the search for knowledge has no taint of evil. The serpent of Eden is here recast as the benign snake of Hippocrates, which confers upon Melampus the gift of understanding the voices of nature and the pattern of causality, one of the main points at issue between nineteenth-century science and faith. However, Melampus remains a symbolic rather than a realistic figure, embodying the author's hope that the discoveries and values of science are not totally alien to those of Romantic poetry. Moreover, there is an uncomfortable suggestion that Meredith is cheating. Either he has carefully tailored the science to fit the aesthetic values of art and poetry or he has failed to understand the full significance of Darwin's theory. He glosses over the terror of nature, partly by ignoring the inanimate universe and partly by taming the internecine struggle for existence that had so appalled Tennyson into a mild and remediable injury for which nature herself provides a panacea if we but search for it in the right spirit. The failure of *Melampus* to convince is an index of how difficult it was to fit such an idealized scientist into realistic literature.

Professional Scientists

The more innovative and memorable portrayals of scientist characters in the Victorian novel result from the authors' efforts to understand the struggles and problems of actual scientists. Elizabeth Gaskell's novel *Wives and Daughters* (1866) is a particularly interesting piece of social history, depicting three scientist characters who span a broad social spectrum. Lord Hollingford, eldest son of an earl, is a wealthy, amateur scientist in the eighteenth-century tradition. Intensely shy and lacking in small talk, he relates easily only to those with whom he can discuss science, namely, his scientific friends from

London and Paris, and the local doctor Gibson, who also finds time to contribute to scientific medical journals. Gaskell is the first novelist to depict the bond of scientific interest as a social leveler, for there were few other grounds on which a lord and the local doctor could have met as equals. This scientific egalitarianism, which could override class distinctions, was one of the attractions of the scientific societies.

The third scientist in the novel, Roger Hamley, is the most interesting, since he represents the first example in literature of a professional scientist in the modern sense. Roger is regarded by his parents as a mere plodder and contrasted unfavorably with his elder brother Osborne, who seems ready to distinguish himself in the expected classical education at Cambridge. Even his mother considered that Roger "was so little likely to distinguish himself in intellectual pursuits; anything practical—such as a civil engineer—would be more the kind of life for him."[16] (We may note the nineteenth-century distinction between *intellectual*, meaning classical, and *practical*, meaning scientific.) However, the unlikely Roger emerges as a new kind of hero. Modeled on Charles Darwin, whom Gaskell knew well, his scientific enthusiasm and careful observation outweigh his social awkwardness. He becomes senior wrangler, wins acclaim for his papers on comparative anatomy and osteology, and, like Darwin, embarks on an exploratory voyage as a naturalist, laying the foundations for a distinguished career as "professor of some great scientific institution" (707).[17]

In her portrayal of the two Hamley brothers, Gaskell contrasts not only two distinct personalities but also the results of two different approaches to education. The classical-humanist criteria favored Osborne with his wit and his ability to write poetry and savor intense impressions and discriminated against Roger's shyness and slow thoroughness. A different set of parameters, however—and Gaskell implies that these parameters are increasingly those of the real world—encourages the abilities of Roger, who, it is clearly suggested at the end of the novel, is the man of the future. *Wives and Daughters* in effect foreshadowed the protracted dispute in the 1880s between Matthew Arnold and Thomas Huxley over the relative importance of the classics and science in the educational curriculum,[18] the latter emerging triumphant.

Gaskell also suggests one of the reasons for the slow emergence of the professional scientist in England, namely, the financial precariousness of a career in science in this period. Despite his outstanding university results and his reputation in the scientific journals, Roger is dependent on wealthy amateurs for an income. Lord Hollingford remarks, "Science is not a remunerative profession, if profession it can be called" (676).

The truth of this statement is developed in George Eliot's greatest novel, *Middlemarch: A Study of Provincial Life* (1871–72), which depicts, among much else, the struggles, financial and personal, of a career in medical research.[19] *Middlemarch* is set some seven years later than *Wives and Daughters*, in a country town that is undergoing vastly accelerated social change, owing to the advent of railways, the repercussions of the first reform bill, and the beginnings of medical reform. Against this background, Eliot's study of Tertius Lydgate, the "new, young surgeon" who aspires both to reform general practice and to pursue first-class research in tissue structure, constitutes the preeminent psychological and social portrait of a scientist in nineteenth-century literature, the realistic counterpart of Mary Shelley's *Frankenstein*. The complexity of Lydgate as a character stems, first, from the density of scientific background Eliot researched for *Middlemarch* and, second, from the fact that she perceived the society of Middlemarch and in particular Lydgate's role in it not merely in literary terms but as a kind of scientific experiment through which to examine the social implications of Darwinian theory. Thus Lydgate, while presented with great realism, has a symbolic significance as well.

As Anna Kitchel's "Quarry for *Middlemarch*"[20] and Eliot's own journal[21] conclusively demonstrate, Eliot read exhaustively not only all the relevant literature of medical and scientific research in the years preceding 1829—the year in which Lydgate is said to have arrived in Middlemarch—but also the background controversy surrounding attempts at medical and sanitary reform in England, as it was reported and argued in the columns of the *Lancet*.[22] Thus, unlike earlier novelists, Eliot does not merely *tell* us that Lydgate is a medical researcher; she places him in the milieu of convincing medical research so that his actions and assumptions, his motives and behavior, result from, and interact with, this background. Eliot also shows the resistance Lydgate experiences from colleagues and the petty jealousies of social intercourse, thereby situating research within the wider social picture.

At the time Eliot was writing *Middlemarch*, medicine had only recently been accorded scientific status. Owing to the system of medical training in England, experimental research in medicine during the first half of the century was almost nonexistent.[23] "About 1829," Eliot remarks in *Middlemarch*, "the dark territories of Pathology were a fine America for a spirited young adventurer."[24] Having studied anatomy and pathology in Edinburgh and Paris, Lydgate is far ahead of his provincial colleagues in his knowledge of the causes and treatment of fevers, including cholera and typhoid,[25] in his use of the stethoscope[26] and the new achromatic microscope, and in his insistence on conducting postmortem examinations when the cause of death is not adequately known. We are also shown several instances of his superiority in diagnosis, which, understandably, affront the medical fraternity of Middlemarch. Thus Lydgate realizes that Fred Vincy is in the pink-skinned stage of typhoid fever, whereas the well-established Dr. Wrench, believing his patient to be suffering merely from "a slight derangement," has been prescribing drugs harmful for Fred's actual condition. Lydgate also rediagnoses Nancy Nash's "tumour" as cramp, and his knowledge of the new methods of treating pneumonia (by permitting the patient's own resistance to combat the disease rather than bleeding and thus further weakening him) allows Trumbull to recover. Perhaps the most impressive indication of Eliot's scrupulous research is seen in Lydgate's treatment of Raffles's delirium tremens, which follows the method proposed only the preceding year by the American John Ware.[27] He also refuses to dispense his own medicines (so avoiding the temptation to overprescribe), an example that further aggrieves his more mercenary colleagues.[28]

These incidents indicate not only a sense of authenticity about the character of Lydgate but also the new attitudes emerging within the discipline of medical science. The gradual change from the mechanistic model that had characterized medicine since the time of Harvey to the idea of process and change implicit in evolutionary theory is mirrored in Lydgate's own attitudes.[29] Lydgate began as a mechanist, fascinated by "his first vivid notion of finely-adjusted mechanism in the human frame" (1:124–25). Then in Paris he came under the influence of the great histologist Marie-François-Xavier Bichat, who rejected the reductionist philosophy, asserting that living bodies "are not associations of organs which can be understood by studying them first apart, and then as it were federally" (1:128).[30] Ultimately Lydgate comes to believe that subjective elements are a necessary part of scientific procedure.

Lydgate's contact with the work of Bichat, who had shown that all the apparently diverse organs were composed of comparatively few kinds of tissues, has filled him with a neo-Romantic desire to discover the fundamental or primitive tissue, of which Bichat's tissues are but modifications. In other words, Lydgate hopes to discover a single common basis for all living structures, a preoccupation parallel to the desires of the alchemists and of Newton, to discover a comprehensive unifying system, and to Frankenstein's obsession with finding the basic secret of life. For Eliot, who had already dealt with a similar theme in her Gothic story "The Lifted Veil," the issue was of immediate, contemporary relevance, since its essentially materialist philosophy was the pivotal point of Darwinism.

The obverse of Lydgate's intellectual passion is a certain arrogance and scorn for the second-rate. Lydgate does not conceal his contempt for the backward medical practices of Middlemarch, thereby arousing the hostility of both his colleagues and his clientele: "Lydgate's conceit was of the arrogant sort, . . . massive in its claims and benevolently contemptuous" (1:130). In Lydgate, this intellectual pride is exacerbated by a residual social snobbery based on his aristocratic family connections, an attitude that sits ill with his progressive notions in science. Eliot makes the point that advanced ideas in one area may fail to modify conventional thinking in another so that even a distinguished mind may be "a little spotted with commonness": "That distinction of mind which belonged to his intellectual ardour, did not penetrate his feeling and judgment about furniture, or women, or the desirability of its being known (without his telling) that he was better born than other country surgeons. . . . Neither biology nor schemes of reform would lift him above the vulgarity of feeling that there would be an incompatibility in his furniture not being of the best" (1:130–31). It is when Lydgate claims greatness that he becomes most ordinary. Despite his superior abilities and motivation, Lydgate, like the protagonist of classical tragedy, is brought low by hubris, including not only intellectual arrogance (to which scientists might be considered especially prone) but also a more mundane snobbery concerning social status and possessions.

Outside his professional interests, then, Lydgate is as conventional and small-minded as the Middlemarch society he so keenly despises. In contrast to his meticulous procedures in science, he leaps, in his private life, from superficial observations to the erroneous conclusions that lead to his disastrous marriage with an incompatible partner and thence, inevitably, to debts and social disgrace. Lydgate thus represents an important development in the representation of the scientist from the Romantic depiction, where the promptings of the heart were depicted as morally superior to those of the mind and submission to the latter invariably issued in evil. It is when Lydgate *fails* to use his mental faculties to the full and succumbs instead to his emotions that he falls. In Lydgate's case this is particularly ironic, because it involves a violation of those very habits of rigorous enquiry that have made him so promising a student of Bichat and Louis. When the latter's new book on fevers arrives, Eliot remarks caustically that Lydgate "read far into the smallest hour, bringing a much more testing vision of details and relations into this pathological study than he had ever thought it necessary to apply to the complexities of love and marriage, these being subjects on which he felt himself amply informed by literature, and that traditional wisdom which is handed down in the genial conversation of men" (1:143).

Eliot also revises the image of the scientist as isolationist. There is considerable irony in the fact that Lydgate, who hopes to extend Bichat's work on tissues and eventually to discover the common basis of all living structures, should endeavor to dissociate himself from society. Bichat had used the old French word *tissu* (derived from the Latin *textere*, to weave) for the structures he discovered, and the study of tissues was subsequently named histology (from the Greek *histos*, web). Yet Lydgate, aware of the intrigues and professional detraction (as well as the expense) contingent on living in London, has come to provincial Middlemarch in order, as he naively thinks, to avoid those very interconnections he so earnestly seeks in the laboratory. He soon discovers that causality operates in society as inevitably as in science. His medical colleagues resent his new ideas and more particularly the arrogance with which he introduces them:[31] "Middlemarch in fact, counted on swallowing Lydgate and assimilating him very comfortably" (1:134). He learns that visiting frequently at the Vincys' house and flirting with their pretty daughter carries social obligations of marriage and that even an honorary position as director of Bulstrode's new hospital carries adverse implications.[32]

Thus each of Lydgate's assumptions—about society, women, and money— not only plays a part in his downfall but runs directly counter to both his scientific training in logic and his search for the fundamental unit of life. If Lydgate fails to find the primary tissue in his laboratory, he encounters it symbolically, and without ever recognizing it, in the woven fabric of the community. Like everyone else in the novel he is an integral part of the tissue of Middlemarch, whatever superficial modifications he may have acquired, and isolation is as impossible in society as in anatomy and histology. The consequence of failing to realize this is that he is evicted from Middlemarch and ends, much to the satisfaction of his wife and his own bitter disgust, as a society doctor "alternating, according to the season, between London and a Continental bathing place; having written a treatise on Gout, a disease which has a good deal of wealth on its side" (2:360).

Eliot also introduces another factor that was to become a perennial difficulty for scientists in the increasingly competitive professional environment of research, namely, the race to establish a discovery before others do so and the difficulty of accomplishing this without private finance.[33] It represents a situation between the time when research was possible only for gentlemen of private means or those with wealthy patrons and the era of publicly funded science, which was not to occur until the end of the century. Lydgate is acutely conscious that "when one has notions in science, every moment is an opportunity" (1:307) and that "Raspail and others are on the same track" (2:23). It was, of course, François-Vincent Raspail, the "father of cell theory," who, in 1833, the year after the novel ends, discovered that very "substance membraneuse des organes Animaux"[34] that Lydgate had been seeking.

There is yet another level of scientific interest in the presentation of Lydgate. Not only does Eliot show the inextricable threads of consequence emanating from both Lydgate's own flaws of character and his interaction with a given society at a given place and time, but she explains them in scientific terms. If Lydgate is unaware of the tightly woven threads of the communal fabric, Eliot is not. With post-Darwinian vision, she interprets it as the social counterpart of the process of natural selection. Lydgate, despite his new intellectual modifications, fails to survive in Middlemarch because he cannot adapt to the environment of that time; nor can Middlemarch tolerate this strange, incompatible organism. When the novel was published, forty years after the time of its setting, increased concern with public health and sanitation, spurred on by several outbreaks of cholera and typhoid in the intervening years, had sharpened public appreciation of preventive medicine and the need for scientific research in medicine. Had Lydgate lived forty years later, he would have been acclaimed.

This time lapse between setting and publication permits a further irony, for as W. J. Harvey has pointed out, Lydgate's failure to find the "primitive tissue" results in part from his asking the wrong question.[35] What is missing from his theoretical consideration of the problem is any reference to cell theory, which had yet to be enunciated. Robert Brown, whose earlier work Lydgate has read without being aware of its importance (he exchanges Brown's book for a preserved sea creature), did his seminal work on the cell nucleus only in 1831, by which time Lydgate was already bedeviled by marital and financial problems, and the crucial work by Schwann and Schneider (the "plodding Germans" whom Lydgate fears as competitors) was completed only in 1838–39. Thus

Eliot suggests that Lydgate's research, like his progressive ideas about medical practice, is premature.[36] Time and place are as essential for the evolution of science itself as for biological evolution.

Middlemarch subverts the Romantic idea of the scientific man of genius who arises in unlikely circumstances and transcends an unfavorable environment to produce a great discovery. Instead, in accordance with the Victorian emphasis on society and her own criterion of responsibility as a preeminent virtue, Eliot depicts a scientist unable to rise above the limitations of self and society. This image of the struggling scientist fighting against financial, personal, and social impediments was to be a recurrent one in the early decades of the twentieth century.

Scientists and Religion

Newtonian cosmology, though indicating in theory the vastness of the solar system, had nevertheless privileged man as the intellectual center of the universe, able to comprehend the divine plan. But by the mid-nineteenth century astronomy and geology had combined to demote him to the status of a local accident in an immense, chaotic system that might at any moment obliterate him by a cataclysmic event such as the geological record displayed with great frequency. If whole species could be rendered extinct, what faith could there be in a loving Creator, in the uniqueness of man, or the value of the individual life?

The most poignant presentation of this theological dilemma facing Victorian intellectuals is to be found in Tennyson's long poem *In Memoriam A.H.H.* (1850), which records the author's attempt to come to terms with the death, at twenty-three, of his close friend Arthur Hallam and to find some meaning in the loss of so promising a life. Although *In Memoriam* was published nine years before Darwin's *Origin of Species* and had already been some eighteen years in the writing, the view of nature expressed in it prefigures the Darwinian problem: nature is fundamentally chaotic, governed only by chance, or if there is any perceptible tendency, it is in the direction of destruction.

> "The stars," she whispers, "blindly run;
> A web is wov'n across the sky;
> From out waste places comes a cry,
> And murmurs from the dying sun;

"And all the phantom, Nature, stands
With all the music in her tone
A hollow echo of my own,—
A hollow form with empty hands."
. . . .
Are God and Nature then at strife,
That Nature lends such evil dreams?
So careful of the type she seems,
So careless of the single life;
. . . .
"So careful of the type"? But no.
From scarped hill and quarried stone
She cries, "A thousand types are gone;
I care for nothing, all shall go."[37]

With the appearance of *The Origin of Species* a storm of controversy swept the nation. Eminent scientists and equally eminent churchmen, statesmen, and other public figures engaged in heated public debates that were eagerly attended. The best-known debate in the nineteenth century was the so-called 1860 Oxford evolution debate between Thomas Henry Huxley, popularly styled "Darwin's bulldog," and Bishop Samuel Wilberforce,[38] but similar confrontations took place throughout the country, more often at the level of an exchange of insults rather than an exploration of the issues involved. The large number of cartoons on the subject of evolution that appeared in *Punch* and the spate of minor novels dealing with the anguish of those who had lost their religious faith through a reading of Darwinism indicate the extent to which the head-on collision between science and religion affected all sections of society (see fig. 15). In such a climate of contention between science and religion, it is not surprising to find a revival of the medieval stereotype of the godless scientist, opposing the authority of the church.

Benjamin Disraeli's response to Darwinism,[39] like that of most nonscientists before about 1870, was conservative and satirical, and the representatives of evolutionary theory in his novels are mere caricatures reflecting this bias. In *Tancred; or, The New Crusade* (1847) Disraeli reaffirmed the doctrine of special creation by ridiculing the new theory. His model gentleman-scientist Tancred remains steadfastly contemptuous in the face of a society lady's gushing exegesis of an evolutionary work, *The Revelations of Chaos*

Figure 15. Charles Robert Darwin, by Linley Sambourne. *Punch* (1882). Wikimedia Commons.

(presumably intended to refer to Robert Chambers's *Vestiges of the Natural History of Creation* [1844]).[40]

A more realistic and sympathetic study of a scientist struggling to maintain his fundamentalist religious faith in the face of Darwinism is Edmund Gosse's depiction of his own father, Philip Gosse, a highly respected amateur biologist who devised and popularized the aquarium as a means of studying living organisms in their own environment. Although this autobiographical novel, *Father and Son*, was not published until 1907, it records the anguish of mid-nineteenth century Christians. Philip Gosse struggled to retain both his intellectual integrity and his fundamentalist faith in creationism. His son records that when he first heard of Darwin's theory, "every instinct of his intelligence went out at first to greet the new light. It had hardly done so, when a recollection of the opening chapter of 'Genesis' checked it at the outset."[41] After long and agonized reflection, Gosse prepared a theory of his own, outlined in his book *Omphalos* (Navel) (1857),[42] "which he fondly hoped would take the wind out of Lyell's sails and justify geology to the godly readers of 'Genesis.'"[43] According to Gosse's thesis, God created the earth complete with the geological strata and their fossils, suggestive of a much longer period of time than the four thousand years calculated from biblical genealogy, in order to test our faith. By such an argument he thought "to bring all the turmoil of scientific speculation to a close, fling geology into the arms of Scripture" (105). Edmund Gosse describes the trauma his father experienced when this "system of intellectual therapeutics," so far from eliciting universal gratitude from reconciled opponents, received only scorn from "atheists and Christians alike [who] looked at it, and laughed, and threw it away" (105). Even Charles Kingsley, from whom Gosse had expected the warmest acclaim, wrote that he could not "give up the painful and slow conclusion of five and twenty years' study of geology, and believe that God had written on the rocks one enormous and superfluous lie" (105).[44]

Gosse diagnosed his father's mental turmoil as arising, in the last analysis, from a lack of humility, from an "obstinate persuasion that he alone knew the mind of God, that he alone could interpret the designs of the Creator" (73). This comment represents a complete inversion of the conventional judgment that scientists who questioned theology were guilty of hubris, while those who clung to religious dogma were walking humbly with their God. Edmund Gosse has flipped the reference point: geology is now the new orthodoxy, and the individual who questions or rejects it is arrogant. This shift in evaluation is symptomatic of the tacit compromise that had been reached by the turn of the century. The rival factions of science and religion made their public peace by an implicit treaty of partition whereby each side agreed not to trespass on the other's designated territory. However, there have continued to be infringements of this unspoken agreement, as in the notorious 1925 "Scopes Monkey Trial" in Tennessee,[45] and in the attempts by militant creationists to have evolution theory expurgated from school and university syllabuses. These incidents provided material for a number of twentieth-century plays, novels, and films and were to reemerge with force in 2009 during the 150th anniversary of the publication of *The Origin of the Species*.

Immensities of Space and Time

For the scientist characters of Thomas Hardy's novels there is no longer any conflict between science and religion, for the latter has ceased to have meaning. Consequently, there is no comfort or psychological bulwark against the immensities of time and space that science was formulating. Instead there is the terrifying vision of a vast, impersonal, uncaring universe as revealed by astronomy and evolutionary theory, in which any appearance of benign purpose is merely a façade; chance and destruction are the characteristic processes of nature. Accepting this harsh perspective, Hardy's scientist characters are typically shown as deficient in human emotions and social consideration.

The two extended portraits of scientists in his novels are those of Henry Knight, an amateur geologist in *A Pair of Blue Eyes* (1873), and Swithin St. Cleeve, an astronomer in *Two on a Tower* (1882). In the tradition of the Romantic stereotype, both are presented initially as observers, detached from any emotional involvement in life. But the reason for this noninvolvement, unlike that of the Romantic scientist, is their awareness of the vastness of geological time and astronomical space that causes them to denigrate the individual affairs of men as trivial and manipulable.

We are introduced to Knight first in his rooms in London, where he has imprisoned nature in an aquarium in order to study it objectively and in isolation. He is the rational scientist as observer, distanced from the object he is studying. However, he is soon to be disabused of the notion that, just because he can explain natural phenomena in terms of physics, an individual may assume such an attitude of detachment and superiority over nature. He performs an experiment to demonstrate his theory concerning updrafts at a cliff face; that is, he attempts to manipulate nature. However, he is rapidly forced to realize that nature is capricious (it sends his hat down the cliff, not up) and that he cannot remain merely a spectator. Obliged to clamber down the slippery cliff face to retrieve his hat, he finds that he cannot easily climb up again and is thus jolted out of the secure role of observer into the precarious position of one involved in a personal struggle for existence. During his uncomfortable ascent, a close visual encounter with a fossilized trilobite embedded in the cliff, the victim of just such a struggle as his own, suddenly renders geology alarmingly personal: "The eyes, dead and turned to stone, were even now regarding him. . . . Separated by millions of years in their lives, Knight and this underling seemed to have met in their place of death."[46] Just as both Knight and the fossilized trilobite have been deposited on the surface of the cliff by natural forces, so human life is as precarious and temporary as that of its extinct ancestor. Knight, a geologist, thus enacts Hardy's own evolutionary pessimism. However, Hardy does not allow Knight to accept death with Darwinian resignation. This scientist, who affects to receive with complacency a theory of nature in which man is infinitesimal, transitory, and unimportant, becomes as fearful and determined to cheat nature of its victim as any of the uneducated country folk he looks down upon when his own existence is threatened.

In his later novel *Two on a Tower* Hardy expands the time scale from the geological to the astronomical. Swithin St. Cleeve is the first astronomer to feature as the protagonist of a full-length novel, and Hardy took some trouble to obtain accurate information for the astronomical details in the novel.[47] As the title suggests (the tower of the title has been fitted out as an astronomical observatory), the novel explores an astronomer's perception of the relationship between the individual and the universe, of which he is both part and victim.[48] Swithin is introduced from the outset as exhibiting two apparently contrary moods contingent upon these two perspectives. On the one hand, he feels intense elation about his future career in astronomy, cherishing an ambition to become another Copernicus. On the other hand, he is prey to deep melancholy and preoccupation with death. Almost his first words in the novel reveal this paradox: "I aim at nothing less than the dignity and office of Astronomer Royal, if I live. Perhaps I shall not live" (9). Hardy suggests from the outset that astronomy necessarily engenders pessimism, as Swithin appropriates the *ars longa vita brevis* theme for this purpose: "Time is short, and science infinite,—how infinite only those who study astronomy fully realize" (10). Demonstrating his telescope to Lady Viviette Constantine, legal owner of the tower, he directs her vision to the sun, which appears as "a whirling mass, in the centre of which the blazing globe seemed to be laid bare to its core. It was a peep into a maelstrom of fire, taking place where nobody had ever been or ever would be" (8). Paradoxically, the telescope, which is intended to make remote objects appear closer, has the opposite effect: the overriding impression is that the magnitude of space annihilates the individual. Swithin tells Viviette, "The actual sky is a horror. . . . You would hardly think, at first, what horrid monsters lie up there. . . . Impersonal monsters, namely Immensities. . . . Those are deep wells for the human mind to let itself down into, leave alone the human body! and think of the side caverns and secondary abysses to right and left as you pass on!" (34–35).

In 1863 a nova had been detected in Scorpio, and Hardy would certainly have read accounts of its decaying brightness, even if he had not observed it personally. The immense wilderness of space was, for Swithin as for Hardy, made yet more terrible by this knowledge of its impermanence: "For all the wonder of these everlasting stars, eternal spheres, and what not, they are not everlasting, they are not eternal; they burn out like candles. . . . Imagine them all extinguished, and your mind feeling its way through a heavens of total darkness, occasionally striking against the black, invisible cinders of those stars. . . . If you are cheerful, and wish to remain so, leave the study of astronomy alone" (35–36).

From perceiving the transience of even the apparently stable heavens (it is stressed that Swithin's interest lies not in the "fixed" stars but in the "variable" stars) it is a small step to an awareness of human mortality and insignificance. The parallel between the stars and human beings is drawn early in the novel in the comparison between the two figures on the tower and the constellation of the Twins (which, significantly, includes a number of "variable" stars). The centrality of chance and accident in human affairs, a common feature of all Hardy's novels but intensified here by the astronomical perspective, implies, in turn, a comment on the limitations of scientific method. Scientists endeavor to explain observed phenomena in terms of universal laws, which presuppose regularity of occurrence and an order underlying the apparent diversity; Swithin is attempting to do for the universe beyond the solar system what Kepler had done for the solar system, and Newton for physics, namely, to formulate the laws of its behavior. Yet, unlike his famous predecessors, Swithin finds not celestial order but chaos—unpredictable stellar outbursts, variable stars, and inexplicable immensities.

That chance and accident are the norm rather than the exception is made explicit when, near the end of the novel, Swithin spends some time observing in South Africa,[49] only to find the same violent disorder in the farthest reaches of the universe: "There were gloomy deserts in those southern skies such as the north shows scarcely an example of; sites set apart for the position of suns which for some unfathomable reason were left uncreated, their places remaining ever since conspicuous by their emptiness" (317–18).

Thus astronomy, like Darwinism, reveals a world of chance and accident at the heart of even the most apparently stable phenomena. Indeed, if Hardy qualifies at all the ubiquity of chance, it is not to posit a benign Providence overruling accidents in order to spare humanity but rather to suggest a wanton, malevolent power stacking the odds even more steeply against us. The maelstrom on the sun extends the terrestrial cataclysms revealed by geology to the celestial bodies, and the repeated references to the animal names for a number of the constellations keep before us the parallel between the heavenly and the earthly domain. It is in this context that Hardy questions the alleged objectivity of scientists, the apparent complacency with which they discuss the immensities of time and space and the transitoriness of man's existence. Swithin is forced to acknowledge his suppressed fears and to espouse a more subjective position.

Swithin also displays the characteristic failings of the Romantic stereotype of the scientist—emotional deficiency and preoccupation with his work rather than with human relationships. Though naive and egocentric, he never overtly becomes the cold-hearted villain deliberately exploiting others, but the effect is much the same. Symbolically, he lives (sleeps and eats) in his elevated observatory, looking down literally and metaphorically on those mortals who view the heavens unscientifically. Moreover, his preoccupation with astronomy leads to a degree of mental rigidity. Although minutely observant of the behavior of the planets, he is strangely insensitive to the emotional states of those around him and interprets everything literally, oblivious of the implicit messages that social convention suppresses. When Swithin leaves Viviette to observe at the Cape of Good Hope, Hardy remarks caustically: "Her unhappy caution to him not to write too soon was a comfortable licence in his present state of tension about sublime scientific things. . . . In truth he was . . . too literal, direct, and uncompromising in nature to understand such a woman" (310). Even in the final encounter between the lovers, Swithin fails to distinguish her real feelings embedded within conventional words. Through the character of his astronomer Hardy explores both the anguish engendered by a scientific understanding of the vastness of the universe and the pain an insensitive scientific viewpoint can produce in the realm of personal relations.

The Exploiters

The mechanistic principles and reductionist philosophy implicit in Darwinism led to a resurgence in literature of the Romantic stereotype of the scientist as emotionally deficient and socially irresponsible, and for parallel reasons. By insisting that organisms could be explained in purely physical terms without reference to emotional or spiritual values, such scientists implied that they themselves were lacking in these attributes.

In her short story "Cousin Phillis" (1865), Elizabeth Gaskell created an otherwise charming character who displays this insensitivity and amorality. Holdsworth, an engineer designing the new provincial railway, is an intelligent, urbane, but basically superficial man who has traveled widely and takes new ideas and values for granted. He trifles with an unsophisticated country girl, Phillis, who assumes from his behavior that he intends marriage; when she hears that he has married in Canada, she falls seriously ill. Much of the moral emphasis and interest of the story hinges on the contrast between Phillis's sincerity and trust and Holdsworth's inability to comprehend such values or to treat her as anything more than a pretty child. The scientifically trained man is associated throughout with the new, with the future, which to him is by simple definition superior to the past. He embraces change—physical (travel, railways), mental (new ideas and theories), and moral (a different view of responsibilities)—and from this sophisticated perspective he sees no harm in paying Phillis elaborate and insincere compliments. In Gaskell's treatment, both the superficiality of Holdsworth and his insensitivity to Phillis's feelings are closely linked to his scientific background.

Similar but more reprehensible is Dr. Edred Fitzpiers of Hardy's novel *The Woodlanders* (1887), a character as cold and amoral as any of Hawthorne's scientists. Although he has set up as a village doctor, Fitzpiers has little interest in general practice and spends more time in reading, study, and scientific experiments. One such experiment involves the examination of the unusually large brain of Grammer Oliver. Fitzpiers tries to induce the old woman to sell him her brain for postmortem analysis, a project clearly intended to indicate his detached view of humanity insofar as it ignores the feelings of the old woman. Like Faustus and Frankenstein, Fitzpiers has studied at a German university, in his case Heidelberg, but unlike his predecessors, he is presented as an intellectual dilettante, flitting between metaphysics, astrology, alchemy, and medical science. Hardy suggests that Fitzpiers has no commitment to the search for truth but merely indulges his idle curiosity.

Godwin Peak, the protagonist of George Gissing's novel *Born in Exile* (1892), one of the first professional geologists in fiction, also displays the dehumanizing effect of scientific study. Gissing stresses that Peak's emotional deficiency and unethical behavior result from his conclusion that there is no moral dimension or meaning in the universe. What is new in this presentation is not only Peak's calm assumption that no scientifically educated person could accept religious beliefs but also his complete lack of regret that this should be so and his failure to see any reason for retaining a Christian ethic after renouncing its creed. Indeed, the implication in this novel, as in Hardy's *A Pair of Blue Eyes*, is that, seen from the perspective of geology, man's life is indeed trivial. On observing the rock strata disclosed by a quarry, Peak reflects that "imagination wrought back through eras of geologic time, held [man] in a vision of the infinitely remote, shrivelled into insignificance all but the one fact of inconceivable duration."[50] Peak instinctively applies the principles of social Darwinism to justify his ruthlessness: "Life is a terrific struggle for all who begin it with no endowments save their brains" (177).

It is apparent from this survey that the realistic novelists of the late nineteenth century were forced to develop new images of scientists and new ways of presenting them. Most of these foreground struggle—the struggle to survive financially, to reconcile scientific knowledge with faith, to find meaning in a universe expanding in space and time. Despite the early favorable depictions of scientists, such as Kingsley's Tom Thurnall and Eliot's relatively sympathetic portrait of Lydgate, the image of scientists in the Victorian novel became increasingly bleak. The discoveries of geology and astronomy and the acceptance of evolutionary theory, especially in its distorted form of social Darwinism, gave birth to the image of a cold, unemotional, and potentially amoral scientist. Such a character links the Romantic stereotype with one of the prevalent twentieth-century images of the scientist, as arrogant, unfeeling, and indifferent to the sufferings that may result from his research.

These attitudes can be seen as arising from two contradictory estimates of science itself. On the one hand, science is assumed to be more powerful than man, since it is invoked to explain him to himself in wholly scientific terms. On the other hand, it is assumed to be so far inferior to man that he believes he can devise and control it. It is the same paradox that underlay Mary Shelley's tragedy of Frankenstein, who also embraced a mechanistic and reductionist view of man (attempting to construct one from dead or inanimate pieces) and found to his horror that the power of the created monster was beyond his control.

CHAPTER 9

The Scientist as Adventurer

> As for my uncle, who never forgot his work, he was carefully examining the nature of the terrain, torch in hand, trying to discover where he was from observation of the strata.... A scientist is always a scientist as long as he retains his composure, and Professor Lidenbrock certainly possessed this quality to an extraordinary degree.
> —*Jules Verne*

The Culture of Progress

The mathematician and philosopher Alfred North Whitehead remarked that "the greatest invention of the nineteenth century was the invention of the method of invention."[1] Certainly the cult of invention and the technology to which it gave rise were to have a major impact on public attitudes to science. It was no longer seen merely, or mainly, as a matter of contentious and divisive theories but rather as bestowing material benefits, the fulfilment of Bacon's promise of a socially beneficial science. Technological progress brought economic prosperity and commercial imperialism in its wake, and the nexus between technology, capitalism, and cultural superiority was firmly in place by the end of the nineteenth century. Science was no longer seen as godless or disruptive; on the contrary, the exploitation of technology to produce more wealth was regarded favorably in moral terms as evidence of due adherence to the Protestant work ethic. There emerged an attitude of complacency, even benign patronage, toward science as an obedient servant that could be relied upon to be both useful and entertaining. Soirees at which the scheduled entertainment was a series of scientific demonstrations were as popular in upper-class society as public lectures on science at the mechanics' institutes were with the working class (see fig. 16). It seemed safe to sport with the technological monster, to regard it as friendly, reliable, even picturesque. It

Figure 16. Eleven O'Clock P.M.: A Scientific Conversazione. From *Twice round the Clock; or, The Hour of the Day and Night in London*, by George Augustus Sala and William McConnell, 1859. Wikimedia Commons.

was, after all, the age when locomotives were called *Pegasus* or *Black Bess* or named after Sir Walter Scott's novels—*Rob Roy, Ivanhoe, Waverley*.

The literary counterpart of this new attitude was the figure of the adventurer-scientist, a resurgence of the Romantic hero, but now allied with science rather than opposed to it. Heirs to the optimism of both the utopian tradition and the wonderful-journey stories, these characters entered into the popular culture of their time as humanity's advance guard, extending the frontiers of experience, whether in space or time, confident of subduing to their will whatever they found there and transcending mankind's former limitations.[2] Their right to dominate nature, the universe, or whatever alien societies they encountered was as unquestioned as the imperialist regimes, of which they were the literary and scientific reflections, to impose their form of civilization on the "primitive" and "benighted" areas of the globe. Technological might was, by definition, morally right. In such a climate, progress was the supreme good. Sociologist Lewis Mumford has observed of this period, "One could not have too much progress, it could not come too rapidly, it could not spread too widely and it could not destroy the 'unprogressive' elements in society too swiftly and ruthlessly: for progress was a good in itself independent of direction or end."[3]

Jules Verne: The Scientist as Optimist

The belief that scientific discovery was the greatest of all adventures and that European man would progressively master nature is reflected preeminently in the early novels of Jules Verne, whose scientist characters are vivacious and optimistic to a degree that only this period could have accepted so uncritically. These debonair heroes of the series collectively (and significantly) entitled *Les voyages extraordinaires* and later subtitled *Les mondes connus et inconnus* are bent on adventure, courageously risking their lives for the delight and the honor of the quest as they explore the last frontiers—the interior of Africa, the interior of the earth, outer space, the depths of the ocean.[4] These myths of conquest, wherein the marvels of science engage with and overcome the marvels of nature, are not merely entertainment but an expression of logical positivism with a strongly didactic subtext, intended to elicit in the youthful readers of Hetzel's *Magasin d'éducation et de récréation* a spirit of bravery, invention, and optimism about the future that he associated with the American national character, identified in particular with inventors such as Edison, Tesla, and Alexander Graham Bell.

Verne's explorer-scientists not only function as intellectual and moral guides to both the known and unknown worlds but, as Roland Barthes has suggested, they also define the familiar, which they symbolically carry with them inside their vehicles, by means of its opposite.[5]

Himself an engineer, Verne was adept at including, with an air of assurance, scientific or quasi-scientific explanations of technological inventions, but his particular focus on various novel means of transport has deeper significance. Motion, especially speed, becomes a metaphor for the assertion of individual freedom and environmental domination, a notion encapsulated in the motto of Verne's most famous character, Captain Nemo, *Mobilis in mobili*—"moving within the moving element." Thus the projectiles shot into space, the balloons, and the submarines are repeatedly characterized by their velocity. However, while these various means of transport catapult the adventurers into the unknown, they also provide a safe shelter, a cocoon within which the intrepid individual may investigate and explore new realms while effectively remaining in the comfort of his home. Barthes comments that Verne "has built a kind of self-sufficient cosmogony, which has its own categories, its own time, space, fulfillment and even existential principle."⁶ Indeed, these vehicles exhibit as much ingenuity in the service of material comfort as in technological efficiency. Their ornamental elegance, displaying the cultural icons of the Second Empire, both removes the threat of the inhuman machine and celebrates the superiority of French artistry (see fig. 17).⁷

Overburdened with their representative function, Verne's individual scientists are rarely developed psychologically, lest they detract from his primary

Figure 17. Nautilus Salon, by Alphonse Marie Adolphe de Neuville and Edouard Riou. From Hetzel edition of *Vingt mille lieues sous les mers*. Wikimedia Commons.

concern, the journey itself. Otto Lidenbrock, Pierre Aronnax, Paganel, and Palmyrin Rosette are distinguished from the explorers, Captains Hatteras, Servadac, and Grant, chiefly by their determination to explain some device or expound a catalog of marvels in scientific terms. Verne's first novel, *Five Weeks in a Balloon* (1863), set the pattern for the majority of its successors. Drawing on a recent actual invention, Nadar's balloon *Géant*, which had crashed during its second flight, he imagined an improved version, the *Victoria*, which, powered by hydrogen gas and equipped with an effective navigational mechanism, was capable of traveling virtually anywhere.[8]

This early novel introduced the trio of characters who, with minor variations, were to feature in nearly all Verne's adventures—the resourceful but eccentric scientist, in this case Dr. Fergusson, obstinately pursuing a dangerous obsession; his foil, the hotheaded but honorable Kennedy; and the phlegmatic servant, Joe, loyal unto death if necessary. Once the *Victoria* is aloft, the main purpose of the balloonists is to afford Verne the opportunity of describing the journey across Africa, geography being the one science in which, as a member of the Société de Géographie, he could claim expert standing. The intrepid Fergusson delivers detailed lectures on all matters of interest in between the adventures, the outcome of which, given his resourcefulness, is never in doubt. Similarly, in *The English at the North Pole* (1864) the details of the journey, the descriptions of life aboard ship, and the difficulties of crossing the Arctic Ocean, scrupulously derived from actual explorers' accounts, are prominent, but Verne has little interest in assessing the value of such a trip or the motivation of the explorers. His most popular novel, *Journey to the Centre of the Earth* (1864), was based largely on his imaginative synthesis of some then-current theories of terrestrial structure. Verne allows his protagonists, Professor Lidenbrock, a geologist, and his nephew Axel, an apprentice mineralogist, to discuss these at some length, Lidenbrock invoking Sir Humphry Davy's theory that the core of the earth could not be liquid, because if it were, the attraction of the moon would cause twice-daily internal tides, with consequent periodic earthquakes.[9] The journey involves a conflation of the idea of John Cleves Symmes that the earth was essentially hollow, containing five concentric spheres, all with openings at the poles, and Edmund Halley's theory of the aurora as emanating from the interior of the earth. Eventually, however, all is resolved; divergent theories are not permitted to threaten the intellectual preeminence of Lidenbrock or the ability of science to resolve theoretical problems.

Characteristically in Verne's stories nature yields up her secrets to the resourcefulness and determination of the scientists. From his first encounter with the Icelandic volcano of Sneffels, Lidenbrock never doubts the successful outcome of his contest with nature: "The Professor never took his eyes off it, gesticulating as if he were challenging it and saying, 'So that is the giant I am going to defeat!'" (85). And at a climactic moment of danger he cries, "Air, fire and water combine to block my way! . . . I won't give in . . . and we shall see whether man or Nature will get the upper hand!" (206). This "defeat" of nature is a dual one: at the physical level it involves the survival of the party against all the violent forces and hardships nature can muster, even that of a volcanic eruption, but it is also an intellectual defeat, signified by the imposing on nature of a scientific explanation for whatever she can produce. When the party encounters a subterranean forest of strange umbrellalike trees, Axel reports:

> I quickened my step, anxious to put a name to these strange objects. Were they outside the 200,000 species of vegetables already known, and had they to be accorded a special place among the lacustrian flora? No; when we arrived under their shade, my surprise turned to admiration. I found myself confronted with the products of the earth, but on a gigantic scale. My uncle promptly called them by their name.
> "It's just a forest of mushrooms," he said.
> And he was right. (166)

In no time Axel has assigned them a Latin classification, *Lycoperdon giganteum*, measured them, compared them with their domestic counterparts, and made the colonizing process all but explicit by declaring their similarity to "the rounded roofs of an African city" (167).[10]

This obsession with definitive explanations continues to the end of the novel. Even after the safe return of the adventure party to a chorus of universal scientific acclaim, Axel and his uncle are mortified by their failure to explain why they were apparently traveling in a direction opposite to that which they had predicted. "One aspect of the journey—the behaviour of the compass—remained a mystery, and for a scientist an unexplained phenomenon is a torture for the mind." When, six months later, Axel notices that the poles of their discarded compass are reversed, he suddenly realizes, "During the storm on the Lidenbrock Sea, that fireball which magnetized all the iron on the raft simply reversed the poles of our compass!" (253–54). Whereupon all is explained and Lidenbrock is overjoyed.

The unacknowledged aim of all the *Voyages extraordinaires* is to convert the bewildering diversity of the infinite universe to a state of tame finitude. Barthes finds the culmination of this aim in Verne's *Mysterious Island*, "in which the man-child re-invents the world, fills it, closes it, shuts himself up in it, and crowns this encyclopaedic effort with the bourgeois posture of appropriation: slippers, pipe and fireside, while outside the storm, that is, the infinite, rages in vain."[11]

The same obsession with order underlies the fanaticism with which Lidenbrock and Axel, in admirable Linnaean fashion, indefatigably name, classify, and codify everything they encounter, intellectually colonizing the hitherto alien universe and making it safe for human domination. Like their predecessors in New Atlantis, Verne's scientists are compulsive collectors of facts and natural objects. Museums and encyclopedias feature repeatedly as both the results and the symbols of this systematizing process. Captain Nemo's collections have converted the *Nautilus* into a vast submarine museum; Professor Lidenbrock's study represents a private museum; and numerous other scientists in the *Voyages extraordinaires*, if they do not actually possess physical collections, are themselves living encyclopedias. The geographer Paganel has the complete classification of South American birds at his fingertips; Axel can recite the catalog of dinosaurs; and Conseil can reel off the taxonomic identification of fish.

Not only are Verne's scientists armed with all available measuring devices —clocks, compasses, sextants, and maps both ancient and modern—but they characteristically stamp the geographic features they encounter with their own names, thereby announcing their appropriation of the hitherto unknown in a form of intellectual as well as geographic imperialism. In naming Hansbach, Axel Island, Port Gräuben, Cape Saknussemm, and the Lidenbrock Sea, Lidenbrock and Axel not only lay claim to enduring fame but establish their continuing possession of nature. Verne seems unaware of the irony involved in this procedure of laying nominal claim to geographic features in a location (inside the earth) that no one else will ever visit. His scientists not only explain objects and events to their own satisfaction but adopt, with almost missionary zeal, the responsibility of enlightening others. They indefatigably embark on a program of lecturing, explaining, and decoding the cryptogram of nature. Axel affirms Verne's article of faith: "However great the wonders of Nature may be, they can always be explained by physical laws."[12] Cyrus Smith, the engineer of *The Mysterious Island* (1875), who has transformed a desert island into a mechanized utopia, perpetually lectures and instructs his companions in the practical applications of science: "The colonists had no library at their disposition; but the engineer himself was like a book—always ready, always open to the exact page they needed, a book that solved all their problems for them and that they leafed through regularly" (292). Professor Palmyrin Rosette has prepared a lengthy treatise on comets, which is included in full for the edification of the reader. In line with Hetzel's didactic intention, these scientist-pedagogues are lionized by their societies. Otto Lidenbrock's book, whose title is also *Journey to the Centre of the Earth*, allegedly creates a sensation throughout the world, being printed and translated into every language, and its author becomes the "corresponding member of all the scientific, geographical, and mineralogical societies in the world" (254).

An interesting aspect of the characterization of Lidenbrock and Axel is their pro-evolutionary stance. Verne gratuitously introduces the recent discovery of 100,000-year-old fossils and in the 1867 edition has Lidenbrock, to his intense joy, stumble on a fossilized specimen of Quaternary man, findings incompatible with the Genesis account of creation. Axel, too, has a hallucinatory dream in which he travels back through the evolutionary history of the earth to the birth of the planet as a gaseous nebula.[13]

Lidenbrock represents one of the earliest examples of the eccentric, irascible scientist. In the twentieth century such characters were to be metamorphosed first into the comic, absentminded professor, out of touch with his surroundings and later into the prototype of the mad and dangerous fanatic, prepared to sacrifice himself and everyone else in the pursuit of science. There are already mild suggestions of this in Verne's character. "Otto Lidenbrock was not, I must admit, a bad man; but, unless he changes in the most unlikely way, he will end up as a terrible eccentric" (8), confides Axel at the outset; and Lidenbrock's willingness to sacrifice his nephew and Icelandic guide along with himself in a maniacal bid for fame is only defused by comedy and by the reader's realization that Axel, as narrator, has survived to tell the tale. Lidenbrock has other characteristics that are potentially more malevolent. He exhibits a furious temper, which he directs against anyone who crosses him, and he is quite unconcerned about his pupils: "His teaching was . . . intended for himself and not for others. He was a selfish scholar" (8). Yet he is presented with a beguiling humor, which renders him nonthreatening. For this purpose Verne recasts as endearing eccentricities several devices used to satirize the virtuosi. Like Shadwell's Gimcrack, Lidenbrock has an attractive young ward in the house but cannot see the obvious fact that his nephew is in love with her. Like the virtuosi, he has a conglomeration of apparently ill-assorted facts at his disposal, although his facts are relevant, even essential, to the occasion. His knowledge of obscure Icelandic history and geography and his acquaintance with the work of the alchemist Arne Saknussemm and the runic alphabet all prove to be vital for the success of the adventure.

Lidenbrock's ruthless determination to carry out his incredible plan is also offset by his frontier-style virtues—courage and optimism when all seems hopeless and the ability to continue his scientific investigations whatever the personal dangers involved. When the hapless Axel asks incredulously, "What! You still think there's a chance of escape?" he replies firmly, "Yes, I do. As long as this heart goes on beating, I can't admit that any creature endowed with will-power should ever despair" (235), and of course his faith is vindicated, whatever the cost to our credulity. Axel affirms: "A scientist is always a scientist as long as he retains his composure, and Professor Lidenbrock certainly possessed this quality to an extraordinary degree" (237). As the party is caught up in the eruption of Mount Etna, which, unknown to them, is finally to fling them to safety on Stromboli, Lidenbrock remains calm and smiling in the midst of his calculations until finally Axel admits: "My uncle was absolutely right; and never had he struck me as bolder or more self-assured than at that moment when he was calmly working out the chances of being involved in an eruption" (240). This sangfroid was to become a trademark not only of Verne's scientists but also of the heroes of twentieth-century pulp science fiction. A similar combination of ingenuity and optimism under extreme circumstances was to reemerge in the character of Mark Watney in Andy Weir's science fiction novel *The Martian* (2011, film 2015), combined with a new ingredient of self-deprecatory humor.

Professor Pierre Aronnax, of the Paris Museum of Natural History, the narrator of *Twenty Thousand Leagues under the Sea* (1870), shares several of Lidenbrock's ambiguous traits, including a readiness to sacrifice his assistant Conseil (the counterpart of Hans in *Journey to the Centre of the Earth*) to his own scientific curiosity and a phlegmatic acceptance of imprisonment aboard the submarine *Nautilus* in return for the opportunity to observe and photograph underwater phenomena for the first time.

Unlike most of Verne's stock scientist characters, the antisocial and enigmatic Captain Nemo, an engineer and the first serious oceanographer in literature, is a complex character whose genesis appears to owe something to a number of Verne's contemporaries, including Colonel Charras and the young Prince Albert of Monaco, one of the earliest oceanographers.[14] Yet Nemo's interest for us depends less on his being a scientist than on his mysterious past, which is presumed to account for his insatiable desire for vengeance, his passion for music, and his hatred of British imperialism.[15] This latter characteristic, while it doubtless chimed in with Verne's own Anglophobia, owes something also to the American Robert Fulton, who had built an actual submarine called the *Nautilus* (first tested in 1801), which he offered to Napoleon for placing powder mines beneath the hulls of unsuspecting British warships. Fulton's motto, *Libertas maris, terrarum felicitas* (Freedom of the seas is the world's happiness), is not dissimilar to Nemo's *Mobilis in mobili* (Moving within the moving element), painted around an *N*. Nemo combines compassion for the oppressed with the ruthlessness of the avenger, who, motivated by a "fierce, implacable defiance towards human society,"[16] can cold-bloodedly drive through a British ship and watch its passengers sink. Nemo is, in effect, an underwater Robin Hood, distributing to suffering races the wealth he extracts from submerged wrecks. Aronnax comments, "His heart still beat for the sufferings of humanity, and his immense charity was for oppressed races as well as individuals" (191).

An extension of this pity for the oppressed is Nemo's advanced environmental attitude toward the oceans and endangered species. He is far ahead of contemporary thinking in his refusal to countenance the killing of the southern whales as urged by the harpoonist Ned Land. "'To what purpose?' replied Captain Nemo; 'only to destroy! We have nothing to do with whale-oil on board. . . . Here it would be killing for killing's sake. I know that is a privilege reserved for men, but I do not approve of such murderous pastime. . . . Leave the unfortunate cetacea alone. They have plenty of natural enemies'" (214).

The enormous popularity of Verne's writings during his lifetime indicates how accurately he reflected current attitudes toward a science associated primarily with technology and engineering marvels. His scientists are neither stupid nor sinister but heroic in their acceptance of danger and invariably resourceful in extricating themselves. Embarked on a physical journey, a metaphor for the intellectual journey they believe will reveal a marvelous future, they embody the limitless nineteenth-century confidence in science to overcome mere physical dangers and to plumb the depths of the hitherto unknowable. In these stories complex technological marvels also guarantee the superiority of the scientist protagonists who manipulate them so calmly, controlling the power currency of the future.[17]

In many of these stories the ethical qualities appropriate to some swashbuckling adventure—bravery, coolness in the face of danger, and self-sacrifice—are manipulated to vindicate more sinister and more socially destructive traits. The success of Verne and others in making scientific ruthlessness acceptable in a dangerous situation, defusing it by an element of noninjurious comedy, was to have dangerous cultural repercussions. It impeded a critical appraisal of the potential danger to society of arrogance and rampant individualism, both of which were made to seem essential for success in the technological and capitalist enterprise.

From the journey through unknown space it seems, in hindsight, only a small step to a journey in time, yet it was not until 1888, when Wells published his early story "The Chronic Argonauts," the forerunner of "The Time Machine," that such an idea was seriously considered. Although, as we shall see in chapter 10, the majority of Wells's scientific romances are more ambiguous in their presentation of scientists, in "The Chronic Argonauts" the adventurer motif is still prominent in Dr. Nebogipfel, the exotic traveler who transcends all the boundaries accepted by others, even the last frontier of time. Wells draws on many traditional features of the alchemist stereotype as a guarantee of his argonaut's strangeness and intellectual obsession: "thin lips, high cheek-ridges, and . . . large, eager-looking grey eyes that gazed forth from under his phenomenally wide and high forehead . . . [and] glowed like lights in some cave at a cliff's foot."[18] Physical comfort and social intercourse are alike irrelevant to Nebogipfel, who appears arrogant and secretive to the point of being reclusive. Nevertheless, we are persuaded to feel sympathy for him, because his motivation is not desire for financial gain but the hope of finding in a future, wiser generation the personal fulfillment his present world cannot provide.

Arthur Conan Doyle: Science and Spiritualism

The Professor Challenger stories of Arthur Conan Doyle, although written a decade and more into the next century, are effectively variations on this late-nineteenth-century stereotype of the scientist as adventurer. The first novel of the series, *The Lost World* (1912), introduces the eminent biologist George Challenger as a domineering caricature, his towering stature and impressive scientific reputation exceeded only by his colossal arrogance. Many of his physical attributes may have been based on those of William Rutherford, Doyle's professor of physiology at Edinburgh University—the booming voice, the powerful figure, the Assyrian black spade beard, and the macabre sense of humor—but Doyle went on to develop Challenger's eccentricities for his own purposes, both stylistic and didactic.

In *The Lost World* Challenger is still mainly comic and presented in a situation reminiscent of Verne's stories. Like Lidenbrock and Nemo, he is equipped with the obligatory character foils—the Axel-like Malone, the admiring narrator who professes objectivity if not skepticism; the intrepid young Lord Roxton; and the irascible Professor Summerlee, Challenger's sworn rival. Like Lidenbrock, Challenger is prepared to risk his own life and that of everyone concerned on an expedition to the upper Amazon to verify the report of an inaccessible plateau, allegedly unchanged since the Jurassic Age. Challenger contrives to place the burden of proof on his opponents, especially Summerlee, on whom he vents his aggressive and sarcastic insults, even physical violence, without even a pretense of scientific objectivity in debate. Doyle characterizes the whole scientific fraternity in much the same way. His set piece is the meeting of the Zoological Institution, where the entire assembly degenerates into pandemonium, leaving "the students rigid with delight at seeing the high gods on Olympus quarreling among themselves."[19]

Once on the expedition, however, Challenger's courage and integrity and his use of scientific method predominate over the bombastic aspects of his character: "All day amid that incessant and mysterious menace, our two professors watched every bird upon the wing, and every shrub upon the bank, with many a sharp contention . . . but with no more sense of danger . . . than if they were seated together in the smoking-room of the Royal Society's Club in St. James's Street" (92–93).

In the next Challenger story, *The Poison Belt* (1913), Doyle capitalized on the rampant speculation and fears that had attended the 1910 appearance of Halley's Comet.[20] Intellectual arrogance is still much in evidence as Challenger predicts that the earth is about to pass through a belt of poisonous ether that will eventually kill the entire population. He alone knows how to take avoiding action—by laying in a store of oxygen sufficient to outlast the results of the poison: "'Alone of all mankind I saw and foretold this catastrophe' said he with a ring of exultation and scientific triumph in his voice," graciously permitting his wife, Malone, Roxton, and Summerlee to share his shelter.[21] Challenger's equanimity concerning the seemingly imminent extinction of humanity and the lack of authorial criticism of this position, suggest that Doyle regards Challenger's fatalistic stance as a proper or at least inevitable outcome of scientific objectivity: "'As to death, the scientific mind dies at its post working in normal and methodic fashion to the end. It disregards so petty a thing as its own physical dissolution as completely as it does all other limitations upon the plane of matter" (101–3). Fortunately, in the event, the effect of the poisonous ether is temporary and people reawaken unaware of having been in an induced stupor.

Challenger's popularity soon suggested to Doyle a way to enlist his character in the defense of spiritualism, a movement of which he was himself a devotee. In *The Land of Mist* (1926), Challenger condemns spiritualism without having examined the evidence and effectively loses the debate with a spiritualist because he has been too arrogant to study the case of his opponent. Here Challenger's scientific integrity is sacrificed on the altar of spiritualism as he is publicly castigated by his adversary in the terms that hurt most: "The fact that a man was a great physiologist and physicist did not itself make him an authority upon psychic research" (473–74). Challenger is readmitted to a state of scientific grace by confessing his error, and upon witnessing an example of psychic empathy between his daughter and his dead wife, he capitulates completely, enrolling in the ranks of the spiritualists to undertake an adventure into the unknown psychic realm.[22]

One of the unusual features of Challenger qua scientist is his highly emotional nature. Unlike the Romantic stereotype of the scientist, and equally unlike Doyle's more famous creation, Sherlock Holmes,[23] Challenger is never a cold, rational figure but a cauldron of emotions violently expressed. His notorious fits of rage are counterbalanced by his deep affection for his daughter and his dead wife.

Having converted Challenger to the spiritualist cause, Doyle could not very well use him again for the same purpose. In *The Maracot Deep* (1928), therefore, he created another scientist of similar attributes—eccentricity, crusty demeanor, even ferocity, and obsessive devotion to his research, whatever the dangers involved for himself or others. Like Verne's heroes and Challenger, the marine biologist Maracot has a supporting cast in his journey to the floor of the Atlantic at a point called the Maracot Deep. When the steel cable holding their diving ball breaks and the ball settles thirty thousand feet down, Maracot, like Challenger, faces what seems certain death with serenity and scientific integrity, busily taking notes to the last. Much of the novel is concerned with the submarine civilization of Atlantis, but more important for Doyle's evangelistic purposes is Maracot's conversion from a position of scientific rationalism to a spiritualist perception. Empowered by Warda, a spirit of good, he faces, on behalf of the citizens of Atlantis, the evil Lord of the Dark Face, Baal-Seepa, and destroys him. Subsequently Maracot, like Challenger, becomes an evangelist for spiritualism.

It is clear that Doyle used his scientist characters for a didactic purpose, paradoxically to further what now seems an unscientific cause but at the time was one among many fringe sciences that hovered on the edge of respectability.[24]

In all these examples the adventurer-scientist is an isolated individual, or at most supported by a small band of assistants necessarily inferior to him in mental ability. (At this stage there is, of course, no question of a female adventurer-scientist.) These adventurer-scientists are the precursors of the heroes of popular twentieth-century pulp science fiction—the space travelers who journey forth intrepidly where no man has gone before, unfailingly noble in their efforts to repel evil Martians and other galactic conspirators bent on invading the good planet Earth or mistreating her well-intentioned ambassadors.

Even before the turn of the century, however, the progress of science was less the work of individuals and increasingly contingent on the backing of socially funded scientific institutions. In turn, scientists were beginning to wield considerable power in society. While some writers believed that this would usher in a Baconian utopia, many of the most perceptive writers in Britain and Europe, including Verne himself in later life, expressed increasing mistrust of technological progress. The atmosphere of Verne's later novels became much darker, more pessimistic, and the character of his scientists changed accordingly.

CHAPTER 10

Efficiency and Power

The Scientist under Scrutiny

> To this day I have never troubled about the ethics of the matter. The study of Nature makes a man at last as remorseless as Nature. —*H. G. Wells*

The Dangers of Mechanization

By material standards, the wealth and power that had continued to accrue to Britain as a result of her Industrial Revolution confirmed the victory of the machine and its methods. Insidiously the criteria of the machine-oriented system—uniformity, competency, material productivity, unfailing "obedience" to orders—became the tacitly accepted values of society, and hitherto neutral terms such as *order, stability,* and *efficiency* acquired moral overtones. But at the same time as the writers considered in chapter 9 were extolling the high-technology adventure, there were also sporadic protests against the culture of efficiency and uniformity. These protests were diverse and often radical. William Morris's utopian *News from Nowhere* (1890), for example, proposed a return to an essentially medieval, socialist system, with an economy based on handicrafts and organized through guilds. In this chapter we examine the image of the scientist in turn-of-the-century literature that attacked the philosophy of scientific materialism and its technological infrastructure.

The principles most commonly disputed included those already condemned by the Romantic writers: materialism, objectivity, mechanism. The last concept had expanded from La Mettrie's animal-machines (see chapter 6) to include biological evidence from Darwin's *Descent of Man* (1871) and Francis Galton's *Natural Inheritance* (1889). The long-standing distinction between living and nonliving had been discarded by chemists after 1828, when Friedrich Wöhler synthesized the organic substance urea from the inorganic compound ammonium cyanate, and the eminent biologist John Tyndall publicly asserted that all mysteries of nature would ultimately be explained in mechanistic terms.[1] Despite the apparent unanimity in the ranks of science,[2] there was considerable criticism of this view from other quarters. The cult of the machine was blamed for weakening man's body, individuality, mind, character, and morale and came under attack on philosophical as well as social and aesthetic grounds. Whereas, a century before, the Romantic writers had opposed scientific materialism as being inimical to the creative process and the natural expression of the emotions, the writers discussed in this chapter raised intellectual objections about the wider social implications of mechanization and scientism.

One of the first literary assaults on the machine was delivered by Samuel Butler. On first reading *The Origin of Species* Butler had been so impressed that he wrote to Darwin expressing his admiration. Later, however, like Darwin himself, he realized that the weakest part of Darwin's theory was its inability to account for the chance variations upon which it hinged,[3] and he thereupon defected to the vitalist camp.

In chapters 21 to 23 of his utopian fantasy *Erewhon* (1872), collectively entitled "The Book of the Machine," Butler describes a land, Erewhon (an anagram of *nowhere*), where machines have been outlawed because they threatened to enslave the populace. The Erewhonian philosophers had argued that, far from making the machines work for them, people would soon become merely tenders of their machines, subservient to their every "wish." Butler's satire rests upon the implicit premise that if animals are *only* machines, it is valid to invert the reductionist line of reasoning and apply the theories associated with animals to machines. Just as simple forms of life had developed into more complex organisms with increased levels of consciousness and purpose, and supplanted earlier forms, so machines would evolve in complexity and consciousness and overthrow their human masters. For example, since calculators could now "do all manner of sums more quickly and correctly than we can,"[4] and do them without tiring or slacking, they were already more successful, that is, more evolved, than humans. Moreover, machines have already taken on many of the characteristics formerly considered unique to living organisms, even reproduction, for machines can produce new machines. Man, by contrast, becomes increasingly reliant upon the machine to do all his hard labor, until his organs atrophy, leaving the machines to become the real rulers of the world. After hearing these arguments from their philosophers, the Erewhonians have decided to destroy all machines invented during the preceding 271 years, that is, since Erewhon's industrial revolution. Even watches are forbidden, since their complex mechanism poses a similar threat. Although presented as a fable, "The Book of the Machine" constitutes one of the most sophisticated arguments of the time against the galloping mechanistic assumptions of Victorian society. It points to what H. L. Sussman has called "the central paradox of Western philosophy, the conflict between the deterministic implications of science and the inward apprehension of volitional freedom."[5]

The proposition that humans would become willing, and eventually helpless, slaves of the machines that answered their every need was expanded by Rudyard Kipling in "With the Night Mail" (1905) and its sequel, "Easy as A.B.C." (1912), and by E. M. Forster in "The Machine Stops" (1909). This latter story describes a world in which virtually the whole population of the earth lives underground, totally dependent on the machine for their survival. "The Machine develops—but not on our lines. The Machine proceeds—but not to our goal. We only exist as the blood corpuscles that course through its arteries, and if it could work without us, it would let us die."[6] This idea was to become the basis for a number of twentieth-century dystopias, including Yevgeny Zamyatin's *We* and Aldous Huxley's *Brave New World*, which will be discussed in chapter 13.

The converse anxiety, that human beings themselves might be no more than mechanisms, either an engineering model of pumps and levers or a complex chemical laboratory, threatened belief in emotions, abstract values, and mind as more than just the brain that housed it. This apprehension was expressed in several ways in literature, one of which involved scientists creating "intelligent" machines without thought of the consequences. An obvious test case was the chess-playing automaton, since chess required a high degree of creative thinking as well as logic. One of the first such stories was Edgar Allan Poe's "Maelzel's Chess Player" (1836). Given Poe's predilection for mystery and the powers of the imagination, it is perhaps not surprising that this automaton is a fake, activated remotely by a concealed master chess player.[7] But in Ambrose Bierce's more interesting story, "Moxon's Master" (1894), the automaton chess player is genuine and, piqued at being beaten at the game, kills its creator. Bierce's story includes a dialogue between the scientist-inventor Moxon and the narrator about what, if anything, distinguishes living systems from machines. Instead of propounding the mechanist view that all organisms are merely complex machines, Moxon proposes the contrary view, namely, that "all matter is sentient, that every atom is a living, feeling, conscious being."[8] He claims to have shown that plants think, and he now believes that even the constituent atoms of minerals "think," since they arrange themselves into mathematically perfect patterns. The irony implicit in the ambiguous title extends that inherent in *Erewhon*: Moxon has made a machine to entertain himself and to demonstrate his theories, but he has not thought through the consequences. In accordance with his own proposition the machine is alive and therefore not content to accept a subservient, machinelike role; experiencing frustration at losing the game and anger against its opponent, it reaches forward and strangles Moxon with its iron hands. This ending looks back to *Frankenstein* and forward to the twentieth-century stereotype of the scientist unable to control his inventions, but "Moxon's Master" is unique in its highly intellectual exploration of the consequences of mechanistic theory.

Psychology and Emotions

The mind was not the only feature to be explored in relation to the machine. The emotions, the soul, and psychological states were also fair stakes in the fashionable game of find the mechanism.[9] Darwin's theory undermined confidence in the nature and superiority of humans over other animals, and in its wake writers explored the notion that the human personality was an inextricable mix of higher and lower natures. Stevenson's classic story *The Strange Case of Dr. Jekyll and Mr. Hyde* (1886), which has inspired some thirty-four film versions, serious and comic, from 1908 to 2008, including two musicals, is an expression of this anxiety that even moral character might be changed by material (in this case chemical) means. Like so many erstwhile fables of science, this once improbable tale was soon to acquire an uncomfortable degree of verisimilitude in a world where drugs and lobotomies were used to modify the personality of social misfits. Stevenson, who had studied engineering and law before turning to literature, was greatly interested in contemporary psychology, and his story is usually considered as a parable about human nature, which of course it is. A story about the splitting of a personality into distinguishable but, as it turns out, inseparable components has universal relevance, and the fact that Jekyll is a scientist may seem incidental. Jekyll's research in chemistry is, however, more than a plot device, and the parallels with Frankenstein are clear and intentional. Stevenson's particular censure falls on the confidence of contemporary science in its ability to subdue and improve upon nature, including human nature, to usher in a utopian future that will supersede the rosiest visions of Eden. The basically good but naive Dr. Jekyll believes that technological progress is at least parallel to, and perhaps even synonymous with, ethical progress and that sin and guilt are mere atavistic remnants from an earlier stage of evolution. He expects, by means of his chemical knowledge, to expunge the less desirable moral elements from his nature and thereby evolve a superior individual (see fig. 18). Stevenson is careful to stress that the procedure Jekyll uses is not at fault. Scientifically considered, Jekyll's experiment, like Frankenstein's, is a brilliant success. The flaw lies in the scientist himself, in his assumption of his own perfectibility and his consequent inability to see that the evil part of his nature cannot be discarded but will, on the contrary, grow stronger without the constraints of its moral counterpart. G. K. Chesterton, who was later to be a trenchant critic of Wells's utopias, pointed out the major irony of the story: "The real stab of the story is not in the discovery that one man is two men, but in the discovery that two men are one man."[10] The inescapable duality of Jekyll and Hyde was a metaphor not only for the nature of man but in particular for the nature of the scientist in his role as creator of another being. Jekyll, in attempting to reconstruct a new self, encounters, as Frankenstein did, his other self, a doppelgänger, through the fission of his own personality.

A very different attitude to the creation of a mechanical "human" is expressed in the symbolist science fiction novel *L'Eve future* (Tomorrow's eve) (1886) by the French writer Auguste Villiers de l'Isle-Adam. Here a fictionalized version of the American inventor Thomas Edison has constructed a lifelike gynoid, Hadaly, which needs only to be given a human likeness. When his British friend Lord Ewald visits and complains that his fiancée, Alicia, while physically beautiful, is emotionally and intellectually vacuous, Edison provides Hadaly with the physical appearance of Alicia and the spirit of a virtuous and emotionally sympathetic woman.[11] Ewald falls in love with her and travels back to England with Hadaly in a coffin. However, a fire on board destroys her, and Ewald is inconsolable. Unlike earlier fictional characters who attempted to create a human substitute, Edison is not punished. On the contrary, Villiers celebrates Hadaly the gynoid as superior to any human counterpart, fulfilling Edison's boast, "I promise to raise from the clay of Human Science as it now exists, a Being made in our own image, and who, accordingly, will be to us WHAT WE ARE TO GOD."[12]

Villiers's philosophic novel focuses largely on the construction of Hadaly and her emotional empathy, but the benumbed gynoid of *La femme endormie* (1899), by Madame B—— is little more than a sexual toy that accommodates

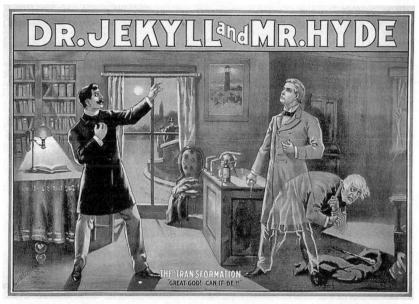

Figure 18. The Strange Case of Dr. Jekyll and Mr. Hyde, poster from the 1880s. Wikimedia Commons.

without resistance the bizarre sexual requirements of her clients. It has been suggested that this pornographic novel is a "carnivalesque mimicry"[13] of *L'Eve future*, insofar as the gynoid, significantly named Mea ("Mine") by the man Molaus, who has purchased her to be his bride, has been reduced to passivity and commodification to serve male sexual fantasies. However, behind the frankly erotic sequences lies the recurrent, problematic issue of equating an artificial human with an actual person and the effect this has on all concerned. Molaus finds Mea perfect in every way, yet she causes him to see and treat every woman as a similar object of sexual desire. As we see in chapter 15, this fear of the consequences of designing perfection continues to be debated in numerous twenty-first-century novels and films, in relation not only to androids but also to the genetic screening of embryos.

Science and Power

From the 1880s on, even Jules Verne, originally so sanguine about technological inventions, became increasingly pessimistic and cynical about science and scientists.[14] Unlike the early *Voyages extraordinaires*, his later writings show a technology shaped no longer for the benefit of society but for the purposes of self-aggrandizement and scientists who have become power-crazed maniacs, obsessed with despoiling nature and destroying humanity. In *The Begum's Fortune* (1879) a French scientist, Dr. Sarrasin, and a German scientist, Herr Schultze, each construct an ideal city, France-Ville and Stahlstadt, respectively. In the aftermath of the Franco-Prussian War Schultze uses his science to revolutionize warfare and initiates a lucrative arms trade to ensure long-term racial superiority for the German people. He is prevented from destroying France-Ville only by an act of Providence such as Verne, disenchanted with human morality, was to invoke with increasing frequency against the evil machinations of successive scientists. Even initially moral scientists such as Robur are not proof against corruption. In *Robur the Conqueror* (1886) the hero acts responsibly, if paternally, like the scientists of Bacon's New Atlantis, refusing to impart his technological knowledge until the society, in this case the United States, is morally prepared for it: "My opinion is, as of now, that nothing should be rushed, not even Progress. Science must not get ahead of social customs. . . . It [my secret] will belong to you the day that you become wise enough to use it constructively and never abuse it."[15] However, in the sequel, *Master of the World* (1904), Robur has deteriorated into a maniacal terrorist (his new vehicle is called *Épouvante*, or "Terror"), wreaking gratuitous violence to prove his superiority until he too is disposed of by the power of Providence in the form of lightning. *The Floating Island* (1895) is a bitter satire of technological materialism, suggesting that Verne had come to realize the sinister potential of the obsession with technological power. Similarly in *Topsy Turvy* (1889) Barbicane, Maston, and the other adventurers of *From the Earth to the Moon* set out to alter the earth's rotational axis in order to melt the polar icecaps and access the mineral deposits beneath. Immune to the pleas and protests from all nations threatened with disaster by this exercise, they are prevented from carrying out their plan only by a mathematical error in Maston's equations. Other irresponsible scientists in Verne's late stories—Thomas Roch in *For the Flag*, Wilhelm Storitz in *The Secret of Wilhelm Storitz*, and Orfanik in *The Carpathian Castle*—are insane and hence even more dangerous because of the unpredictability of their actions. They prefigure the obsessive, immoral scientists discussed in chapter 12 below. Marcel Camaret, of *The Amazing Adventure of the Barsac Mission* (1920), is not actively evil; on the contrary, he is a dreamy, absentminded scientist absorbed in abstract problems and unaware that he is being manipulated by Harry Killer, the evil tyrant of Blackland. Yet Verne makes it clear that the social responsibility of scientists cannot be waived, or their failures pardoned, merely because they had not intended any harm. Camaret is pardoned only after he has destroyed the evil city of Blackland.

Ironically, the most influential and systematic critic of scientists in the last decades of the century was H. G. Wells, the first English novelist to have received formal training in science before he began to write fiction and the writer who was later to be most completely identified with new scientific utopias, the twentieth-century counterparts of Bacon's New Atlantis. Through his experience at the South Kensington Normal School of Science, where he studied biology under T. H. Huxley, Wells was able to expand considerably on the range of scientists depicted in literature, introducing entomologists, geneticists, ecologists, and microscopists. The development of the scientist character in Wells's novels functions as both an index of the increasing influence of actual scientists in his society and a dramatization of his own changing perception of their potential power, which he viewed with varying degrees of apprehension and optimism. Through successive novels and short stories Wells explored the social implications of the impending technological boom and the inadequacy of the earlier code of scientific ethics—strict objectivity and refusal to acknowledge responsibility for the results of one's research, the criteria of so-called value-free research.

Wells's most famous story, "The Time Machine" (1895), a reworking of "The Chronic Argonauts," introduces a range of complex attitudes toward science. It embodies both the optimism he associated with contemporary scientists and the reservations he himself felt at this stage about science and its values. Our first impression of the nameless Time Traveller is of someone open to new ideas. This immediately distinguishes him from his guests, who embody the normal skepticism of society toward anything mentally or physically disturbing. This framing scene inclines us to sympathize with the Time Traveller and, provisionally at least, to accept his reasoning. However, the Time Traveller is no simple hero, and the story he relates carries its own trenchant criticism of the intellect, which, it seems, is a shaky reed to rely on in the future. The main section of the story concerns the Time Traveller's stopover in the year 802,701, when England, at least, has become literally the "Two Nations" that Benjamin Disraeli had symbolically called it.[16] On the one hand there are the effete, childlike Eloi, the beautiful people who inhabit an apparently idyllic paradise in the sun, laughing their way through a vegetarian, close-to-nature existence; on the other, the carnivorous, subterranean Morlocks, the real rulers of the upper world, who prey upon the Eloi as though upon cattle. In Wells's mythology, the Eloi represent one obvious facet of the Romantic ideal; the Morlocks, by contrast, are presented as hideous, ruthless, inhuman, and it is they who are identified with technology. This association, simplistic as it necessarily is in Wells's treatment, generated a powerful image that was visually reinforced in subsequent films of "The Time Machine." Technology is linked with a dark, subterranean existence, an impression that references both the alchemist's cave and the working-class underworld of late Victorian London. It was a powerful symbol for the way the Victorian industrial system preyed upon the lives of its workers who, for virtually their whole working life, were little more than machine fodder, pent up in dark factories, sealed off from the natural rhythms of light and dark.

Set against this visual imagery and the unforgettable picture of the dying planet, the Time Traveller himself also contributes to Wells's characterization of the scientist. Whenever he feels secure in his ability to solve a problem by reasoning, he is invariably on the brink of disaster. Thus he foolishly leaves his time machine unattended, having cleverly, as he thinks, removed the crystal bars essential for driving it through time, only to find on his return that it has disappeared, the Morlocks having simply lifted it up by brute force and transported it through space. He departs from the Palace of Green Porcelain confident in his ability to ward off the heliophobic Morlocks with his matches, camphor, an iron bar, and a bundle of firewood, the equipment suggested by his reason; instead, he unwittingly starts a forest fire, thereby killing Weena, and loses his sense of direction. Witnessing the "decaying vestiges of books" in the deserted museum, he is forced to acknowledge the futility of his "own seventeen papers upon physical optics."[17] Yet, despite his repeated failures, the Time Traveller continues to trust in rationalism. By a series of successive approximations, he evolves a theory to fit the observed facts of the year 802,701; and at the end of the story, still a firm believer in scientific method, he again sets off in his time machine, equipped, as he thinks, with the requisite apparatus to furnish incontrovertible "proof" to his skeptical friends.

The doubtful value of the intellect is reflected and amplified in the Time Traveller's ambiguous attitude toward the future. The narrator maintains that the Time Traveller is fundamentally a pessimist about the future: "He, I know—for the question had been discussed among us long before the time machine was made—thought but cheerlessly of the Advancement of Mankind, and saw in the growing pile of civilization only a foolish heaping that must inevitably fall back upon and destroy its makers in the end" (83). Certainly, his experiences in 802,701 and, even more dramatically, at the demise of life on earth confirm this view. Yet he sets off with as much enthusiasm as ever on his next voyage into time. In other words, he, like the narrator, acts "as though it were not so."

The Time Traveller in fact is no simple hero but the victim of considerable irony on Wells's part. Having spent years constructing a time machine and having endured the extreme discomfort of time traveling in order "to get into the future age," he arrives in 802,701 only to find himself terrified at the "possibility of losing my own age" (34). Yet, even knowing the risks involved, he remains determined to endure the whole procedure again. The ambiguity underlying the story reflects Wells's own pessimism about the future, partly in response to the prevailing fin-de-siècle atmosphere,[18] but more immediately reflecting the ambivalence of T. H. Huxley, who had painted a relatively pessimistic view of humanity's future in his recent Romanes lecture, "Evolution and Ethics."[19]

Despite these ambiguities, the Time Traveller emerges as a sympathetic character, especially in the superb scene of the dying planet, where he experiences total isolation and the most extreme helplessness. He may struggle, albeit with difficulty, against the Morlocks; he cannot struggle against the second law of thermodynamics. In this scene he emerges as the protagonist of a cosmic tragedy, defeated by an insuperable fate. Thus the Time Traveller, Wells's first developed fictional scientist, is a sympathetic but fundamentally powerless figure who, despite his ambitious plan to transcend the limitations of time, is unable to overcome either the forces of evil in society or the inevitability of nature's laws; on the contrary, he errs and suffers, always within the limitations of the human condition.

In his volume of short stories *The Stolen Bacillus and Other Incidents* (1895) Wells explored several examples of scientists, often themselves ruthless, caught up in circumstances beyond their control. "The Diamond Maker" and the title story are simple studies of the scientist exploiting his discovery to gain power, a figure Wells was later to develop in Griffin, the Invisible Man. But two stories, "The Moth" and "The Lord of the Dynamos," are more complex and deserve closer scrutiny. Beneath the comedy of "The Moth" Wells explores the motivation of an apparently dedicated entomologist. We discover that Hapley's twenty-year devotion to his research sprang exclusively from hatred of his rival Pawkins, whose death leaves a void in his life. In Hapley's deranged mind Pawkins returns to taunt him in the form of a rare moth whose capture becomes the sole purpose of his life. In the superficially comic character of Hapley, therefore, Wells suggests the obsessive hypocrisy whereby, on the pretext of scientific argument, a scientist may vent his own subjective passions and obsessions.

"The Lord of the Dynamos," in the same volume, is one of Wells's most effective short stories, depicting, among other things, various prevailing attitudes toward science and technology. The shed housing the dynamos that supply the electric railway at Camberwell symbolizes the technological society, its chief and most obvious attribute being power—in both senses of the word. To it come representatives of three diverse attitudes toward technology. Holroyd, the "practical electrician," reveres the dynamo, partly because it is more powerful than him and partly because it allows him to exercise physical power over his subordinates. He embodies the attitudes of those social Darwinists who professed to find in evolutionary struggle and the technological superiority of one race or group a justification for laissez-faire policies and exploitation of workers. "To James Holroyd, bullying was a labour of love,"[20] and he despises and bullies Azuma-zi, an immigrant worker from Southeast Asia who also worships the dynamo, but for the powerful sense impressions it evokes. Eventually, Azuma-zi, goaded beyond endurance, propels Holroyd against the live terminals, as his sacrifice to the Lord Dynamo. When he is unsuccessful in similarly sacrificing a replacement manager, Azuma-zi flings himself onto the live terminals as a sacrifice to, and mystic communion with, his lord. The parallel nature of the two deaths underlines the fundamental similarity between these apparently disparate worshipers of technological power, Azuma-zi parodying Holroyd and, by extension, the worship of technological power that Wells observed in his contemporaries. If these two characters were the only representatives of science in this fable, we might assume that Wells's attitude toward the technological society was entirely pessimistic. However, there is a third character, the scientific manager, who unites efficiency and feeling within a framework of social responsibility, providing the first example of a combination Wells was later to develop in the "noble" scientists of his utopias.

Wells's Ruthless Scientists

The Island of Doctor Moreau (1896), Wells's most complex and provocative critique of scientism, is one of the great modern myths in the tradition of *Faust* and *Frankenstein*. Moreau himself is at first presented realistically enough as a somewhat fanatical biologist. Exiled from Britain after a scandal involving vivisection, still a topical issue in the 1890s,[21] he now lives, with his assistant Montgomery, in almost complete isolation on an island of uncertain coordinates. Significantly, Moreau, like Frankenstein, seems at first a charismatic figure. Not only is his physical appearance impressive (Prendick remarks upon "his serenity, the touch almost of beauty that resulted from his set tranquillity, and from his magnificent build"),[22] but there is evidence of that noble dedication to research and contempt for ease and social rewards that were to distinguish the noble scientist stereotype of the early twentieth century.

Yet Moreau also functions on a number of symbolic levels, which universalize the meaning of the novel. It is clear, for instance, that he embodies several characteristics of the alchemist. Like Frankenstein, he conducts his secret research into the creation of human beings, finding any human interruption almost intolerable. Just as Frankenstein paradoxically creates his living monster from the materials of death, collecting bones and sinews from graves and charnel houses, so Moreau's experiments, too, are intimately associated with death—with the literal death of the "failures" and with the baptismal bath of pain through which the animals must pass, as though through death, to be made "human." There is little doubt that Wells intended the obsessive Moreau to represent the new image of the mad scientist as genius linked to insanity. His name is almost certainly borrowed from the French psychiatrist Jacques Moreau, whose *Morbid Psychology* (1859) was an early statement of the similarities of genius and idiocy and both as a disease.[23]

Moreau further resembles Frankenstein in cutting himself off from the structures and values of society, dismissing as stupid those who do not approve of his methods. He thus reflects in his own character the profound hubris implicit in his experiment and in the science he typifies. As Frankenstein embodied the idea of the Creator current in Mary Shelley's generation, Moreau represents just such a creator as might be derived from the evolutionary process. The almost arbitrary succession of beasts passing through Moreau's hands reenacts the evolutionary process as nature's giant experiment, wherein much "material" must necessarily be lost for the sake of a few "successes." Moreau, who uses these same terms when describing his work, is as ruthlessly amoral as the Darwinian picture of nature expressed by Tennyson: "So careful of the type she seems / So careless of the single life."[24] Indeed Moreau enacts almost exactly the dangers Huxley had warned would result from an "imitation" of the cosmic process and from any attempt to derive a social ethic from it: "Cosmic evolution may teach us how the good and evil tendencies of man came about; but in itself it is incompetent to furnish any better reason why what we call good is preferable to what we call evil than what we had before. . . . Let us understand once and for all, that the ethical progress of society depends not on imitating the cosmic process, still less on running away from it, but in combatting it."[25]

Moreau justifies his relentless pursuit of his work by just such an appeal to "natural" philosophy: "I am a religious man, Prendick, as every sane man must be. It may be, I fancy, I have seen more of the ways of this world's Maker than you—for I have sought His laws, in *my* way, all my life" (107). The "laws" of nature that Moreau exemplifies are those that underlie the evolutionary process—chance, waste, and pain—and it is repeatedly insisted that Moreau's creations involve all these aspects.[26] Thus Moreau functions as both a parody of the Old Testament Creator and an allegory of evolution itself. In this latter role, he implicitly exposes some of the problems raised by Darwinism not only for religious orthodoxy but, equally, for a humanist belief in the essential nobility and goodness of man. Moreau, who appears to have transcended human limitations, uncovers the innate bestiality of man and finds that evolution operates in reverse, a parody of the title of Darwin's book, *The Descent of Man* (1871).

Moreau also embodies both the arrogance of science in the social context and the ruthlessness of the social Darwinists, who translated the same ideology from biology to a system of economic rationalism. Contemptuous of social condemnation, Moreau insists on his right to inflict pain on his experimental creatures: "Each time I dip a living creature into the bath of pain, I say, 'This time I will burn out all the animal, this time I will make a rational creature of my own'" (112), justifying his procedures by appeal to the time frames of geology and biology, cosmic dimensions of space, and eons of time: "A mind truly opened to what science has to teach must see that it [pain] is a little thing" (106). "After all, what is ten years? Man has been a hundred thousand in the making" (113).

Typically, Moreau's research criteria are impersonal, excluding a moral code: "To this day I have never troubled about the ethics of the matter. The study of Nature makes a man at last as remorseless as Nature" (108). His only guidelines are those of the experiment itself: "I went on with this research just the way it led me. That is the only way I ever heard of research going. The thing before you is no longer an animal, a fellow-creature, but a problem" (107).

However, Moreau does not escape Wells's irony. Just as the Time Traveller, having invented a time machine in order to overcome the restrictions of time, finds himself still the prisoner of time, in each case a time more inimical than his own age, so Moreau, obsessed with achieving mastery over evolution and the limitations of human creativity, falls victim to the very plasticity he sought to achieve, being killed by one of his hybrid creatures. Similarly, just as the Time Traveller finds that the further he travels into the future, the more that future resembles the remote past, so Moreau finds that the more he attempts to overcome the limitations of the human condition, the more he uncovers those limitations; the more he determines to humanize his beasts, the more he demonstrates the bestiality of man.

Despite the dominance of Moreau, both on the island and in the novel, the two other scientists of the novel, Moreau's assistant Montgomery and the narrator Prendick, also contribute to Wells's composite picture of the scientist. Montgomery is judged by Prendick to be a fundamentally weak character, assisting Moreau even while disapproving of his methods. He represents the scientist caught up in a morally repugnant situation from which he feels unable to escape. It is he who, taking over where Moreau's surgery left off, is attempting to teach the Beast Folk to speak. Unlike the research biologists Moreau and Prendick, Montgomery has been a doctor and retains something of the compassion associated with his calling. It is one of the many somber considerations of the novel that in seeking to retain humane feelings while working on Moreau's project Montgomery has virtually become insane. Like his increasingly frequent bouts of drunkenness, his final words express his inability to cope with a world of chance events: "The last of this silly universe. What a mess" (162); but they comment equally trenchantly on the Darwinian picture of a universe governed by chance and accident. In this Montgomery reflects the response of many late-nineteenth-century minds to Darwinism. In their diverse ways, Tennyson, Arnold, Clough, and Hardy all made a similar comment at some stage.

The third scientist, Prendick, is an important and subtly drawn figure whose attitudes change considerably during his stay on the ironically named Noble's Island. After overcoming his initial fear and disgust of the Beast Folk, he pities them, but after Moreau's death he takes over his godlike role on the island, employing both physical and moral force (symbolized by the whip) to keep the Beast Folk under control. On his return to England, he buries himself as effectively in his private research as Moreau had done on his island. Significantly, he takes up chemistry and astronomy: "There it must be, I think, in the vast eternal laws of matter, and not in the daily cares and sins and troubles of men, that whatever is more than animal within us must find its solace and its hope" (191–92). The parallel with Gulliver's pretensions to be above the rest of humanity, to live, like the Houyhnhnms, on a higher plane of pure reason, remote from physical considerations, is strikingly clear.

Thus *The Island of Doctor Moreau* contains not just the element of "theological grotesque," which Wells himself identified,[27] but equally a trenchant satire on scientism, on the isolationism of scientists and their contempt for the layman and, ultimately, for mankind. By portraying the same fundamental attitude, not only in the mythological figure of the fanatical Moreau but also in the apparently ordinary, decent Prendick, Wells extends his implied criticism to include even the respectable gentleman scientist who has "taken to natural history as a relief from the dullness of [his] comfortable independence" (15) and who has "done some research in biology under Huxley" (41–42).

The first film based on *Dr. Moreau* was *The Island of Lost Souls* (1933). In this Moreau explains to Parker (Prendick in the novel), the castaway visitor to his island, that his experiments are intended to accelerate evolution, turning animals into humans through surgery and psychological conditioning. He plans to mate his most successful creation, the Panther Woman, with Parker. However, the Beast Folk revert to type and eviscerate Moreau before starting a fire that engulfs the island. This is a graphic enactment of Huxley's warning that the progress of mankind consists not in emulating evolution and still less in running away from it but in combating it. The second film, *The Island of Dr. Moreau* (1977), emphasized bidirectional transmutation. Moreau injects Braddock (Prendick of the novel) with a serum to bestialize him, but he escapes and the effect wears off. In the 1996 version, supposedly set in 2010, Moreau uses a more contemporary method, genetic engineering, to transform his animals into humans.

In Wells's next novel, *The Invisible Man* (1897), the dramatic interest revolves around the deterioration of character when great power is suddenly placed within its grasp; it is significant therefore that Wells uses a scientist for such a role. Like Faust and Frankenstein, Griffin began his researches in optical density idealistically, for the sake of knowledge, but became corrupted with two countervalues: secrecy and the desire for fame. His account of his accidental discovery whereby he could make an animal transparent and hence invisible shows that he was immediately aware of the power factor: "I beheld, unclouded by doubt, a magnificent vision of all that invisibility might mean to a man,—the mystery, the power, the freedom."[28]

One of Griffin's most chilling aspects is his emotional deficiency, which is explicitly presented as an extension of his scientific objectivity. Despising affection as a sentimental illusion and society as stupid and obstructionist,

Griffin does not scruple to dispose of these and embarks on a reign of terror. The novel is thus essentially a tale of scientific hubris in which Griffin inevitably brings about his own nemesis. Much of the dramatic irony results from the paradoxical inversions whereby Griffin, like the Time Traveller and Moreau, achieves his desires only to find that they restrict rather than increase his power. Once he has achieved invisibility, Griffin's voluntary isolation becomes involuntary, and the arrogant scientist who believes he has progressed beyond "common humanity" is reduced to the most basic survival activities. His brilliant intellect, directed in the pursuit of one goal, becomes monomania, and having become invisible, he now desires most of all to become visible again. This too he finally achieves, but only after being battered to death like an animal by the villagers, rendered brutal by the fear he has instilled.

Like Moreau, Griffin also functions as a symbol of science itself. In an invisible man, who seemingly represents a disembodied intellect, Wells has parodied the primary claim of scientific method, namely, that the observer is invisible. Hence Griffin's career, from the pursuit of pure knowledge to the instigation of his reign of terror, is an implicit comment on the progress of science and a prediction of the likely consequences of adopting its values without question. Griffin represents the culmination of Wells's sequence of unmistakably reprehensible scientists, but before turning to the portrayal of those noble scientists who were to usher in his utopian society, Wells depicted two others, Cavor and George Ponderevo, whose interpretation remains ambiguous.

Cavor, the chemist of *The First Men in the Moon* (1901), who produces a gravity-screening substance called cavorite, remaining blissfully intent on his discovery while disasters threaten, at first appears to be merely another absentminded scientist. But despite the considerable element of comedy, there is a serious moral implicit in the complex portrayal of Cavor, which was entirely lacking from Verne's story of a voyage to the moon. Compared with Moreau and Griffin, Cavor is untainted by the desire for power or personal gain, or even for scientific fame.[29] His assistant, Bedford, remarks incredulously, "It wasn't that he intended to make any use of these things, he simply wanted to know them."[30] Cavor's physical journey to the moon is also a symbol of his readiness to undertake adventurous mental journeys, and like Verne's scientists, he displays in the cause of science an impressive unconcern about his own safety. When imprisoned by the Selenites, the lunar inhabitants, he meditates calmly on the abstract correlations between the earth and the moon, and just before his recapture by them, and the certain death this implies, he is busily scribbling notes about the various forms of Selenite.

But despite these engaging qualities, Cavor is not presented as a hero; on the contrary he is morally suspect. If Bedford, the imperialist and entrepreneur, hopes to colonize the Selenites for the sake of wealth (and the parallel with the genocide of the Tasmanian aborigines under British colonial rule is made explicit at the beginning of the novel), Cavor is prepared to sacrifice not only himself but anyone else in the cause of his research. When his first attempt to make cavorite results in the death of his three assistants, he merely comments, "That is a detail. If they have [perished], it is no great loss; they were more zealous than able" (31). He then proposes to pass the disaster off as a cyclone and collect a considerable share of any compensation. Fundamentally, then, the disarmingly cheerful Cavor is as amoral as Moreau, a fact Bedford perceives early in the novel, commenting: "He had troubled no more about the application of the stuff he was going to turn out than if he had been a machine that makes guns" (21). When, at the end of the novel, Cavor radios back to earth his impressions of lunar society, it is clear that he admires the highly rationalized system whereby the Selenites condition their offspring for the jobs they are predestined to do, so that "each is a perfect unit in a world machine" (258). He regrets his initial instinctive distaste at the means of adaptation as an unfortunate lapse from scientific objectivity and expresses a hope that he will soon learn to appreciate these methods fully, that is, without any emotional involvement. In effect, the Selenite society parodies Cavor's belief that the scientific end justifies the psychological means and that efficiency is the preeminent criterion of morality.

Like the Time Traveller, Moreau, and Griffin, Cavor receives the retribution he deserves. Having unreservedly admired the strict rationalism of the Selenites, he becomes its victim. It would seem that at this stage Wells himself was ambivalent about the rationalist ideals of Cavor and the Selenites, for the satire underlying the comparison between earthly and lunar societies is clearly double-edged. Each falls short of a utopian ideal that would permit both freedom for the individual and an efficient organization of society. Indeed, Wells was the first to enunciate this fundamental logistical problem, which has been echoed in a plethora of twentieth-century dystopias, beginning with his own *When the Sleeper Awakes*: how to combine freedom and efficiency. Our own highly computerized society still has not resolved this question, as recurrent concerns about privacy and security of personal identity data indicate.

Wells's other ambivalent treatment of a scientist in this decade was George Ponderevo, narrator of *Tono-Bungay* (1909). Although this novel has been valued chiefly for its commentary on the social and intellectual climate of late Victorian and Edwardian England, George Ponderevo also marks Wells's first treatment of a scientist as protagonist of a realistic social novel. Ponderevo does not begin as a scientist, but much of the novel is concerned with his rejection of the institutions and attitudes that obstruct science. These incidents themselves serve to define the role of the scientist by contrast. Ponderevo writes, "I have called it *Tono-Bungay* but I had far better have called it *Waste*,"[31] and his story presents one instance after another of waste and confusion at both the individual and the social level. Science emerges as the only viable alternative to this all-pervading chaos against which Ponderevo struggles, and which Wells himself was to struggle with all his life:[32] "Through the confusion something drives, something that is at once human achievement and the most inhuman of all existing things. . . . Sometimes I call this reality Science, sometimes I call it Truth" (335).

The qualities Ponderevo possesses or acquires that enable him to overcome this almost universal disorder, at least in his own life, include a certain detachment, both emotional and mental, in his perception of others, a desire for an overview, and a pervading skepticism, which prevents him from giving his allegiance to any cause or relationship. Temporarily, he participates in the grandiose marketing plans of his uncle and in romantic relationships (with Marion and Beatrice), but his enthusiasm is always qualified, distanced. Love is viewed as a distraction, a rival for the scientist's attention, and after an unsatisfactory marriage and an enervating affair Ponderevo achieves a serene detachment from relationships. On the other hand, science is personified in terms of a mistress who, unlike her rivals, fulfills her promises: "Scientific truth is the remotest of mistresses, she hides in strange places, she is attained by tortuous and laborious roads, but she is *always there*! Win her and she will not fail you; she is yours and mankind's for ever. She is reality, the one reality I have found in this strange disorder of existence" (239). The pursuit of science, which demands both physical and mental discipline, is contrasted with the self-indulgence evident in Ponderevo's uncle and in British society generally.

Ponderevo is also skeptical of conventional religion. All the representatives of religion in the novel are treated with irony as reactionaries, identified with traditions that are seen as irrelevant and powerless to adjust to the necessary changes in society. Instead, Ponderevo substitutes science as a secular religion akin to a revelation and pursues it with the zeal of a convert, regarding all other systems as mere illusion. In later writings, Wells was to make the connection more explicitly, even mystically. In the essay "Religion and Science" he specifically identifies the scientist with a religious pursuit: "The scientific worker, whatever his upbringing may have been and whatever sectarian label he may still be wearing, does in fact believe in Truth—which is his God—in a God who is first and foremost Truth and mental courage. His life business is unfolding the divinity in things, and the real conflict is between the Truth as he unfolds it and the priests and exploiters of the false Gods who still dominate most men's lives."[33]

For Wells, scientists were the only members of society associated with an acceptance of change, but change almost inevitably involves destruction, and this, it seems, the scientist must accept without regret. Ponderevo registers no nostalgia for the passing of tradition, just as he has no hesitation about shooting the native who threatens to interfere with the "quap" project. It is symptomatic that the final symbol of science in the novel is Ponderevo's destroyer, X_2, which he has sold without any patriotic scruples to a non-European power. Hence Ponderevo, in the final analysis, shows a ruthlessness emanating from his belief in the principles of science that differs only in degree from that of Moreau and Griffin. The ambiguity with which he is presented, compared with the clear authorial condemnation leveled against the earlier characters, is indicative of Wells's own changing attitudes toward science.

Trenchant as they are, these skeptical treatments of scientists and technologists in the late nineteenth century are little more than a thoughtful pause between the unthinking exuberance of Verne's adventurer-scientists and the frank glorification, in the first decades of the twentieth century, of the new scientist hero, either as inventor or as leader-elect of a future utopia. These characters are the subject of the next chapter.

CHAPTER 11

The Scientist as Hero

Scientia redemptor mundi.
—Martin Atlas

Despite the misgivings voiced by the writers discussed in chapter 10, the beginning of the twentieth century was characterized by a wave of optimism. Fin-de-siècle weariness and pessimism were swept aside by a renewed belief in progress that mushroomed along with the material triumphs of technology. Carl Becker has commented that to nineteenth-century Europeans, "progress was not so much a theory to be defended as a fact to be observed."[1] This was even more true in the first decades of the twentieth century. The almost unquestioning faith in progress was inextricably bound up with the belief that the products of science would inevitably lead to a better world with better human beings, better government, and only as much departure from perfect happiness as might be necessary to avoid universal boredom. This confidence in progress as unfailingly good was most pronounced in America, where scientific progress was itself a "conservative" idea insofar as it was an extension of the beliefs and values that had characterized American culture from the beginning of European settlement.[2]

It was almost inevitable that the cult of progress should be associated with science, for they shared the same values, namely, a focus on modernity and the future, the ideal of objectivity, rejection of external authority, belief in the increase of knowledge, and the corollary that such knowledge would inevitably be for the betterment of mankind—all Francis Bacon's ideals, decked out even more attractively and forcefully in modern dress. During the 1870s and 1880s T. H. Huxley, the eminent biologist, and Norman Lockyer, editor of the premier science journal *Nature*, used their public positions to insist that good government and sound politics should be based on the principles of science and carried on by those with scientific training.[3] Scientists were encouraged to pontificate on all matters social, political, religious, educational, and moral as well as technological and became the new cult heroes. This uncritical acceptance was reflected in a large number of literary works spanning the period from the 1880s to the Second World War and later. Most of the authors had little knowledge of science beyond what was reported in the popular press, but since this was equally true of their readers, there was little if any pressure on them to research the areas they wrote about so glibly. Scientist characters in these stories reflect the popular credulity about the limitless power of science for good, and few questions, if any, are asked. Biographies of scientists in the 1930s and 1940s were even more popular and uncritically adulatory, and books for young readers such as *The Story of Louis Pasteur* (1936), *The Insect Man* (1936), *The Story of Alexander Graham Bell* (1939), and *Madame Curie* (1943) depicted these saints of science battling ignorance to bring the world the treasures of their researches.

The portrayals of heroism vary considerably, from simple stories of inventors adventuring into space to psychological dramas in which the scientist hero agonizes over what is the best government for utopia. These heroic scientists will be discussed under four headings—the scientist as inventor, the scientist as world savior, the scientist as detective, and the scientist as utopian ruler. Because of the enormous proliferation of science fiction and films in the twentieth century, it is impossible to attempt a comprehensive survey and only representative examples have been chosen. Again, because of the ephemeral nature of much science fiction and the films based on it, in some cases a brief plot summary is necessary to comment on the scientist character.

The Scientist as Inventor

The inventor whose discoveries prove to be of surpassing benefit to mankind is a feature of American rather than of European literature, where theoretical or "pure" science continued to embody the tradition. In the New World, still mindful of frontier values, the inventor who exhibited resourcefulness, optimism, and adventure was perceived as a vitally important member of society, while the pure scientist was regarded as not only less useful but potentially less trustworthy, even sinister, still tainted with the moral flaws of the European alchemist.

America's relatively brief history of science already boasted such prominent inventors as Benjamin Franklin, Alexander Graham Bell, George Eastman, Nikola Tesla, and Thomas Edison, all of whom had attained the status of national heroes. Edison, in particular, became a cult figure, his life story of progress from poverty to fame carrying considerable appeal in the egalitarian climate of the time. Although he had received only three months' formal schooling, Edison eventually established the first industrial research laboratory, his "invention factory," as he called it, and by the end of his life had more than one thousand patents to his name, including those for the electric light bulb; the phonograph; the carbon-resistance telephone transmitter, which greatly improved on the audibility of Bell's model; and the kinetograph, the first fully effective motion picture camera.[4] All these adjuncts to daily living kept Edison's name and fame before the general public to a degree that was unique at the time and has scarcely occurred since.[5]

American writers at the turn of the century incorporated these household names as characters in their fiction, resulting in a subgenre that John Clute has described as "Edisonade."[6] In *Edison's Conquest of Mars* (1898), for example, Garrett P. Serviss enlists not only Edison but Lord Kelvin and Wilhelm Röntgen. Authority was assumed to emanate from the mere mention of such characters. These fictional inventors quickly became associated with another turn-of-the-century preoccupation, Martians. The hypothesis that there was a Martian civilization had achieved popularity as a result of Giovanni Schiaparelli's discovery of the grooves, or *canali*, of Mars from the Milan observatory in 1877. Schiaparelli's word *canali* was mistranslated as "canals," implying a construction by intelligent beings. From this speculation the astronomers Camille Flammarion in France and Percival Lowell in America developed elaborate theories about the purpose of such canals, suggesting that they might be an irrigation system for conveying water from the polar ice caps of Mars to the arid regions around the Martian equator. They also speculated about the nature of the hypothetical Martians and the state of Mars itself. Lowell devised a scenario, widely believed to be factual, that Mars was an aging planet whose inhabitants would soon have to flee and colonize another planet. What place more likely than their neighbor, Earth? Within a short time two streams of Martian literature had sprung up, one an extension of the journey to the moon and one in which the Martians invaded Earth, the latter swelling the general torrent of "invasion" novels precipitated by George Chesney's *Battle of Dorking* (1871), a seemingly realistic account of the invasion of Britain by a German-speaking nation.[7]

The journey-to-Mars narratives assumed that man was still supreme in the universe; therefore in these stories the inventors of earth spaceships could afford to be gracious galactic ambassadors. One of the most popular was Robert Cromie's adventure story *A Plunge into Space* (1890), describing a journey to Mars in a spaceship driven by differential use of an antigravity shield. Cromie's hero, Henry Barnett, is a scientist who after twenty years of seclusion has discovered "the origin of force." Although he is introduced as somewhat comic in the manner of the absentminded professor, he soon emerges as an intrepid physical hero as well as an intellectual giant, directing the space flight with consummate bravery and dignity. "Some of those below drew a deep breath and half muttered a prayer. The engineer himself neither sighed for earth nor prayed to heaven—he was a man of science. His long white fingers did not quiver, his pulse did not increase a beat."[8] Barnett is also a moral hero, able to understand and appreciate the ethically advanced social system of the Martians, who have outgrown aggression and competitiveness. He not only brings his party safely back to Earth but heroically grapples with one member who has become insane, saving the others at the cost of his own life. In this story Barnett is portrayed as a neo-Renaissance gentleman of science, comparable to the socially advanced and ethical Martians.

In Garrett P. Serviss's romance *Edison's Conquest of Mars* (1898) Edison, too, is unfailingly chivalrous to his Martian enemies. When the commander of the expedition threatens the emperor of the Martians, Edison counsels mercy, and the "venerable Lord Kelvin," who is also aboard, behaves "with the courage and coolness of a veteran in every crisis."[9] Edison's fictional nephew Frank features as the young hero of Weldon Cobb's *Trip to Mars* (1901), serving his apprenticeship as assistant to Nikola Tesla, another real-life popular inventor hero.[10]

By contrast, in the stories of Martians arriving on Earth, the confrontation was naturally seen as intrinsically more threatening and the invaders less virtuous. In H. G. Wells's *War of the Worlds* (1898) the Martians fleeing from their dying planet are as ruthlessly bent on colonizing Earth as the Europeans had been in establishing an empire (see fig. 19). They easily overpower the disorganized English but are finally defeated by a simple bacterial infection to which humans have built up an immunity but which kills the bloodsucking Martians. Wells's training as a scientist is apparent in any comparison of *The War of the Worlds* with other contemporary Martian stories. Given the premise of the existence of intelligent life on Mars, the sequence of events in his story is logically impeccable.

The prolific fiction of Martian confrontation after Wells was to have a

Figure 19. Cover illustration by Frank R. Paul for *Amazing Stories*, August 1927, depicting a scene from H. G. Wells's *War of the Worlds*. Wikimedia Commons.

long-lasting effect on the character of the scientist in space. Fear of invasion spelled the demise of the gentleman ambassador desirous of learning from his hosts. The adolescent readers of pulp fiction wanted human heroes with whom they could identify, and the lurid cover illustrations of Hugo Gernsback's magazine *Amazing Stories* and its later rival *Astounding Stories*, with their depictions of space cowboys slaying galactic Indians, both fulfilled and promoted this expectation.[11] The heroes of most of these pulp stories are scientists who invent either novel means of traveling to Mars or new and exotic weapons with which to combat the Martians.

Although scientists and engineers had been employed as weapons designers for centuries, this was the first time they had featured in this way in literature. It paved the way to the expectation prevalent during the First World War (and even more pronounced in the Second) that a scientist's patriotic duty was to direct his expertise toward the construction of ever more effective forms of killing. Certainly in these early works no shadow of blame attaches to the scientists, whether they initiate an attack or recommend mercy.

In retrospect, these scientist-inventors seem crudely conceived, preoccupied with a relatively narrow range of devices—interplanetary travel and communication, new sources of energy (an emphasis triggered first by the Curies' discovery of radium in 1902 and later by accounts of the potential of Rutherford's splitting of the atom), and new kinds of weapons to defeat evil invaders, whether earthly or alien. But they answered the deeply ingrained desire of their (mostly young) readers to transcend the limitations of the material world, for, in a time of relative political peace, scientists replaced military heroes as the conquerors of a new kingdom.

This was also an important factor in the phenomenal success during the 1930s of science fiction pulp magazines, whose appeal resulted largely from their ability to convey a heady sense of the scope and grandeur of the world that science might unlock. Science fiction told its readers that they too could transcend seeming physical limitations of space and time through the power of science. In the tradition of H. G. Wells, the science fiction writers of the 1930s invented the world that less than half a century later would become reality, though at the time only their fans believed them. They wrote about the increasing dependence of society on technology, nuclear and chemical weapons, mind-controlling drugs, the population explosion and its probable social consequences, and instantaneous global communications. It was not coincidental that the avid teenage fans of those magazines (almost exclusively male) became the astrophysicists and rocket designers of the sixties and seventies.

The scientist characters of the pulp magazines were chiefly adventurers, the successors of Verne's heroes, voyaging through the galaxy, doing battle with the forces of evil and breaking through spatial, temporal, and psychological barriers. Almost without exception they were aggressively male and represented a society of male elitism. If women featured at all in these pulp stories, they had a minor part, irrelevant to the real action and subservient to male authority. Their role was usually that of the scantily clad victim menaced by some alien monster, while the human hero, temporarily reduced to the role of voyeur, watched as the monster prepared to enact his own suppressed desires before rescuing her (see fig. 20). On examination this supposedly futuristic genre proves to be highly conservative, even reactionary, in its presuppositions about society and human relations. Its strongly sexist and hegemonic structure was to become the prototype for American science fiction for half a century, leading Ursula Le Guin, one of the great science fiction writers of the next generation, to describe it as "a perfect baboon patriarchy." "From a social point of view most SF has been incredibly regressive and unimaginative. All those Galactic Empires, taken straight from the British Empire of 1880. All those planets—with 80 trillion miles between them!—conceived of as warring nation-states, or as colonies to be exploited, or to be nudged by the benevolent Imperium of Earth towards self-development—the White Man's Burden all over again. The Rotary Club of Alpha Centauri, that's the size of it."[12] The same imperialistic, might-is-right attitude is still evident in the film *Avatar* (2009), in which humans are colonizing Pandora, a moon in the Alpha Centauri system, in order to mine for the rare mineral unobtanium, desirable as a superconductor. Determined to gain access to the mineral, the Resources Development Administration is prepared to attack the native Na'vi and destroy their sacred Hometree.

The first significant break with these simplistic stereotypes came in 1969 with the *Star Trek* TV series. The five-year mission of the starship *Enterprise* was to explore new worlds, to seek out new life and new civilizations, rather than to combat or colonize. In these stories, science remains both the metaphor and the rationale for adventuring through space, "the final frontier," and resourceful astronauts, now including female scientists, retained their hero status, as evidenced by the continuing *Star Trek* series and films and by the *Stargate* series and movie (1994). Both *Star Trek* and *Stargate* emphasize a peaceful purpose—to explore, understand, and communicate. More recently Andy Weir's novel *The Martian* (2011) and the highly successful film (2015) based on it affirm similar values. This mission to Mars is an exploratory and investigative one, not a colonizing or combative one, and the "Martian" is not an alien but an astronaut, Mark Watney, who, lost and presumed dead after an intense storm, is left behind on Mars, when his manned space mission is aborted. Watney, a botanist, is endowed with the same qualities as the earlier scientist heroes—resourcefulness in solving the most daunting problems, optimism, and endurance—and of course, against all odds, he returns safely to Earth; but we should note some important cultural differences. Watney

Figure 20. Uncredited illustration in the first issue of *Marvel*, October 1938, showing an extraterrestrial creature attacking a woman while the hero watches, temporarily powerless to help.

is not a traditional superhero but modest, self-deprecating, and humorous, and instead of rescuing a female victim, he is himself rescued by the female mission commander.

The Scientist as World Savior

More frequently the scientist hero of early-twentieth-century popular fiction not only produced some beneficial invention but used it to save the world—or at least the part of it that was regarded as important by the author and his reading public—from an evil enemy. The nature of the enemy varied greatly in different periods, from an interplanetary alien or hostile European, Middle Eastern, or Asian power to atomic war or a natural disaster affecting the entire planet. It was this wide spectrum of hypothetical hostile situations that permitted the scientist-savior figure (always male) to endure as long as he did.

The rapid social and technological changes and large-scale historical events of the first half of the century, including two world wars, offered continual scope for new kinds of heroism to emerge. The scientist locked in combat with the champion of evil, whether terrestrial or intergalactic, gave way to the Wellsian scientist saving society from endemic inefficiency, waste, and wars. This figure, in turn, was overtaken by the purveyor of nuclear power guaranteed to drive the world's industry and usher in "atomic" peace and prosperity for all. All these characters are essentially similar in their conception, representing both Bacon's dictum that knowledge is power and his optimistic assumption that knowledge is, by definition, good.

One of the first novels to feature a scientist-savior was Bram Stoker's Gothic horror story *Dracula* (1897), in which Professor Van Helsing, although somewhat upstaged by his more famous opponent, saves England and, by extension, Europe from invasion by Count Dracula (significantly of Eastern European origin) and his vampire hordes. Van Helsing is an interesting and unusual character in that, while he is presented as both a medical doctor and a famous scientist, he is also critical of the narrow rationalist methodology of Victorian science. Another character introduces him as "a philosopher and a metaphysician, and one of the most advanced scientists of his day; and he has ... an absolutely open mind."[13] His "absolutely open mind" allows Van Helsing to employ the talismans of religion along with the remedies suggested by science, both old and new. He invokes the powers of hypnotism, blood transfusion, and the herbal remedy of garlic, along with the crucifix and the consecrated Host. Far from surrounding himself in the clinical atmosphere of the laboratory, he identifies his mission with a chivalric and essentially religious past: "We go out as the old knights of the Cross ... to set the world free" (156). Such an alliance of science and religious tradition is atypical, especially in conjunction with a scientific savior such as Van Helsing ultimately proves to be, and reflects Stoker's own predilections rather than contemporary scientific attitudes.

The scientist hero who demolished Martians and vampires soon evolved into the scientist as military hero, ensuring that the right side won in a world war and bringing peace through scientifically generated power. These heroes are almost exclusively American in origin, and the stories in which they appear frequently end with the establishment of an American world empire—a strange ideal for a republic so proud of its revolutionary past, but suppressed imperialist dreams could be justified on the pretext of ushering in a new world order and a Pax Americana. An early example of such stories was Stanley Waterloo's *Armageddon* (1898), in which an American inventor produces a dirigible bomber capable of destroying the navies of an evil European coalition massed against the "good" Anglo-American forces. The American scientist, "working hard to perfect a deadly machine, destructive beyond all others invented," enunciates in effect the doctrine of mutual assured destruction, which was to become a linchpin of US policy during the Cold War: "To have a world at peace there must be massed in the controlling nations such power of destruction as may not even be questioned.... When it [war] means death to all or the vast majority of all who participate in it, there will be peace."[14]

The dangers of this late-nineteenth-century saber rattling were to become fully apparent when these same sentiments, dressed up as the doctrine of deterrence, were enthusiastically embraced as self-evident justification for the full-scale involvement of science in the arms race. It is illuminating to compare the opinions concerning world politics held by many of the scientists involved in "Star Wars" research of the 1980s[15] with those voiced by Waterloo's American victors at the conference of nations: "We consider ourselves the approved of Providence in directing most of the affairs of the world" (242). From this position of self-righteousness it was only a small step to the belief that possession of ultimate power justified world domination by the scientists in order to keep the peace. Most of these fictions about scientifically based world states rely on the development of some new form of energy, which, it seems, has first to be used in a destructive mode in order to clear the ground for the good society.

In Simon Newcomb's *His Wisdom, the Defender* (1900), Campbell, a Harvard professor of molecular physics, discovers an antigravitational substance, "etherine," and a "thermic engine" (the forerunner of a nuclear power plant) that powers cars on minimal fuel, thereby precipitating a new industrial revolution. Such a monopoly on the source of power was frequently used by evil fictional scientists to hold the world to ransom, but Newcomb, himself a distinguished astronomer and professor of mathematics in the US Navy, assumes that his physicist will necessarily be unfailingly benign and paternal. He also believes that absolute authority, vested in a noble scientist somewhat resembling himself, is essential to ensure that the world is kept in order. With no suggestion of authorial irony Campbell announces to the nations, "It became evident to me that if I could retain in my own hands the power to guide the revolution, I could bring about its benefits without its attendant evils. To do this my power must be absolute."[16] Having achieved his plan for a federated world state under his own sovereignty, Campbell proclaims himself "Defender of the Peace" and proceeds to abolish war and colonialism. In this sense, his world federation is utopian and in the forefront of many similar treatments in which scientists not only furnish the vision and the means to usher in utopia but also proceed to govern it with wisdom and benevolence. Like Wells, Newcomb found it necessary to resort to force in order to instigate his utopia but assumed that once it was set up, the scientific lion would lie down happily enough with the democratic lamb in a state of permanent efficiency and bliss.

No such alliance occurs in J. Stewart Barney's once popular novel *L.P.M.: The End of the Great War* (1915). Like Newcomb's Defender, his scientist protagonist Edestone, a thinly disguised Edison, perfects an ultimate weapon and thereby forces the nations of the world to the conference table. However, his world state has no place for democracy, which he ridicules for its inefficiency. Instead he sets up a limited Aristocracy of Intelligence, with an absolute dictator at its head. The apparent incompatibility of democracy, perceived as inefficient, with the goals of science had troubled H. G. Wells, but it is clear that Barney, at least, had no qualms about an autocratic stance.

In numerous stories and novels of the early twentieth century, a world crisis, often precipitated by a "mad" scientist (usually a central European who discovers ultimate powers of destruction and holds the nations to ransom), is averted by a noble scientist, who outwits his evil counterpart. Hollis Godfrey's novel *The Man Who Ended War* (1908) is typical of these in featuring the almost obligatory mad scientist, who develops weapons of mass destruction and starts a world war that bids fair to end life on the planet until a heroic American scientist intervenes just in time.

The much-publicized approach of Halley's Comet in 1910 inspired a flood of catastrophe novels and stories, some of which made use of scientists to save the world, or part of it, from imminent natural disasters. In Garrett P. Serviss's "The Second Deluge" (1912) a scientific genius, modestly named Cosmos Versal, predicts that a watery nebula will engulf the earth and builds an ark to save a select band who will repopulate the earth after the danger has passed. In choosing those he will save, Versal, apparently without irony, declares that he will "begin with the men of science. They are the true leaders." Versal remains the ruler of the new society, to which he "taught the principles of eugenics and implanted deep the germs of science."[17]

Only occasionally are the evil scientist and the noble hero combined in the one person, as in the German novelist Bernhard Kellermann's *Der Tunnel*

(1913), which tempers the heroic scientist with many of the elements of the fin-de-siècle skepticism about science examined in chapter 10. Kellermann's protagonist is an American engineer, MacAllan, who idealistically determines to build a sub-Atlantic tunnel in order to foster worldwide understanding. Although his ruthless obsession with the project takes an enormous toll of money and lives, including those of his wife and daughter, MacAllan persists, and the tunnel, once completed, is hailed as the greatest human feat of all time. The Faustian elements of MacAllan, his determination, arrogance, and willingness to sacrifice others to his project, are finally subsumed in his triumph as "der Odysseus der modernen Technik."[18]

During the First World War the heroic scientist was pressed into service for the war effort on the home side, while his evil counterpart was invariably located in the enemy camp. Thus the age-old tournament between good and evil was briefly entrusted to scientific champions. But with the cessation of war, the scientist was immediately recast as the architect of a new and peaceful society. This had already been foreshadowed in many of Wells's utopian stories. A common Wellsian scenario is one in which scientists end the wars started by others and usher in a reign of uncontested peace; the grateful public is then only too happy to embrace the principles of pure science, rendering a military autocracy redundant. Indeed, in *The World Set Free* (1914) Wells even went to the unprecedented length of having his scientist, Holsten, who had just worked out the mechanism of a nuclear chain reaction, decide to suppress the knowledge in case it might be used for "atomic bombs" (a phrase coined by Wells). Despite his efforts, a world war does break out in the fictional world of the novel (set forty years ahead, in the 1950s), and atomic bombs dropped from planes lead to the near extinction of civilization.[19] Holsten, however, survives to advise the few remaining rulers of the nations about the setting up of a utopian world state based on the principles of science and fueled by the same energy source that had produced the bombs. The message was unequivocal: scientists alone could turn potentially evil power to good.

In *The Man Who Rocked the Earth* (1915), Arthur Train and Robert Williams Wood also have recourse to a peace-loving inventor, who signs himself Pax and uses his power to stop the world war that, appropriately enough for the date of publication, is raging at the beginning of the novel. Pax, who has discovered how to concentrate and direct nuclear power of such magnitude that, as a mere test case, he can flatten the Atlas Mountains, divert the Mediterranean into the Sahara Desert, and shift the earth's axis of rotation, is depicted as being inspired only by a noble motive—to stop the war.

Predictably, there is a strong nationalist bias on the part of the authors: the Germans are seen as untrustworthy, reneging on the armistice in order to test their own new weapon, while the Americans are unfailingly honorable and are selected by Pax to arbitrate on the future of the world.[20] On Pax's death, another American scientist is appointed Dictator of Human Destiny to cooperate with the US president in preserving world peace. There is no suggestion that he or his successors might prove unequal to the task of being morally responsible for the world.

The interwar period also produced a spate of novels in which scientists, either individually or in collaboration, were depicted as establishing lasting world peace. These noble scientific heroes have an epic stature little short of godlike as they unselfconsciously assume the mantle of saviors of mankind. The means of effecting world peace range from a show of lethal weapons, with which the noble scientists smartly bring the rest of the world under control,[21] to plans for reforming governments by the setting up of an international brain trust of scientific advisers.[22]

In line with this latter action, British and American scientists after the First World War were quick to point to their contribution in ending the war successfully and to capitalize on their strategic importance by requesting finance for scientific research. The governments of the major powers funded military-oriented science to a degree hitherto unknown, and this practice found vigorous support in novels and stories of the period. It has been calculated that whereas before 1914 two-thirds of fictional apocalypses were assigned to natural causes, after that time two-thirds were attributed to humans, and of these, three-quarters came in the form of world wars involving scientific weapons.[23]

Not surprisingly, in most of these stories written between 1915 and the early 1930s the villains are terrestrial and the danger is a world war, but before it was apparent that another major war was looming, the dangers from which scientist-saviors were required to save the earth were often extraterrestrial in origin, either physical disasters or evil aliens. One of the best known was the stereotypical eccentric scientist Hans Zarkov in Alex Raymond's space opera comic strip *Flash Gordon*. Zarkov featured as Flash's confederate in fighting the evil Ming the Merciless from the planet Mongo.

In two classic novels of the 1930s by the science fiction team Edwin Balmer and Philip Wylie, scientists are shown as not only predicting a natural catastrophe but preserving at least a section of mankind from destruction. *When Worlds Collide* (1932) begins with the discovery by an astronomer,

Bronson, of a dead sun, Bronson Alpha, which, together with its planet Bronson Beta, is on a collision course with Earth. Although there is no chance of saving the planet Earth, a group of scientists, the League of the Last Days, led by "the world's greatest physicist," Cole Hendron, plans to save a remnant of the human race by transferring five hundred of the best brains and fittest bodies—scientists feature largely in the former category, if not the second—to Bronson Beta as it passes close to Earth. Despite natural disasters and attacks by the millions of desperate people not selected for salvation, Hendron coolly guides his two rocket-powered spaceships to safety on Bronson Beta, having witnessed en route the catastrophic collision of Bronson Alpha and the Earth. The sequel, *After Worlds Collide* (1934), describes the setting up of a stable society on Bronson Beta governed by wise and incorruptible scientists.

These heroic scientists are characterized by their imperturbability in the face of impending disaster and especially by their refusal to be swayed from rational decisions by emotional considerations. This suggests that because, at this period, only scientists were perceived as being able to confront and modify impending natural disasters (apparently even disasters on an astronomical scale), the characteristics associated in the popular mind with their ability to do this (namely, their rationality and their cool-headed, even ruthless, decision-making abilities) were reassessed as noble, elevated qualities, while emotional involvement and humane considerations were correspondingly downplayed. This was a dangerous message to impart to the impressionable teenage readers who would become the powerful scientists of the next generation. Some of its repercussions may be seen in the attitudes of the nuclear club, the scientists who worked on the Manhattan Project during World War II and played with the production of a weapon that had a 10 percent chance (considered acceptable) of exploding the earth's atmosphere. William Broad's interviews with the predominantly young scientists involved in "Star Wars" research indicated the perpetuation of these attitudes into the next generation of atomic scientists.[24]

An interesting and more humane variation of the world-savior model is to be found in Rudolf Daumann's *Protuberanzen* (Prominences) (1940). Here the impending fictional disaster is a terrestrial-based one: the onset of another ice age, which threatens to engulf all existing civilizations. Albin Hegar, Daumann's chemist-hero, predicts the imminent catastrophe but is disbelieved by his fellow scientists. He therefore has to persevere against prejudice and ridicule before he can carry out his plan to melt the glaciers. Daumann's hero takes his righteous stand against reactionary self-interest and limited understanding, declaring that judgment about his proposals can rest only with an "authority higher than that of men, Nature itself."[25] Hegar the noble scientist accepts the responsibility conferred on him and finally succeeds in saving the world by means of his cool-energy system, seemingly a precursor of "cold fusion."

During the 1930s a new source of potential heroism became available to the scientist—atomic power. In these early years there was no suspicion that the benefits might be flawed. On the contrary, the extravagant miracles with which atomic power was credited were endorsed by many of the most respected scientists, including the Nobel laureates Robert Millikan[26] and Frédéric Joliot-Curie, so it is hardly surprising that many novels and stories of the period reflect this euphoria. "Atomic Power" (1934), by Don A. Stuart (a pseudonym for John Campbell Jr., editor of the most successful science fiction series, *Astounding Stories*), epitomizes this optimism. In this story the earth is undergoing entropic death, expanding into quiescence, with consequent panic on the part of its inhabitants. Enter a brilliant physicist who discovers atomic power and recharges the energy of the planet, restoring life, health, and happiness to its frozen peoples. Malcolm Jameson's novel *The Giant Atom* (1944) was possibly even more naive in its combination of the scientist as world savior and intrepid space traveler, saving earth from a "reintegrator" that devours all in its path.

By the 1950s, however, the development of atomic power was viewed with more skepticism. The side effects of radioactivity were becoming more widely known, and the benevolence of nuclear physicists in promoting nuclear power was being questioned. More importantly, there was Hiroshima. After the bombing of Hiroshima and Nagasaki, it became increasingly difficult to portray scientists as having the moral qualities to become world saviors and lead humanity to a glorious future. In America, moreover, there was the awkward fact that former German scientists, some of whom had been prominent Nazis, were welcomed into the national pantheon of missiles-research scientists to boost the US Cold War effort. Wernher von Braun, who during the 1950s played an important role in developing American rocket technology, had been a leader in the group that had developed V-2 rockets for Hitler at Peenemünde. The moral implications of this did not bear too much scrutiny and were conveniently glossed over by the media; but although some scientists, notably the Chicago group, which included Leo Szilard, campaigned to reinstate scientists in the public confidence, arguing that they alone possessed the formula for universal brotherhood, they were never to recover more than a pragmatic respect.

It is significant that at least two of the (few) post–Second World War scientist-saviors in literature have recourse to spiritual assistance. In *A Case of Conscience* (1958), set in the twenty-first century, James Blish, a former biologist and medical technician, takes as his protagonist a Jesuit biologist, Ramon Ruiz-Sanchez, who is sent with three other scientists to report on the civilization of the planet Lithia, apparently a totally rational, nonreligious utopian society in which everyone is both free *and* good. Unlike his fellow scientists, the priest concludes that Lithia is a contrivance of the devil to destroy mankind. He performs the ritual of exorcism on the planet of Lithia at the very moment when another scientist is performing an experiment in nuclear fission. The planet vanishes—because of "An error in Equation Sixteen," says a mathematician observer, but Ruiz-Sanchez and the author believe otherwise. Even more startling than the scientist as priest is the scientist as pope, but Oskar Maria Graf's novel *Die Erben des Untergangs. Roman einer Zukunft* (The heirs of doom: A novel of the future) (1959) features just such a protagonist, who resurrects the world after the devastation of war. Again, the miracles of science are shown to be consistent with the miracles of faith, and Graf's brilliant scientist-pope, whose research into strains of wheat solves the problem of world hunger (the first reference in fiction to the Green Revolution), remains a humble, idealized figure able to establish peace and harmony in his utopian society. Although this may seem an idealized fantasy, Einstein regarded Graf's novel very highly, as suggesting a possible path to world peace and prosperity.

Increasingly, though, after Hiroshima, heroism came to be associated with the suppression, rather than implementation, of new knowledge. Many writers suggested that even if the scientists themselves were ethical, they could not avoid being manipulated by power-hungry governments and the military-industrial complex. In such a situation, heroism consists in concealing potentially harmful knowledge from the authorities, a theme that was to provide the basis for the large number of novels and plays to be examined in chapter 16, in which the ethical scientist is in conflict with social forces and usually fails to protect his discovery from misuse. One example of this is James P. Hogan's *The Genesis Machine* (1978), in which a scientist hero of 2005 secretly subverts the government-military machine to produce a version that can be used only in defense, never for offense.

In all these stories and novels featuring a scientist-savior, the propagandist element dominates the characterization. Rarely is there evidence of a moral struggle within the protagonist. As with so much science fiction, the interest of the story lies in the intellectual game of inventing a new idea, and the only struggle depicted is the external one of individual heroes battling a monolithic authority.

The Scientist as Detective

With the increased importance in judicial enquiry of forensic evidence there emerged a new social role for science and a new popular hero, the scientist-detective who solved crimes baffling to the police (crimes often perpetrated by an evil scientist). The prototype was, of course, Sir Arthur Conan Doyle's best-known character, Sherlock Holmes. Doyle daringly included in his problematic hero many aspects of the traditional villain: not only a dedication to analysis and objectivity but also coolness and avoidance of social involvement. In the first Holmes story, "A Study in Scarlet" (1888), Holmes is introduced as having a comprehensive knowledge of chemistry and as having discovered an infallible test for microscopic amounts of blood. Doyle repeatedly stresses, however, that it is not merely the knowledge and application of chemistry that constitutes scientific enquiry but the logical processes of analysis and deduction. With his analytical cast of mind, Holmes epitomizes the distanced observer, coldly surveying humanity's weakness and stupidity, and although the reader is intended to be dazzled by his intellectual brilliance, it requires all the warmth and unfailing admiration of his assistant Dr. Watson to make him even tolerably human. Significantly the only person whose intellect Holmes respects is his opponent, Professor Moriarty, a stereotype of the evil scientist, yet described in terms that suggest Holmes's alter ego: "He is the Napoleon of crime, Watson.... He is a genius, a philosopher, an abstract thinker. He has a brain of the first order.... I was forced to confess that I had at last met an antagonist who was my intellectual equal. My horror at his crimes was lost in my admiration at his skill."[27]

Richard Austin Freeman's popular detective character, Dr. Thorndyke, a doctor of medicine as well as a lawyer, was another forensic scientist who resolved hitherto impenetrable mysteries by means of the same skills that Doyle had popularized. Although Thorndyke is a less chilling personality than Holmes, the retrospective explanations in terms of logical deduction form an important part of the story and the characterization. Stuart White's novels *The Mystery* (1907) and *The Sign at Six* (1912) are also of this kind, but White offers three scientist types—the cold rationalist; the evil inventor; and the seemingly indolent hero, a latter-day Scarlet Pimpernel, who finally solves the mystery by nonrational means. "A man must have imagination and human sympathy to get next to this sort of thing."[28]

Another interesting forensic character was Edwin Balmer and William MacHarg's Luther Trant, a psychologist who, in order to track down and convict criminals, allegedly developed the lie detector. The authors stressed the reality of the hero's techniques: "The methods which the fictitious Trant . . . here uses to solve the mysteries . . . are real methods; the tests he employs are real tests. . . . such as are being used daily in the psychological laboratories of the great universities . . . by means of which modern men of science are at last disclosing and defining the workings of the oldest of world-mysteries . . . the human mind."[29]

In general there is little attempt to characterize these fictional forensic scientists beyond the pattern developed by Doyle. Their connection with laboratory scientists usually resides in their recourse to deductive logic and their use of strategic chemical tests, but overall there is little to distinguish them from nonscientist crime solvers such as Agatha Christie's Hercule Poirot. Nevertheless, they are some of the few specialist scientists who function as social benefactors, promoting law and order and ensuring that evil does not triumph.

The Scientist as Utopian Ruler

An extension of the scientist who saves the world from crime or disaster is the scientist who ushers in a new society where such evils could never exist. Despite the questioning of technological progress by the writers discussed in chapter 10, there were still optimists who envisaged a benevolent technological society where all physical and social problems would be solved by the application of science. The genetic improvement of the race, the cessation of war, a limitless source of power, the socialization of science, and world economic planning to eliminate poverty feature prominently on the agenda of these heroes. Around the turn of the century the utopian Kurd Lasswitz[30] argued, both in fiction and in two essays, "Über Zukunftsträume I u. II" (Dreams of the future I and II) (1899), for unlimited confidence in technology and the overcoming of all social problems by the combination of technology and nature.

The basis for many of these utopias is the development of some new source of power from which all other social benefits are alleged to flow, a reflection of the optimism that attended the discovery, first, of electricity and, later, of atomic power. These writers are less interested in the inventions per se than in the utopian society they facilitate, offering ample riches and leisure for all. Characteristically in such utopias it is scientists who introduce a better quality of life, both physical and social. Politically these utopias are envisaged as a world state governed by an intellectual-scientific oligarchy whose members are agents of technological progress, providing a universally benevolent society.

One of the earliest of these utopias was Kurd Lasswitz's depiction of a future society controlled by benevolent scientists and technicians in *Gegen das Weltgesetz: Erzählung aus dem Jahre 3877* (Against world law: A story of the year 3877) (1878). Synthetic food production feeds the world's population, and techniques for speed learning by thought transfer overcome the problem of coping with the information explosion. Lasswitz's tales are almost diagrammatic in their propagandist zeal, the idealized characters—Cotyledo the botanist, Funktionata the mathematician, Atom the physicist—representing disciplines rather than individuals. Like virtually all the novelists discussed in this chapter, his interest is in social theory rather than psychological exploration. The simple didactic message of these works is that a world governed by scientists would be as free from greed, inequality, and other moral imperfections, as it would be from hunger, war, and pestilence. Despite the literary deficiencies of most of this fiction, the image of the scientist as an exemplar of social responsibility ushering in a new society retained its popularity until the end of the Second World War.

The best-known creator of science-based utopias was H. G. Wells, who, as we saw in chapter 10, had been one of the most trenchant critics of scientism during the last decades of the nineteenth century. Yet in 1901, as though to welcome the nascent century with a new message of hope, Wells published *Anticipations*, an enormously popular social prophecy about the changing lifestyles to be expected in the new century. With this book Wells effectively became, for the British public, "Mr. Future," and much of his subsequent prolific utopian writing may have been an attempt to live up to this reputation. *Anticipations* marked a turning point in Wells's thought, for although its ostensible method is one of extrapolation from existing trends (and it was this illusion that accounted for its plausibility), the success of his social system in fact hinges on a significant new factor, namely, the emergence of an altruistic scientific elite, the New Republicans, who, while understanding the dangers of technology, are capable of mastering them and who exercise a benevolent, paternalistic control over the masses.

Because he believed that he now understood the means for securing utopia, Wells discarded his former pessimism and embraced the regimen of Bacon's House of Salomon, suitably modified for the twentieth century, as the basis for a revised system of government, education, and social order. This new optimism was to inform nearly all Wells's subsequent fiction, both the utopias and the novels, requiring him to recast his former morally dubious scientists as champions against the destructive, reactionary forces of disorder, inefficiency, and waste. He became the literary spokesperson for those contemporary scientists who were crusading vigorously either for greater political influence or for all politicians to have a scientific education.[31]

The first fictional indication of Wells's changed perspective is his short story "The Land Ironclads" (1903), which is concerned with an alleged military clash between English anti-intellectualism and non-English scientific expertise. The latter is embodied in a group of intelligent and efficient young engineers who, more in sorrow than in anger, speedily defeat the disorganized English patriots. The parallel with the Selenite society of *The First Men in the Moon* is clear, but the authorial position has shifted markedly. Technological progress is now shown as inevitable, and Wells's contemporaries are counseled to accept it—out of necessity, if for no other reason. Wells was here providing fictional support for the major political campaign being waged by many contemporary scientists (including Wells's own hero, T. H. Huxley) who exploited the threat, both economic and military, posed by a technologically superior Germany in order to gain political influence and research funding.[32]

A year later, in *The Food of the Gods* (1904), technological change, symbolized as "bigness," is represented as not merely strategic but wholly desirable. Here, as in the subsequent series of blueprints for a better society, thinly disguised as novels, Wells endowed his scientist rulers with moral as well as technical supremacy. Like Bacon, he assumes that they will be permanently ennobled by the qualities both writers regarded as intrinsic to the pursuit of science—honesty, internationalism, altruism, and compassion—but Wells adds a new factor, efficiency, which he invests with moral virtue. *A Modern Utopia* (1908), in many ways the prototype of all his utopian novels, contains Wells's manifesto on the primacy of technological efficiency in solving the world's problems:

The plain message which physical science has for the world at large is this, that were our political and social and moral devices only as well contrived to their ends as the linotype machine, an antiseptic operating plant or an electric tramcar, there need now at the present moment be no appreciable toil in the world and only the smallest fraction of the pain, the fear and the anxiety that now make human life so doubtful in its value. . . . Science stands, a too-competent servant behind her wrangling under-bred masters, holding out resources, devices and remedies they are too stupid to use.[33]

Ironically, this cult of efficiency embodied in his ruling scientific elite, whom he calls Samurai, is the same criterion that Wells had viewed so skeptically in *The Invisible Man* and *The First Men in the Moon*. But gradually, through his own repeated insistence that the principles of science were the only universally accepted beliefs, Wells came to regard science not merely as rationalism and efficiency but in quasi-religious terms as a mystical power, an aspect of the collective unconscious capable of uniting mankind, as "something that floats about us and through us. It is that common impersonal will and sense of necessity of which Science is the best understood and most typical aspect. It is the mind of the race."[34] As Wells's prophecies became more mystical, his scientists became shadowy, remote figures, high priests of the new religion and increasingly remote from reality.

Wells was not the only writer of the period to pin his faith on scientist-rulers. A similar science-based utopian future was envisaged by the German writer Martin Atlas, whose novel *Die Befreiung* (Liberation) (1910) describes a technological utopia, Penon, ruled by Siler, a benevolent, peace-loving scientist. Siler repeatedly stresses that social reforms can emanate only from the wisdom and ability of the noble scientists, who, like Wells's Samurai, constitute the intellectual aristocracy of the world state. Significantly, the Penonite motto is *Scientia redemptor mundi* (Science, redeemer of the world).

Indeed, despite the involvement of scientists during the First World War in developing ever more sophisticated means of destruction, including chemical weapons, several German writers of the interwar period continued to trust in scientific progress as the necessary precondition for utopia. All these utopias were based on the assumption of a new energy source that would solve the world's economic problems and, by extension, all others as well. Hans Dominik's popular science fiction novel *Die Macht der Drei* (The power of three) (1922) discusses the matter almost diagrammatically, with three scientists representing three different attitudes toward the source of power they have discovered in the *Strahler*, a precursor of the laser.[35]

In his later work *Atomgewicht 500* (Atomic weight 500) (1935), a novel strongly reminiscent of Wells's *World Set Free* (1914), Dominik went on to elaborate on what might be involved in such a utopian society. Again the impetus for social revolution comes from the discovery of a new energy source, this time a new heavy element of atomic weight 500, which has the potential to be used either benevolently or as a means of mass destruction. Realizing this and determined to initiate a new age of beneficial technology, the discoverer, Wandel, steadfastly refuses to be intimidated by his scientific peers or manipulated by financial interests, and eventually he succeeds in gaining international cooperation for the production of energy for peaceful purposes.

Ironically, in these technological utopias, instigated and governed by scientists, the very features that were most enthusiastically propounded were to form the basis of the highly critical antiutopias that characterized the following decades. The authors of *We* (1924), *Brave New World* (1932), and *1984* (1949) regarded with horror the devices and systems that had been hailed so optimistically as the hope of the future, because they saw efficiency and individualism as mutually exclusive and evaluated them in inverse order to their predecessors. Yet there was one more attempt to promulgate the Wellsian-style utopia based on efficiency, this time induced by psychological conditioning, namely, Burrhus Frederic Skinner's *Walden Two* (1948), written specifically to counter Aldous Huxley's stringent critique of behaviorist practices in *Brave New World*. Skinner, himself a psychologist and one of the main proponents of behaviorism, had a frankly propagandist motive in writing his fictional account of an ideal, nonpolitical society where behaviorist principles ensured that all its members would work for the happiness and fulfillment of the group and achieve those rewards for themselves in the process. In the standard pattern of utopian fiction, Skinner's narrator, Burris (an obvious echo of Burrhus), leads a party to inspect the community founded by a former academic, Frazier, another convinced behaviorist. Against his will, Burris is forced to assent to the superiority of Walden Two over all other societies, to admire the orderly but varied and stimulating life possible when every citizen voluntarily accepts and acts for the overall good of the community. Frazier claims the right to be judged as a scientist, but he is also the most powerful ruler imaginable, since his control of each individual's behavior is largely unrecognized. Although inevitably he remains a flat character as a result of the weight of propaganda he bears, Frazier represents an interesting attempt, unique in the twentieth century, to justify intellectually the possession of ultimate power, the right to decide what is "the best course for mankind forever": "I've admitted neither power nor despotism. But you're quite right in saying that I've exerted an influence and in one sense will continue to exert it forever. . . . I did plan Walden Two—not as an architect plans a building, but as a scientist plans a long-term experiment, uncertain of the conditions he will meet but knowing how he will deal with them when they arise."[36]

Like *Brave New World*, *Walden Two* is based on a denial of the concept of personal freedom (epitomized in Thoreau's *Walden* [1854]), but unlike Huxley, Skinner justifies this. Frazier asserts, "By using the principle of positive reinforcement—carefully avoiding force or the threat of force—we can preserve a personal sense of freedom" (220). Notwithstanding the vigorous defense of his theories, Skinner fails to answer the objections raised by Huxley, and it is finally disclosed that Skinner's primary goal is not to make people happy and productive but to create a large-scale laboratory for the study of human behavior: "We can study [these things] only in a living culture, and yet a culture which is under experimental control. Nothing short of Walden Two will suffice" (242). Thus, despite his overt role as idealist and world leader, Frazier represents a reincarnation of the scientist who subordinates humanity to his obsession with science.

In all these high-technology utopias nature is regarded much as Bacon had regarded it, as an endless source of material and energy for exploitation. As this assumption has come to be questioned and the disastrous results of such disregard for environmental integrity have been recognized, the utopias based on technological domination over nature have come to seem naive. In the first quarter of the twentieth century, however, there was little hint of ecological fragility, and these utopian schemes were accepted uncritically as the way of the future.

In British and European literature disenchantment with the heroic scientist occurred, with few exceptions, almost immediately after Hiroshima. There are two principal reasons for this. First, in the face of nuclear war, unlike any other war, heroism becomes irrelevant. None of the traditional attributes of heroism—courage, self-sacrifice, even cunning—has any effect on an ICBM. Retaliation, even if possible, is at best only a Pyrrhic victory: there are no winners of a nuclear war. Thus in modern warfare the scientist hero is redundant. The real action is carried out by machines, not human beings, and machines are ultimately controlled, if at all, by politicians and generals, not scientists. The best that scientists can do is attempt to outwit the military-industrial complex by employing its own strategies of military deterrence, bluff, and deception. Second, and perhaps more interestingly, scientists came increasingly to be judged as responsible for the nuclear menace. Robert Oppenheimer's memorable statement that "the scientists have known sin" was to echo through the corridors of memory and fiction for decades.

In America the idealism persisted longer than in Europe. As the physical chemist Eugene Rabinowitch wrote in 1956, "In 1945 after the revelation of the atomic bomb, American scientists enjoyed a brief spell of popular respect and acclaim. It was they, most people realized, who had dramatically ended the war in the Far East; it was they, many believed, who held the keys to the future power and prosperity of America. Some went so far as to look up to scientists as men destined to rule the world in the 'atomic age.'"[37] Among American science fiction writers, in particular, there was a mood of self-congratulation over the apparent fulfillment of their prophecies that atomic weapons would bring the nations to their senses and usher in a reign of peace and prosperity based on atomic power. This chimed in well with the generally euphoric mood of a country that saw itself as the real victor of the war, having ended the conflict when its European allies had failed to do so and having emerged not only unaffected by the privations of war but as the richest nation in the world. The defeat of Japan, regarded in the United States as an evil, militaristic power, was taken as a mandate for noble American scientists to re-create the world in their own image. Atomic power, provided it remained in the safekeeping of Americans, would never be used for evil purposes but only to promote universal happiness and prosperity.

In the United States this euphoria persisted until well into the 1950s. Then, with the launching of Sputniks I and II in 1957, Americans were shocked by the realization that the Russians not only had caught up with them in the development of atomic weapons but had actually preceded them into space.[38] Suddenly scientists, and in particular physicists, were perceived as morally suspect. Either they were seen as deficient in national loyalty and even likely, in an excess of misguided internationalism, to give away their secrets to the Soviet bloc, as Klaus Fuchs and other atom spies had done, or they were retrospectively blamed both for their own part in the development of the weapons and for being unable to render the enemy's weapons harmless. In popular

Western films, the role of the heroic savior in matters nuclear passed from the scientists to government agents (a curious but strategic inversion of traditional moral attitudes toward spying). Only another spy, it seemed, could be trusted to outwit the atom spies.[39] Scientists, by contrast, were increasingly portrayed as either irresponsible, caring only for science, regardless of social consequences, or helpless to put back in the bottle the evil genie they had so carelessly released. It was to be half a century before scientists were again cast as heroes in novels or popular films.

CHAPTER 12

Mad, Bad, and Dangerous to Know
Reality Overtakes Fiction

> I'm a scientist... who's given his life to pure knowledge.... I have a chance of performing the last and greatest experiment known to science. To release the earth's energy to destroy—I hope in a flash--the life on it.... One last triumphant stroke... leaving the mindless cosmos to its own damned dance of blind energies for ever. —*J. B. Priestley*

The scientists' reign as heroes of twentieth-century literature was always open to question, and even the promise of a limitless power source was not sufficient to promote the idea beyond the 1940s. But long before this, and indeed concurrently with much of the scientist-as-hero literature considered in chapter 11, mad and evil scientists holding the human race to ransom had become a commonplace of pulp fiction, horror films, and cartoons and were not infrequently found in mainstream novels and plays. Probably no other profession has provided modern literature and films with so many villains. I have noted that, from his extensive survey of the horror films produced between 1931 and 1960, Andrew Tudor concluded that one-third present science as the main cause of disaster, and of that third a large proportion portray the scientists behind the science as evil.[1] And more recently Peter Weingart and colleagues, after analyzing 222 movies, established that even initially "good" scientists were often subsequently corrupted by powerful, evil interests.[2] Certainly mad scientists feature as one of the four most common sources of terror in horror films, the only other contenders being natural disasters (often precipitated by scientists), the supernatural, and psychic disorder.

The mad scientist has been a feature of cinema from its inception. The very first silent movies were based on fictional examples. In 1897 Georges

Méliès adapted the Faust legend for *Le cabinet de Méphistophélès* and the following year produced *Damnation de Faust* and *Faust et Marguerite*. He also photographed pantomime clips of linked tableaux, his "ludicrous expeditions," showing comic mad scientists, inventors, astronomers, and alchemists, professing to be from the Institute of Incoherent Geography and in 1902 *Le Voyage dans la lune*,[3] a satirical combination of Jules Verne's *From the Earth to the Moon* (1865) and H. G. Wells's *First Men in the Moon* (1901). Méliès was clearly fascinated by the cinematic potential of the mad scientist. He produced short films of an absentminded lecturer; a manic inventor, Crazybrains, in an airship; an astronomer so engrossed in his observations that he falls off the roof of his laboratory. In America Edison Studios dominated the market, and here, too, crazy inventors and scientists featured largely. In its four-minute trick film *A Trip to Mars* (1910), an inventor-professor unintentionally sprinkles two antigravity powders into his waistcoat and flies off to Mars.[4] In the same year, the Edison Manufacturing Company produced the first cinema adaptation of *Frankenstein*, a sixteen-minute horror film, which professed to focus on the psychological elements of Shelley's novel.[5]

Why does a society that is so dependent on its scientists for infrastructure, wealth, lifestyle, medical breakthroughs, and repair of environmental damage and that provides vast sums of money for science research continue to condemn scientists in its most popular mode of entertainment? Why has it perpetuated stereotypes so outmoded, so implausible, almost parodic?

One reason is that the speculative fantasies concerning the evil scientist that we have traced from the medieval alchemist, and that existed in oral tradition long before that in the figure of the sorcerer or prehistoric shaman, were finally overtaken by reality. By the first decades of the twentieth century the power and influence wielded by actual scientists as a result of their technological expertise was perceived as equivalent to, if not surpassing, whatever supernatural efficacy had been attributed to their magic-dependent fictional forebears. The writers to be considered in this chapter expressed the fears of their contemporaries that science and its products had the physical power to crush individuals and whole societies, even the entire human race. Partly because it still retained an aura of mystery, this power was presented as having an unstoppable momentum, beyond the knowledge and control of ordinary citizens, and thus it inevitably raised the specter of what would happen if it should fall into evil hands. However, the reasons why it was more commonly represented at this time as being used for malevolent rather than benign purposes are more complex.

At the most basic level, evil characters in fiction are simply more interesting than morally impeccable ones, but a network of other reasons became particularly pertinent at this time. As we saw in chapter 8, one response to the immensity of space and time revealed by nineteenth-century astronomy and geology and popular accounts of the heat death of the universe through entropy was a sense of powerlessness, a loss of religious faith, even despair. As the century drew to a close, however, and the next generation grew familiar with these ideas, such responses were most often replaced by a cynicism that, discarding ethical perspectives along with religion, resorted to exploiting the here and now without fear of future retribution.

The frequent association of the evil scientists of this period with acts of terrorism whereby they hold a society to ransom suggests an implicit connection between the power of wealthy industrialists and the science-derived technology that enabled them to amass their wealth. Many of the mad and evil scientists in the first decades of the century are represented as extortionists demanding money in return for the use (or nonuse) of their inventions. These characters, who reenact on a world scale the tyranny practiced by H. G. Wells's Invisible Man, are the laboratory counterparts of ruthless capitalists exploiting their workers to gain yet more wealth, confident that they will not be disciplined. These figures feature largely in suspense tales, in which the focus is less on character than on the means whereby some savior, often another scientist, saves society from the evil machinations of the maniac. Although these stories suggest that society's obsession with wealth and technology are ultimately responsible, they characteristically have a happy ending. In this period writers and audiences alike retained a residual belief in the long-term benefits of science and technology and the ultimate survival of humanity.

Rather more interesting than these power- and money-obsessed characters are those whose malevolence and determination to destroy the world allegedly arise from their frustration and despair that the universe is apparently nothing more than a chance eventuality or a flawed mechanism. Given access to sources of apparently limitless power, their potential for destruction offers scope for scenarios of chilling horror.

Also emerging in the literature of this period is the first explicit connection between scientists and the machinery of war. Previously, evil scientists were depicted as individuals bent solely on their own advancement, but in the early twentieth century, by linking technology with new and more effective military weapons, writers signaled the increasing reliance of society on its scientists for the nationalistic posturings that were to lead to two world wars and the Cold War. Some also refer obliquely to the manipulation of the war machine by scientists in order to obtain funding for their research by arguing that it is of strategic significance to defense. Thus many of the writers considered in this chapter foreshadow, albeit often in simplistic form, the later, more complex discussions regarding the changing power balance between science and society.[6]

In modern fiction the rise of the evil scientist is directly contingent on some new technology, usually a source of physical power, over which they hold a monopoly. Dynamite was followed by electricity, which was not only finding an increasing number of useful applications but was rumored to be capable of annihilating a whole city from a distance. Just when electricity was being demystified, radioactivity, a power more strange than any before, was being publicized. Wilhelm Röntgen had discovered X-rays in 1895, and the following year Henri Becquerel had discovered radioactivity in radium. Two years later Marie and Pierre Curie had observed the effect in both radium and polonium. Radioactivity seemed mysterious, uncanny, and closely related to the whole alchemist tradition. It glowed in the dark, it was associated with the transmutation of elements, it was capable of killing animals (Pierre Curie had killed a mouse with a dab of radium and told reporters that he would not wish to share a room with a kilogram of the substance), and its power was rumored to surpass that of electricity. Contemporary scientists did nothing to dispel this aura of mystery surrounding radioactivity. In 1903 Sir William Crookes, a British physicist who understood very well how to exploit nationalistic fervor and the power of the press, chose to describe the energy locked in one gram of radium as being able to "blow the British navy sky high."[7] This graphic image, combining both power and destruction, made it seem highly desirable to have a monopoly on such energy. In his Nobel Prize address in 1905 Pierre Curie had even hinted briefly at the possibility that it might be better not to know the secret of such power: "One may suppose how radium could become very dangerous in criminal hands, and here we might ask ourselves if it is to mankind's advantage to know the secrets of nature, if we are mature enough to profit by them, or if that knowledge will harm us." But like most scientists of the period, he immediately rejected the pessimistic alternative: "I am one of those who believe with Nobel that mankind will derive more good than harm from such discoveries."[8]

The mysterious power of radium was soon overshadowed in the popular mind by the awesome implications of the one equation that everyone, even the least mathematical, could remember: $E = mc^2$. The possibility of converting matter into energy was adapted to military uses in popular fiction long before it became a reality, the first example being in H. G. Wells's novel *The World Set Free* (1915). Again, the scientists, for reasons as diverse as the impulse to divest themselves of sole responsibility for the outcome of their work and the need for research funding, were quick to publicize the enormous power of their discovery. Frederick Soddy, a British chemist who had worked with Rutherford on radioactive decay, wrote of scientists in 1932, "If ever they are in a position to transmute the elements at will—it would put into the hands of men physical powers as much greater than those they now possess as these are greater than the forces at the command of the savage," adding, "but of this particular contingency there is, perhaps fortunately, as yet no sign whatever."[9] This ambivalent message of warning and optimism was still current in the 1930s, when Sir Ernest Rutherford called his popular book on atomic physics *The Newer Alchemy*, introducing the notion that the quest of the alchemists had been realized, since elements could now be transmuted into others. It was in this climate that Flash Gordon, hero of space opera comics, could sabotage an "atom furnace" to outwit his archenemy Ming, who boasted that "Radioactivity will make me Emperor of the Universe."

Radioactivity and its possible effects, both known and speculated upon, were not the only source of power and fear associated with science. During the First World War, scientists were called upon to devise newer and more lethal military hardware and chemical weapons. Fighter aircraft, new explosives, bombs, submarines and tanks, and the use of chlorine and mustard gases to kill thousands at a time turned even the most sinister predictions of Wells's *War in the Air* and *War of the Worlds* into reality. It became apparent that the power of science could be used for mass evil as well as for good, and the optimistic ideas put forward by H. G. Wells, J. B. S. Haldane, and others that scientists could be safely entrusted with such potential received a severe setback.

The dropping of the atomic bombs on Hiroshima and Nagasaki and Oppenheimer's memorable statement that "the scientists have known sin" revived and amplified the long-standing archetype of the evil scientist from Dr. Faustus onward. Spencer Weart remarks, "When the United States dropped atomic bombs on Hiroshima and Nagasaki, the news had no immediate impact on the shape of the stereotype of nuclear scientists. Rather the stereotype already formed leapt into new prominence, its emotional force redoubled.... The name of Frankenstein was invoked everywhere from street corners to the US Senate."[10] To those who were appalled by the destruction it did indeed seem that power-crazed physicists capable of destroying humanity were walking the earth, respected, courted, and jealously guarded by their governments as a national defense weapon.

The "Frankenstein" tag referred not only to the horror engendered by the bomb itself but to the unassailable power of the scientists as possessors of unique knowledge who denied any social responsibility for its consequences. The American physicist Freeman Dyson, who had remarked, "The Faustian bargain is when you sell your soul to the devil in exchange for knowledge and power. That, of course, in a way is what Oppenheimer did," also acknowledged in himself the terrible attraction of such power:

> The glitter of nuclear weapons. It is irresistible if you come to them as a scientist. To feel it's there in your hands—to release this energy that fuels the stars, to let it do your bidding. To perform these miracles—to lift a million tons of rock into the sky. It is something that gives people the illusion of illimitable power and it is, in some ways, responsible for all our troubles, I would say—this, what you would call technical arrogance that overcomes people when they see what they can do with their minds.[11]

The spate of novels, plays, and films dealing explicitly with the scientists working on the development of the bomb at Los Alamos indicates the fascination this theme had for the twentieth century.

Because of their potential for instigating disaster on a cataclysmic, even planetwide scale, evil scientists quickly superseded the scientist hero in films and comic strips. These mad, bad, and dangerous scientists fall into two main categories: (1) the power maniacs, usually physicists, obsessed with the potential of some new energy source and bent on world domination or destruction on the grand scale; and (2) those whose basic philosophy is inherently evil in some more subtle and sinister way. The latter group includes a significant component of biologists, depicted, according to the perennial Faustian stereotype, as attempting to usurp the province of the Creator by creating, changing, or extending life. These characters were to prove no less prophetic than the atom-smashing destroyers; they are the fictional precursors of researchers into genetic engineering, in vitro fertilization, and ecological manipulation.

Power Maniacs and World Destroyers

The belief that scientists were on the brink of discovering a new, immensely powerful form of energy suggested a source of imminent disaster that lent itself readily to literary treatment and produced a flood of stories depicting mad scientists unleashing their newfound powers of destruction upon the world. These paranoiacs often appeal to their catastrophic reading of nature (as revealed by geology and astronomy) as justification for destruction, and it is apparent that all of them, for one reason or another, actively hate life. A number of such scenarios appeared at the turn of the century and continued to form a large component of the offerings in the pulp science fiction magazines until 1945, when such fictitious horrors were overtaken and rendered obsolete by factual accounts of the effects of the atomic bomb.

The earliest treatments were necessarily nonspecific about the details of the threat posed by the mad, bad scientist. Where there is some attempt to account for his motivation, Darwinism is frequently invoked, either on the grounds that from the evolutionary perspective the extinction of humanity is of little significance, or because, like Doctor Moreau, the evil scientists have adopted what they interpret as the ruthless ethics of Darwinian nature. In most of the stories dealing with such events, character is entirely subservient to the plot, but there are two early and interesting examples where the motivation of the scientist is examined in some detail.

In W. H. Rhodes's story "The Case of Summerfield" (1871), the mad chemist Summerfield threatens that unless the citizens of San Francisco pay him the sum of one million dollars, he will throw into the Pacific Ocean a pill that will instantly convert it, and subsequently all the oceans of the world, into a blazing inferno, destroying every living thing.[12] Rhodes insists that Summerfield is both insane ("I thought I could detect in his eye the gleam of madness" [19]) and brilliant (his facial features resemble those of both Newton and Alexander von Humboldt), and suggests that from his post-Darwinian perspective Summerfield sees only evil and destruction in nature. Like Swithin St. Cleeve in Thomas Hardy's *Two on a Tower*, he points to the history of catastrophes throughout the universe, quoting von Humboldt's account of the destruction of a star in Cassiopeia and the deterioration of Sirius. Appealing to the violent and inanimate aspects of the universe and using a nonhuman time scale, he argues that ethical criteria are irrelevant.

For Summerfield the inducement underlying the threats is solely hope of profit. A more interesting study is provided by Robert Cromie's classic novel *The Crack of Doom* (1895), in which a mad scientist, Herbert Brande, is motivated by a warped idealism. Believing, like Summerfield, that "the universe is a mistake" and animate nature a regression, he determines to reduce the solar system, or at least the planet Earth, to "pure elemental ether." Post-Darwinian nature is the villain of his universe. "She has no system . . . she is not wise. . . . The theory of evolution—her gospel—reeks with ruffianism, nature-patented and promoted. The whole scheme of the universe, all material existence . . . is founded upon and begotten of a system of everlasting suffering. . . . Wholesale murder is Nature's first law."[13] Brande also voices a doctrine that was to become almost a cliché in post–World War II discussions of science, namely, that science *cannot* be stopped: "No man can say to science, 'thus far and no farther.' No man ever has been able to do so. No man ever shall!" (20).

After the publication of Rutherford's *Radioactivity* in 1904, a new source of power, and with it new and sinister aspects of behavior, became available for literary embellishment, and a wave of novels appeared envisaging the misuse to which the release of such enormous reserves of energy might be put. Among the early atom-smashing devices, miraculous rays are almost obligatory, and the writers' nationalist prejudices decided which devices were morally good and which were evil. In George Griffith's *The Lord of Abour* (1911), for example, a characteristic example of contemporary "invasion" literature, the Germans invent a ray that can "demagnetize" metal so that it crumbles to dust, and the British fleet is destroyed by rays from the enemy's wooden ships, a fictional enactment of Sir William Crookes's provocative statement quoted above. However, the British retaliate with the products of their "good" science—helium-radium bullets—which soon demolish the enemy. In Arthur Conan Doyle's story "The Disintegration Machine" (1929), which also purports to derive from Rutherford's work, Latvian scientist Theodor Nemor has invented a machine "capable of disintegrating any object placed within its sphere of influence . . . [into] its molecular or atomic condition."[14] The amoral Nemor is intent only on selling his invention to the highest bidder, but he is persuaded by the British scientific hero, Professor Challenger, to demonstrate his own machine, whereupon he is disintegrated—permanently.

Less optimistic about the possibility of controlling the evil was Upton Sinclair's play *The Millennium: A Comedy of the Year 2000* (1907), in which a new radioactive element, radiumite, is let loose by a mad professor, killing all life on earth except for eleven plane passengers flying above the contamination.

The following year Anatole France was exploiting the possibilities of atomic rays in his novel *Penguin Island* (1908), in which terrorists destroy whole cities with pocket-sized atomic explosives.

The Czech writer Karel Čapek, better known for his play *R.U.R.* (see chapter 14 below), also drew a connection between destructiveness on a grand scale and the irresponsible pursuit of what is scientifically interesting. In his novel *Krakatit* (1923), an engineer, Prokop, has invented an explosive, krakatit, and believes that there is within the earth an incalculable force of explosiveness (atomic fission). At first Prokop cares only about the excitement of discovery but is gradually converted to a sense of social responsibility. His "solution" is not to reject his discovery but to look forward to its peaceful use.

Murray Leinster's two ingenious stories "The Storm That Had to Be Stopped" and "The Man Who Put Out the Sun" (both 1930) also revolve around the efforts of a mad, power-crazed scientist, Preston, to make himself a world dictator by holding the world to ransom over his transmission of the sun's rays through the atmosphere, leading to catastrophic climatic changes and all the symptoms of nuclear winter. Happily, his evil plans are thwarted by another scientist, Schaaf, but, as in most of these suspense-centered stories, the rescuing scientist is a mere deus ex machina, and there is no explanation of why it is any more safe to trust his intentions.

Underlying all these stories is the premise that scientists have access to an unprecedented level of technological power, far beyond that of any other group in society, and are therefore capable of affecting the whole planet, if not the solar system. While, as we saw in chapter 11, it was sometimes postulated that this power would be used to abolish war and usher in an era of plenty for all, the scenario in twentieth-century fiction is more frequently one in which scientists, despite their intellectual brilliance, are not to be trusted. Their habitual preoccupation with abstract thought and statistical parameters renders them ready to sacrifice whole populations if this should appear necessary to further their research.

With the development of increasingly powerful weapons of mass destruction, speculative fantasy was superseded by fact. Georg Kaiser's play *Gas I* (1918) was one of the first to exploit the fear of gas attacks during the war to intensify the picture of a ruthless scientist/engineer determined to develop industrial gas, whatever the cost to the workers. "Our gas feeds the industry of the entire world," he proclaims.[15] When a dangerous fault in the process is discovered, the engineer denies that there could be a flaw in his calculations, and even after an explosion kills hundreds of workers, he urges the survivors on with promises of wealth and power in terms reminiscent of wartime rhetoric: "There is your rule, your mastery—the empire you have established" (623). Kaiser's engineer personifies the dehumanization of the industrial age, in which desire for material wealth and the power of science outweigh humane considerations. In the sequel, *Gas II* (1920), a world war breaks out and the same engineer modifies the gas factory to produce nerve gas. "The gas that kills!" he screams maniacally, holding aloft a glass "bomb" of deadly gas in order to rally the workers. "Ours the power! Ours the world!"[16] Kaiser, himself a witness of the horrors of actual gas warfare, makes no attempt to devise a happy ending; his play ends with the extermination of the entire population.

By the mid-1930s, as news of the race to produce the first atomic weapon leaked out, the atomic bomb superseded all other scientific horrors, both in the public imagination and in literature and film. The first horror movie to refer to the nuclear menace was *The Invisible Ray* (1936), featuring an initially visionary astronomer, Dr. Janos Rukh, played, significantly, by Boris Karloff, fresh from his triumph as Frankenstein's monster. Rukh, having been exposed to strong radiation X from a meteor, can kill with his touch and slowly loses his mind. Crazed by his wife's infidelity he proceeds to eliminate his colleagues but cannot bring himself to destroy his wife. He is finally annihilated in flames from the radiation. In the 1940 horror film *Dr. Cyclops* the alchemist-like Dr. Alexander Thorkel has a laboratory in a remote Peruvian jungle where he experiments with shrinking living creatures using radium from a nearby deposit. When his secret weapon is discovered, Cyclops ruthlessly turns it on the interlopers, shrinking them to the size of mice, intending to kill them. In all these cases the suggestion is clear that scientists are not to be trusted with any such extraordinary powers.

More realistic, though certainly no less pessimistic, is J. B. Priestley's novel *The Doomsday Men* (1938), concerning a plot to explode an atomic device in the Mojave Desert. The three MacMichael brothers responsible for the project are mad fanatics who, for diverse reasons, desire to end the world. The novel was to acquire a grim retrospective irony when, seven years later, a group of apparently sane men was to carry out just such a project in another American desert at Los Alamos. Of the three brothers in the novel, only one, Paul, is a scientist. Priestley explains his fanaticism as a combination of pride in his own preeminence as an atomic physicist and despair, like that of Cromie's character Brande, at the centrality of chance and accident in the universe. Priestley implies that for physicists bent on discovering order in the universe, the realization that chance and accident predominate is devastating.[17] Paul MacMichael explains to another scientist his plan to set off an atomic chain reaction that will peel the skin off the earth, "like an orange, only faster!":

> I'm a scientist. A good one, . . . who's given his life to pure knowledge. . . . I have a chance of performing the last and greatest experiment known to science. To release the earth's energy to destroy—I hope in a flash—the life on it. That life, in my opinion, was an accident. . . . Out of the eternal dance and changing patterns of light and energy . . . mind has somehow emerged, to acquire knowledge but also to understand its own noble despair. But it can still use that knowledge for one last triumphant stroke, one supreme act of defiance, . . . grandly destroying itself, leaving the mindless cosmos to its own damned dance of blind energies for ever.[18]

Given his premise, MacMichael's conclusion, like Brande's in *The Crack of Doom*, is entirely logical.

The actual deployment of atomic bombs on Hiroshima and Nagasaki, at first widely accepted by the Allies as a necessary means of ending the horrors of war, was soon reassessed by writers who perceived that such weapons represented not an end but a beginning. One of the earliest treatments of this theme was F. H. Rose's novel *The Maniac's Dream* (1946), which effectively recasts Conrad's problematic character Kurtz from *Heart of Darkness* (1902) as a maniacal physicist who forces his natives to construct the ultimate weapon. In his characterization of the physicist who sees himself as the agent of vengeance to destroy the destroyers, Rose presciently suggests both the endless proliferation of weapons of mass destruction that the Cold War was to produce and the insanity of believing that the chain of destruction can be halted at will: "The bombs which the Americans made for the destruction of their enemies in Japan, are as nothing compared with the developments that have followed. . . . It has often happened in history that the destroyer of others has himself been destroyed by his own device. America and Britain, now that they have unloosed upon humanity this awful power, talk glib nonsense about controlling the uncontrollable."[19]

The subsequent Cold War produced its own particular brand of fictional scientists, who by humanitarian criteria must be considered either evil or deranged in their obsession with testing the latest nuclear weapons and their use of every subterfuge to escalate the arms race and effect a confrontation between the superpowers. The new and more dreadful feature of these characters is their apparent normality. They are perceived by their colleagues as dedicated and patriotic Americans, motivated by essentially normal psychological reactions. L. Sprague de Camp's story "Judgement Day" (1955) focuses on such an atomic scientist working at Los Alamos, whose indifference to humanity is traced back to his treatment at school, where, in response to being ostracized for his cleverness, he developed a disaffection for people that has gradually reached psychopathic proportions. At the moment of his research triumph he reflects: "I find myself getting more and more indifferent to everything but physics, and even that is becoming a bore." Once he has the opportunity of destroying most of humanity, including his former tormentors, he scarcely hesitates: "I'd kill them by slow torture if I could. If I can't, blowing up the earth will do."[20] The British film *Seven Days to Noon* (1950), the story of a formerly naive and compliant physicist, Dr. John Willingdon, who has been involved in the manufacture of atomic bombs, explored this question further. Willingdon's mind cracks when he grasps the enormity of what he has done; he steals a bomb and announces that he will blow himself up together with the whole of London unless the British government promises to renounce nuclear weapons.

The most influential of all nuclear war films featuring the maniacal scientist was Stanley Kubrick's *Dr. Strangelove, or: How I Learned to Stop Worrying and Love the Bomb* (1964), based loosely on the novel *Two Hours to Doom* (1958), by British author Peter George. The central figure, Dr. Strangelove, a scientist-strategist in a motorized wheelchair, is, on the one hand, a reissue of the Romantic stereotype of the emotionless, mechanized scientist who has lost his humanity (the wheelchair suggests a parallel with another dehumanized character, D. H. Lawrence's Clifford Chatterley). But as a composite figure, comprising elements of Otto Hahn, Edward Teller, and Henry Kissinger, he also represents a macabre combination of mad doctor and state scientist.[21]

Christopher Frayling has traced the connections between *Dr. Strangelove, or: How I Learned to Stop Worrying and Love the Bomb* (1964) and *Metropolis*, seeing the wheelchair-bound Strangelove as "the Rotwang of the nuclear age."[22] He describes Strangelove as one of the great archetypes of the scientist in film: "a combination of the mad scientist (Frankenstein-style), prosthetic scientist (who has lost touch with his humanity), corporate scientist (detached from the consequences) and genius specialist working for the military."[23] Rotwang's mechanical right hand has morphed into Strangelove's black-gloved right hand, with its propensity to jerk into the motion of "*Sieg Heil*" reinforcing Strangelove's repeated reference to the American president as "Mein Führer." As a former Nazi physicist, who had worked on the German V-2 rockets, he has only admiration for the Soviet's Doomsday Machine, which will inevitably detonate if there is an attack on the Soviet Union, destroying life on earth, and he coolly calculates the likely number of megadeaths in such an event. His back-up plan is to relocate in deep mine shafts out of reach of radiation several hundred thousand people whose descendants could repopulate the earth a century later. The director, Stanley Kubrick, had become fixated on the notion that a nuclear war could be started by accident or through the actions of a mad scientist, and Strangelove, with his white hair falling over his forehead, is clearly in that tradition. More immediately he also recalls the German aerospace engineer Wernher von Braun, who had invented the V-2 rocket for Germany before defecting to the United States just prior to the Nazi surrender and was placed in charge of the American Space Flight Center to build the Saturn V rocket.

The exaggerated symbolism of medieval sorcery and references to Frankenstein, alongside the emphasis on Strangelove's sinister debility from which he miraculously "recovers" at the climactic ending, shocked audiences into considering the complex reasons behind the arms race. Not least among these was the motivation of the scientists themselves, especially the physicists, who were powerful figures in the military-industrial complex as long as they continued to produce ever more ingenious weapons of mass destruction. The film raised such questions as Who else would employ these specialist scientists in such numbers, on such funding levels? and Was it not, therefore, in the physicists' own interest to further the escalation of the Cold War?

Less grotesque but, for that reason perhaps more sinister, is Professor Groeteschele of *Fail-Safe*, also released in 1964, based on the 1962 novel by Eugene Burdick and Harvey Wheeler. If Strangelove references von Braun, Groeteschele seems based on Edward Teller, who worked on the Manhattan Project. As the scientist advising the Pentagon during the Cold War he is wholly devoid of human considerations. "60 million dead is the highest price we should be prepared to pay in a war," he declares, having completed a profit-and-loss account that to him appears a wholly reasonable risk calculation.

A later and more sophisticated treatment of the mad scientist seeking to destroy the world is Philip Dick's allegorical novel *Dr. Bloodmoney; or, How*

We Got Along after the Bomb (1965). It depicts a postapocalyptic world after a nuclear holocaust, set off when the paranoiac physicist Dr. Bluthgeld incorrectly certified that a high-atmosphere bomb test would be quite safe. Bluthgeld believes that he personally precipitated the destruction by calling down nuclear thunderbolts through his own psychic powers, because in his clinical state of deluded reference he confuses secondary and final causes. Bluthgeld, therefore, represents quite literally the mad scientist; he has even sought treatment from a psychiatrist. On the one hand, his delusions of grandeur lead him to believe that he is the chosen instrument of God for the punishment of mankind: "I am the center. God willed it to be that way."[24] Yet he also, in his way, regrets this. Thus Bluthgeld stands as the parodic personification of the complex characteristics Dick attributed to the mad scientist archetype—paranoia, delusions of grandeur, obsessive behavior, and belief that he is the instrument of God. Although we at first assume that Bluthgeld's belief in his ability to effect a nuclear cataclysm through psychic powers is further evidence of his insanity, Dick progressively undermines any distinction between "real," or external, events and those that occur within the psyche. If Bluthgeld invented the atomic bomb, he is indeed responsible, as he himself believes, for the 1972 fallout and for the imminent Third World War; whether the actual means are psychic powers or a causal chain of events is, in Dick's analysis, ultimately irrelevant.

Bluthgeld is finally removed from the action by an equally mad technologist, Hoppy, but such a character, endowed (as a result of mutation caused by radioactive fallout) with increased psychic powers, proves no less dangerous than Bluthgeld. Dick suggests, in response to the escalation of the arms race, that no one can be trusted with the power necessary to arbitrate and control those already in power.[25] If we compare this no-win situation with the easy solutions proposed by writers earlier in the century, involving some allegedly benevolent dictator who enforces world peace, we can see the full extent of Dick's pessimism, articulating a general cynicism during the Cold War about the coexistence of power and benevolence.

Philosophies of Evil: Biologists and Psychologists

More subtle but no less dangerous on a smaller scale than the insane and evil physicists seeking to destroy the world are those scientists intent on exploiting their influence over living beings to gain wealth, fame, or power. The early years of the century produced a variety of such characters in literature, many of them modeled closely on the Frankenstein archetype, and they too were quickly incorporated into films and pulp fiction. With few exceptions, these sinister characters were biologists, with, later, a sprinkling of psychologists; indeed, until the onset of the Cold War these disciplines surpassed physics as the breeding ground of scientific mania. Fictional biologists incurred the further occupational hazard of falling prey not only to insanity but also to blasphemy. While a reductionist methodology was acceptable in the physical sciences, in biology it was censured not only as leading to emotional retardation on the part of the scientist but as engendering a desire to assume divine control over life and the mind, the last bastion of spiritual and moral free will. The public utterances of eminent biologists such as T. H. Huxley and John Tyndall had done little to discourage this idea. Tyndall had asserted that although there was as yet no clear proof of a link between consciousness and molecular activity, nevertheless, all the mysteries of nature would ultimately be explained in mechanistic terms,[26] and Huxley had claimed that consciousness was reducible to a mere reflection of molecular movement, psychic events in the mind being caused solely by physical and chemical events in the nervous system. The immoral biologists of fiction invariably adopt this reductionist stance.[27] While the image of the evil physicist was tempered by the American cult of the inventor, biology had no such "champions" and continued to languish under the imputation of godlessness.

An early-twentieth-century example of this stereotype occurs in Maurice Renard's once popular novel *Le docteur Lerne, sous-dieu* (1908), a horror story concerning an evil German scientist, Klotz (nationalist prejudices resulting from the Franco-Prussian War are clearly operative here), who has "killed" a kindly French doctor, Frédéric Lerne, and transplanted his own brain (and hence his immoral personality) into Lerne's body. Klotz has then embarked upon a series of grafting and transmutation experiments based on those of Dr. Moreau (Renard dedicated the novel to Wells), intending eventually to sell new, virile bodies to geriatric millionaires (hence the title of the English translation, *New Bodies for Old*) and ultimately to render himself immortal by transferring his own soul to a piece of machinery. Klotz thus reenacts the alchemist's search for the elixir of life, albeit now pursued through surgery, and is accordingly decked out in all the trappings of the alchemist— clandestine activities, soul transference, a desire for godlike powers, and a concomitant lust for gold—but his bizarre belief that immortality resides in machinery makes explicit the connection with reductionist biology. This concept has reemerged in twenty-first-century narratives in which immortality is associated with transferring the neural content of a brain to a supercomputer to create a sentient machine, as in the film *Transcendence* (2014).

In the same year Somerset Maugham also revived the evil alchemist model, in *The Magician* (1908), a novel strangely prescient of in vitro fertilization and embryo culture. Like his medieval precursors, Maugham's young scientist, Oliver Haddo, is obsessed with a desire to vie with God by creating homunculi and thus studies occult lore, including Arabic and Hebrew formulas, in order to create his embryos. When it appears that these require virgin blood for their nourishment, Haddo has no scruples about sacrificing his new wife, Margaret, for the purpose, a detail that would hardly have surprised medieval readers but proved less convincing in the twentieth century.

Klotz and Haddo are representative of many other characters of the period who are censured for their hubris in attempting to improve upon, or reverse, the course of nature. André Gide also chose an amateur biologist as the type of the inhumane scientist. Before his miraculous conversion to Catholicism, Anthime Armand-Dubois features in *Les Caves du Vatican* (1914), translated as *The Vatican Cellars*, as an "unbeliever and freemason," intent on jeering at friends, family, and children. As evidence of his inhumanity, he conducts experiments on rats, blinding, deafening, castrating, skinning, and starving them, before moving on to deal with human beings. His aim is "to reduce all animal activities to what he termed 'tropisms.' . . . An entire category of psychologists would admit nothing in the world but tropisms."[28] Gide takes care to indicate that Armand-Dubois's procedures in his work for the Académie des Sciences are typical of the science fraternity of his time. His rat mazes, with their many variations to keep the animals from their food, are described as "diabolical instruments, which in a later age became the rage in Germany under the name of *Vexierkasten*, and were of great use in helping the new school of psycho-physiologists to take another step forward in the path of unbelief" (6).

In most science fiction of this period there is a strong (albeit simplistic) moral message, reinforced by the retribution meted out to the presumptuous biologist for activities that continued to reference the goals of alchemy. Thus in A. Hyatt Verrill's representative story "The Plague of the Living Dead" (1927) a biologist, Farnham, develops a serum that inhibits the aging process and even proves capable of restoring dead creatures to life. However, the revived animals and erstwhile dead people behave "immorally," and even

Farnham is made to see the blasphemy of his "interference with the laws of a most wise and divine Creator."[29] A similar line is followed in "The Ultra-Elixir of Youth" (1927), another moralistic story by the prolific Verrill, in which a scientist attempts to reverse the effects of aging and the author concludes that this "has brought home to us the terrible consequence of attempting to interfere with the plan of the Creator."[30] Perhaps the most extended moralizing of this type occurs in Alexander Snyder's "Blasphemers' Plateau" (1926), in which Dr. Santurn, another decadent biologist, determines to prove that the so-called spirit is "merely the vibrations which stimulate the electrons in their orbits."[31] Predictably, Santurn proves to be as ruthless and evil as he is irreligious and meets a suitably Faustian, if logically confusing, end.[32]

Such clear-cut moral statements, combined with the exaggerated descriptions of research on living organisms, made these characters eminently suitable material for the horror movie. The advent of this genre can be dated from the 1931 version of *Frankenstein*, in which Boris Karloff rocketed to fame as the nameless monster produced, as the introduction clearly states, by the "man of science who sought to create a man after his own image, without reckoning upon God." The huge success of *Frankenstein* and its numerous successors detailed in chapter 7 above, says much about the climate of popular paranoia at the time. It was followed in the same year by the equally successful *Dr. X*, which made use of a similar formula, a group of scientists working in an isolated old Gothic mansion to create synthetic flesh. These mad, morally doubtful, and certainly irreligious scientists were to set the trend for the next fifty years, for in the celluloid world scientists are far more often mad than sane. Such films focus on the Frankenstein stereotype of the obsessed biologist isolating himself from the natural affections of the living to create artificial life from the materials of death. Yet, interestingly, science itself, as distinct from its practitioners, is rarely denigrated in these films, and even the evil scientist is sometimes presented as more sinned against than sinning. *Dr. Jekyll and Mr. Hyde* (1931), *Dr. X* (1931), *The Murders in the Rue Morgue* (1932), *The Vampire Bat* (1933), *The Walking Dead* (1936), and *Dr. Renault's Secret* (1946) are all concerned with biologists who have played God by attempting to create or restore life, but there remains a residual, if vague, admiration for science itself. The films can thus be seen as contemporary documents of Western society's own confused cost-benefit analysis of science, its risks, and its promises, the same ambivalence that had attached to alchemy.

The abhorrence at transgressing species boundaries, and especially that between humans and other animals, was vividly captured in the film *The Fly* (1958), in which a scientist, André Delambre, constructs a matter transporter device, the disintegrator/integrator through which living organisms are transposed. Sure of success, he enters one of the chambers, unaware that a fly is in the other. When he transports himself, his atoms become mixed with those of the fly so that he now has the head and arm of a fly, although he retains his own mind, while the fly acquires a miniature human head and arm. In a series of horrific consequences both are crushed to death, but prior to this Delambre provides an explicit moral directed at scientists: "The more I know, the more I'm aware of, the more I'm sure I know so little." In a 1986 remake the scientist, now called Seth Brundle, has a very different personality and more contemporary methodology: gene transfer. His teleportation project between telepods is still only partially successful with baboons when he decides to enter one pod and emerge from the receiving one. It gradually becomes apparent that he has merged at the molecular-genetic level with a fly trapped in the telepod and is becoming more insect-like. Desperate to dilute his fly genes with more pure human DNA, he plans to fuse himself, his girlfriend, and her unborn child into one entity but instead becomes fused with the metal of the pod. Although intended as a general analogy for disease, this horror film was seen by some critics as a cultural metaphor for the AIDS epidemic.[33]

An attempt to come to terms with the potential for social development in the light of the failings of individual scientists is explored in some depth in Alfred Döblin's monolithic *Berge, Meere und Giganten* (Mountains, seas and giants) (1924), which anticipates the adverse ecological consequences of scientists' meddling with nature. In this vast novel, spanning several centuries in the future, industrial scientists and research groups have effectively assumed total power over society, converting it into a huge technological experiment about which no questions may be asked. Insisting on scientific progress as an end in itself, Döblin's scientists, having already devised machines to do all the necessary labor, turn to the production of devices for which there are as yet no applications. In a remarkable prefiguring of the escalation of the Cold War, they perfect more and more sophisticated weapons, which in turn require wars for their justification, so that the scientists are portrayed as being directly responsible for the fomenting of another war.

One of the most interesting of Döblin's scientists, and one of the few women scientists in the literature of this period, is the biologist Alice Layard, leader of the North American *Frauenkamaraderie*. Originally motivated by the desire to make a contribution to society, she announces the discovery of a synthetic food whereby entire populations can be fed without the need for arable land or even sunshine. This discovery immediately becomes a political weapon that the women seek to keep for themselves, and Layard deliberately contaminates the food, causing an epidemic, in order to discredit her (genuine) discovery rather than surrender it. Döblin thus suggests that research can change from an altruistic activity to a means of acquiring prestige and power until, paradoxically, the research itself is sacrificed to the latter end.

Still obsessed with circumventing nature and striving for the seemingly impossible, Döblin's Promethean scientists embark on another project, which becomes a major disaster: melting the ice sheets of Greenland. In the process, primitive animals and plants, long extinct, are thawed out, producing grotesque monsters; and the scientists, in self-defense, propose to assume similar proportions, a procedure reminiscent of Boom Food in H. G. Wells's *Food of the Gods*. Here, however, the inflated size of the scientists is symbolic of their pretensions and their hubris rather than, as in Wells's treatment, of their moral stature. Eventually all the giants die, leaving a happy, pastoral, antitechnological society with the motto *Wissend und demütig* (Wise and humble). Thus Döblin's solution, like Ruskin's, is a reactionary one: return to a preindustrial system. Scientists are either too ignorant of consequences or too morally flawed to be entrusted with power; hence their attempts to transform nature are doomed to disaster. The enormous box office success of the film *Jurassic Park* (1993) and its successors, *The Lost World: Jurassic Park* (1997), *Jurassic Park III* (2001), and *Jurassic World* (2015), suggests the continuing attraction of this view.

A more sophisticated philosophical attack on immoral biologists is to be found in the very different premises of Aldous Huxley and C. S. Lewis. Huxley's attitudes toward science and scientists varied considerably during his long life as a writer. In *Along the Road* (1925), written early in his career, Aldous Huxley, grandson of the eminent Thomas Henry Huxley and brother of another distinguished biologist, Julian, reflected with regret on his own failure to become a scientist, concluding, "Even if I could be Shakespeare, I think I should still choose to be Faraday."[34] Again at the end of his life, as is evident from his last novel, *Island* (1962), he came to see in science possibilities both for solving the world's ecological problems and for helping humanity to achieve its spiritual potential. However, in the period between, he was intensely critical of the limitations of science, regarding it, first, as inadequate for the evaluation of ethics and aesthetics and leading to a loss of creativity, individuality, and liberty by its insistence on a narrow range of objective criteria. Already in his early nonfictional work *Jesting Pilate* (1926) Huxley blamed scientific paradigms for the disappearance of values from Western civilization, arguing that scientific materialism is a closed intellectual system that precludes the relevance, or even the possibility, of a value system: "If men have doubted the real existence of values, that is because they have not trusted their own immediate and intuitive conviction. They have required an intellectual, a logical and 'scientific' proof of their existence. . . . When you start your argumentation from the premises laid down by scientific materialism . . . [you] must infallibly end in a denial of the real existence of values."[35]

Huxley came to regard science as a root cause of both loss of individual liberty and the rise of nationalistic totalitarianism. In *Science, Liberty, and Peace* (1946) he outlined the reasons why scientists must bear a large degree of political responsibility. First, by helping to create the powerful weapons of modern warfare, scientists have endowed political leaders with the means of holding the masses in fear and subjection; second, they have acquiesced in the destructive uses to which their discoveries have been put.

It was out of this largely negative assessment of science in the real world that Huxley created his fictional scientists. The best-known expression of his belief that science stunts man's development and insulates him from coming to terms with the reality of his existence is, of course, *Brave New World* (1932), with its description of a society founded exclusively on the principles that Huxley associated with technology: efficiency, pragmatism, standardization, restricted thought, and the consequent dehumanization of the individual. Along with E. M Forster's "The Machine Stops" (1909), Yevgeny Zamyatin's *We* (1920), and George Orwell's *1984* (1949), it stands as a rejection of the Wellsian utopias grounded in an unquestioning faith in the beneficence of science.[36] The more immediate cause for Huxley's and Orwell's grim picture of a technologically determined society was their reaction against the dehumanizing theories of the behaviorist school; hence research into psychological conditioning is depicted as the most insidious and sinister aspect of science, surpassing even the destructive power unleashed by the physical sciences and the pretensions of biologists peddling eternal life. For both these writers psychologists are the consummate villains of the whole scientific pantheon.

Huxley presents no individual scientists in *Brave New World* (since loss of individuality is part of the point he is making), but overall the novel subverts and parodies the scientist-as-hero stereotype. Indeed, the whole ethos of this post-Fordian society, with its complex system of genetic engineering and its redefining of personal relations and emotions as obscene, is a reductio ad absurdum of what Huxley believed were the inevitable implications of scientific method linked to assembly-line efficiency. *Brave New World* became a cult novel, eliciting many derivative treatments of the dystopian theme. Gerald Heard's *Doppelgängers* (1948), for example, is set in 1997, after the "Psychological Revolution," when Earth is ruled by a benevolent dictator who uses popular science to keep the masses unthinking and happy. Others follow Huxley less closely, but the message is much the same—rejection of a mechanistic society where humanity has been reduced to an accepting automaton status.[37]

The Christian apologist C. S. Lewis, who from his position as a conservative classicist engaged in a prolonged public controversy with the socialist biologist J. B. S. Haldane over the degree of influence science should exert in society,[38] also produced a popular trilogy castigating biologists in general. Masquerading as science fiction, it was a carefully contrived propagandist attack on scientism, which in Lewis's Neoplatonist view must inevitably lead to the demise of religious values. He regards the two basic propositions of modern science—reductionist materialism and chance—as more insidious and dangerous than any overt destruction wreaked by mad, bad scientists, because while the latter will inevitably be abhorrent to society and elicit a counterattack, the rationale of science appears unexceptional, even attractive, and can thus subvert a society's ethical values without resistance. Lewis sets out to denounce this philosophy, embodied in the stereotyped character of Weston (Western man), a physicist, who wholeheartedly embraces accepted scientific values and who is therefore, in Lewis's treatment, both ruthless in his attempt to exploit society, and indeed the universe, and simultaneously limited in his ability to understand the beauty and complexity of that universe.

Since his main concern is to condemn materialism, Lewis, unlike the majority of science fiction writers, has little interest in the technical side of his story. In the first volume, *Out of the Silent Planet* (1938), we are given no details about the spaceship in which Weston and his companions travel to Malacandra (Mars). Indeed, Lewis decries pulp science fiction of the *Amazing Stories* kind for its "scientification" and describes the Machiavellian Weston as "a man obsessed with the idea which at this moment was circulating all over our planet in obscure works of 'scientification' in little Interplanetary Societies and Rocketry Clubs, and between the covers of monstrous magazines."³⁹ Being a materialist, Weston is, Lewis assumes, necessarily amoral and therefore has no scruples about dabbling in vivisection, supporting eugenics, or experimenting by force on a hapless local village boy, since in his view the boy is "incapable of serving humanity and only too likely to propagate idiocy. He was the sort of boy who in a civilised community would automatically be handed over to a state laboratory for experimental purposes."⁴⁰

To oppose this view, Lewis introduces his champion, Ransom, a philologist and Christian humanist whom Weston regards as "insufferably narrow and individualistic," since he fails to subscribe to the doctrine that the ends justify the means, that the individual should be sacrificed to the race and the present generation to the hypothetical future. Weston asserts, "Life is greater than any system of morality; her claims are absolute. It is not by tribal taboos and copy-book maxims that she has pursued her relentless march from the amoeba to man and from man to civilisation" (*Out of the Silent Planet*, 154).

Unlike most of the mad, bad scientists reviewed in this chapter, Weston is at one level an idealist in that he is interested not in personal gain but in a cause outside himself. His apparent altruism is therefore more dangerous than obvious self-interest. Lewis spends considerable time at the end of *Out of the Silent Planet* making Weston appear morally stupid through the standard satirical device of having him explain his position to someone of a completely different moral perspective, in this case the all-wise Oyarsa of Malacandra.⁴¹ Weston affirms his Darwinian and racist beliefs in the innate superiority of the human species over the individual and, plunging deeper into the moral mire at every step, deduces from this a mandate for man to spread to other worlds, killing the inhabitants if necessary, in order to survive the death of this and other planets. "Humanity, having now sufficiently corrupted the planet where it arose, must at all costs contrive to seed itself over a larger area. . . . Planet after planet, system after system, in the end galaxy after galaxy, can be forced to sustain, everywhere and for ever, the sort of life which is contained in the loins of our own species"⁴²—a clear allusion to H. G. Wells's repeated statements about "the beings now latent in our thoughts and hidden in our loins [who] shall . . . reach out their hands amidst the stars."⁴³

In *Perelandra* (1943), the second volume of the trilogy, Weston's philosophy is modified somewhat in order to allow Lewis to attack another "heresy" of science, the pantheistic view he associated with Olaf Stapledon, namely, that the movement of all life is toward a purposeful life force, which has directed evolution. Such a doctrine is based on what Lewis calls "chronological snobbery," the belief that the new is, ipso facto, superior to the old. This view is, of course, diametrically opposed to Lewis's classical humanism, which locates the golden age in the past, and to his affirmation of the Christian doctrine of the Fall. For good measure, Weston also espouses the Baconian ideal of scientific utilitarianism and makes his textbook statement on this point too in order to be knocked down by Lewis's champion Ransom.

In the third novel of the trilogy, *That Hideous Strength* (1945), Lewis turns his attack upon psychologists, in particular behaviorists of the Skinner school. Their spokesperson announces Lewis's version of their intentions: "If science is really given a free hand, it can now take over the human race and recondition it; make man a really efficient animal."[44] The organization set up to implement this program of efficiency is the National Institute of Coordinated Experiments (N.I.C.E.), a sinister scientific foundation run by power-seeking bureaucrats who employ techniques of media manipulation to brainwash individuals and eventually society. Its goals include the fusion of state and applied scientific research, control of the environment, eugenics, and the remedial treatment of criminals, all of which Lewis identified with the subjection of nature and the surrender of individual freedom. The goals of N.I.C.E. are listed as "quite simple and obvious things, at first sterilisation of the unfit, liquidation of backward races (we don't want any dead weights), selective breeding. Then real education, including pre-natal education. . . . Of course it'll have to be mainly psychological at first. But we'll get on to biochemical conditioning in the end and direct manipulation of the brain" (47). Lewis's characterization of Weston forms an interesting contrast to the other evil scientists discussed in this chapter because of the detail and ferocity of the attack and the author's intense personal involvement in the argument, which, ironically, renders his case as biased as the one he is bent on destroying.

In Ira Levin's novel *The Boys from Brazil* (1976, film 1978), the evil scientist is allegedly the actual Dr. Josef Mengele, the medical doctor who performed horrific experiments on the prisoners of Nazi concentration camps. Still a devotee of Nazi philosophy, Mengele is obsessed with creating clones of Hitler, one of whom will become the new führer. *The Boys from Brazil* will be discussed in more detail in chapter 16, but Mengele deserves mention here as epitomizing a twentieth-century version of the mad, bad, and highly dangerous scientist, obsessed with an evil political philosophy that, in his mind, justifies any means to his end.

In this analysis of evil scientists in twentieth-century literature it is apparent that many of their characteristics are identical with those depicted in previous centuries—arrogance, desire to usurp divine authority (particularly as creator), materialism, reductionism, lust for political power. The essentially new element is the resentment voiced by some of these fictional scientists that the classical Newtonian picture of a predictable, orderly system has been overturned by the revelation that chance, accident, and disorder prevail in nature. In several of the works discussed in this chapter this resentment leads to a manic obsession with removing life from the system altogether in order to purge it of its disorder, suggesting that their authors had not yet come to terms with the neo-Gothic horror raised by nineteenth-century astronomy and biology. Most of these writers were still unaware that contemporary physics was positing a parallel model of chance and unpredictability at the subatomic level.

All the characters explored in this chapter are the creations of male writers, whose diagnostic methodology involved searching for particular and individual causes for the violence or obsession they depicted. Later in the century, however, feminist writers and critics were to propose an alternative interpretation. They see the characteristics of these mad, evil, and dangerous scientists as an extension of the values of their society, and these, in turn, they explain as consequent on the masculine role models perpetrated by that society. Christa Wolf, one of the most outspoken women writers of the (former) Eastern bloc countries, expressed this view pointedly in 1981, at the height of the Cold War:

> Rockets and bombs are after all not by-products of this culture; they are the consistent manifestation of thousands of years of expansionist behaviour; they are the inevitable embodiments of the syndrome of industrial societies which, with their more! faster! more accurately! more efficiently! have subordinated all other values, many of which were measured by more human norms. . . . I am plagued by the thought that our culture, which could only attain what it calls "progress" through force, through domestic repression, through the annihilation and exploitation of foreign cultures, which has narrowed its sense of reality by pursuing its material interests, which has become instrumental and efficient—that such a culture necessarily has to reach the point it has now reached. . . . For three thousand years . . . women have not counted and do not count. Half the population of a culture has *by its very nature* no part in those phenomena through which that culture recognises itself. . . . It occurs to me

that they [women] therefore have no part in the experiments in thought and production that concern its destruction.[45]

In Wolf's analysis, then, the mad, bad, and dangerous scientists represent a consequence of the excessive concentration on reason, efficiency, and objectivity that a patriarchal hegemony extols. The scientists to be discussed in the next two chapters represent different aspects of this impersonal approach. These fictional presentations are essentially neo-Romantic in condemning the emotionless, inhumane, and amoral scientist, but these analyses are based on a more sophisticated understanding of cosmology, make use of different metaphors (the computer features largely), and perceive the problem as much more pervasive throughout the whole cultural ethos than their nineteenth- and early-twentieth-century predecessors could have envisaged.

CHAPTER 13

The Impersonal Scientist

> I have only one relationship and that's with my work. . . . I don't want relationships, I don't want involvements.
>
> —*Stephen Sewell*

Of all the charges against the scientist in literature, none has been leveled more frequently or more vigorously than that of aloofness and emotional inadequacy. This was not an attribute associated with the early representatives of science—the alchemists, the Faustus stereotype, or the Restoration virtuosi. On the contrary, these precursors of the modern scientist were depicted as passionate, inspired, even religious in their zeal. Emotional deficiency appeared as a characteristic of fictional scientists only after Newton's celestial mechanics eliminated much of the awe and mystery from man's perception of the universe and Enlightenment philosophy elevated the cultivation of objectivity as the precondition of scientific method. The English Romantics certainly ascribed this diminished respect for the emotions and subconscious states to the culture of scientific rationalism, and twentieth-century writers have extended this Romantic stereotype of the scientist who has sold his emotional soul for scientific prestige.

While most of the evil scientists considered in chapter 12 were drawn from the ranks of physicists, biologists, and psychologists, it is mathematicians and computer scientists who feature prominently among those characters who have reneged on personal relationships. This suggests that whereas physics and biology are popularly identified with power and activity, mathematics and its more recent offshoot, computer science, are more closely associated with abstraction and dehumanization. Why is this?

The exponential rate of expansion of the computer into so many areas of communication appears, at one level, to have transformed Descartes's dream of the mathematization of the world into reality. Although the computer has by no means solved all of humanity's problems, it has significantly changed the prevailing view of what is possible and the value systems espoused by technologically advanced societies. The criteria for evaluating computing systems are primarily those of efficiency, consistency, and simplicity, that is, the qualities associated with the reification of experience, and it is these criteria that are increasingly invoked for the evaluation of success in both the social and the personal spheres. The fear that computers would take over the world was seen by later writers as being already realized in a more insidious way through this change in values.

In technological societies, communication has been largely reduced to the transmission of facts through minimalist and impersonal media such as email, Facebook, smartphone aps, tweets, blogs, hashtags, Flipboards, messaging, and YouTube. These social media standardize the message in the interests of speed of transmission and minimal time requirement for input and reception of the message. Tweets, for example, are limited to 140 characters. The ultimate impersonal message is the binary code of the computer, and growing numbers of computer-literate people regard the world, both animate and inanimate, as a series of problems soluble by algorithms. So many daily activities are already based on numbers and computation that we hardly notice the insidious implications of reducing human beings to numbers, a practice that, before the advent of vast computer networks, was reserved for the army and the prison, where individuality was deemed to have been forfeited. Even the social sciences, in endeavoring to mimic the physical sciences, have had recourse to a statistics-based methodology for the assessment of everything from intelligence to social change.

In the broader community, this trend did not go unchallenged. There have been various attacks on the edifice of scientific materialism. Various anti-intellectual movements, which included the hippie cult of the 1960s and renewed interest in the occult, Eastern religions, meditation, astrology, and various forms of mysticism and alternative medicine were a protest against the perceived sterility and inhumanity of utilitarian rationalism initiated by science. Writers, too, registered their dissent. Kitchen-sink drama, the preoccupation with mental stress and nervous breakdown as a result of social pressures and alienation, and the boom in fantasy fiction and horror films were all expressions of anomie and a pervasive discontent with a society dominated by scientific materialism.

Another response was the frequent appearance in literature of emotionless scientist characters, especially psychologists, mathematicians, and computer scientists. These impersonal figures are presented not merely as emotionally deficient individuals but as representatives of a deeply depersonalized society. Sometimes they are less vilified than pitied as victims of the system, but more frequently they themselves are too obtuse or too deeply flawed to realize their inadequacies.

Scientists as Representatives of the Depersonalized Society

Scientists are depicted as fitting representatives of the Western technological society insofar as they embody the vision of a utopian future and the potential to produce the state-of-the-art technology that accumulates wealth and power. The constellations of competition operating in all sectors of society—economic, military, political, as well as within the sciences themselves —drive the "scientification race,"[1] since scientific and technological knowledge is seen as fundamental to success in all these spheres. This powerful knowledge is identified with

1. cultivation of rationalist skills and corresponding suppression of the emotions;
2. an objective perspective;
3. efficiency elevated to a moral value;
4. reification of individuals to statistical units; and
5. integration of technological and economic systems so that the former receives further justification, because it secures wealth, and hence political dominance, for the society that possesses such expertise.

Scientific writing reinforced this claim for objective truth through its restricted, wholly factual vocabulary, its rejection of rhetorical and poetic figures and topoi, and exclusion of personal pronouns by substitution of the passive voice. The real author/experimenter is thereby removed from the alleged sequence of events in favor of a supposed direct object reference—"letting the facts speak for themselves."[2] This is the stylistic correlative of the character of the impersonal scientist.

Most of the above criteria were included to varying degrees in the Romantic attack on science and scientists, but more recently feminist critics have introduced a new perspective by identifying these factors with the cultural suppression of male emotions within Western patriarchal society. As we shall see in chapter 18, and in Christa Wolf's "Selbstversuch: Traktat zu einem Protokoll" discussed at the end of this chapter, feminist criticism from the 1980s alleged a correlation between male-dominated hegemonic systems and the unspoken criteria, prejudiced against feminine values, for evaluating behavior in those societies.

An early depiction of an impersonal mathematician appeared in Louis Macneice's poem "The Kingdom" (1943), where it is implied that the scientist, who is, significantly, nameless, has lost his humanity and substituted his graphs for a normal human relationship with a living child.

> he lived by measuring things
> And died like a recurring decimal
> Run off the page, refusing to be curtailed;
> Died as they say in harness, still believing
> In science, reason, progress,
> ....
> ... plotting points
> On graph paper he felt the emerging curve
> Like the first flutterings of an embryo
> In somebody's first pregnancy; resembled
> A pregnant woman too in that his logic
> Yet made that hidden child the centre of the world
> And almost a messiah; ...
> ....
> ... Patiently
> As Stone Age man he flaked himself away
> By blocked-out pattern on a core of flint
> So that the core which was himself diminished
> Until his friends complained that he had lost
> Something of charm or interest.[3]

Initially scientific materialism was also linked with industrialism, nearly always a topic for disapproval in literature. D. H. Lawrence, whose neo-Luddite attitude toward industrialism and its deleterious effects on both the English countryside and society could best be described as late Romantic, created a female character, Winifred Inger, in *The Rainbow* (1915), in whom the connection between science and the machine is explicitly argued. Winifred, a science teacher, represents both the new independent woman and the emotionless aspects of industrial society that Lawrence most abhorred. Winifred, the product of a "scientific education," is tainted with many of the same faults as Lawrence's industrialist characters, Clifford Chatterley and Gerald

Crich: "The real mistress of Winifred was the machine. She too, Winifred, worshipped the impure abstraction, the mechanisms of matter. There, there in the machine, in the service of the machine, was she free from the clog and degradation of human feeling. There, in the monstrous mechanism that held all matter, living or dead, in its service, did she achieve her consummation and her perfect unison, her immortality."[4] Winifred appears only briefly in *The Rainbow*, as vying for the allegiance of the main character, Ursula, who rejects her. But before long the complex connections between scientific reductionism, a depersonalized society, and their political equivalent, totalitarianism, were explored in the three great dystopias of the twentieth century—Yevgeny Zamyatin's *We*, Aldous Huxley's *Brave New World*, and George Orwell's *1984*. All three writers conclude that conformity is accepted by populations as the price of security.

Inspired by Wells's ambiguous dystopian novel *When the Sleeper Wakes* (1899), the Russian writer Yevgeny Zamyatin, who had been trained as an engineer, created in *We* (1924) the paradigmatic scientific dystopia.[5] In *We* a whole society, the One State, is managed, down to the most intimate details, on wholly mechanistic principles. The villain of Wells's novel was capitalism, which ruthlessly suppressed the workers. For capitalism, Zamyatin substitutes scientific obsession with efficiency, showing how this reduces the individual to a nameless number, a robot-like unit, forced to conform to a "mathematically perfect life."[6] "Every morning, with six-wheeled precision, at the same hour, at the same minute, we wake up, millions of us at once. At the very same hour, millions like one, we begin our work, and millions like one, we finish" (13). Every action, every movement, is part of a stylized, "unfree" dance pattern, every part prearranged according to the Tables of Hours. Even sexual desire has been dealt with in a quantified and scientific way by the Lex Sexualis: "A Number may obtain a licence to use any other Number as a sexual product" (30), and the "product" must acquiesce. Dreams, emotions, imagination, and spontaneity are strictly suppressed.

The narrator, number D-503, is an engineer, Chief Builder of a new spaceship, the *Integral*, designed to bring "mathematically faultless happiness," by force if necessary, to aliens in the farthest reaches of the universe. He therefore begins a diary, a guide to life in his society, which he intends to send as part of this galactic crusade, for at first he is in complete accord with the values of the state. Indeed, his conditioning in rational thought has been almost flawless: the great terror of his childhood was his first confrontation with irrational numbers. "I wept and banged the table with my fist and cried, 'I do not want that square root of minus one.' . . . This irrational root grew into me as something strange, foreign, terrible; . . . it could not be defeated because it was beyond reason" (37). It is this cultivated fear of the irrational that induces D-503 and the rest of his society to conform: there is security in a system where everyone is an element in the pattern, a configuration in Euclidean geometry.

In the process of writing his diary, however, D-503 looks afresh at his world and becomes increasingly critical of its enforced harmony. This process is initiated by his disturbing encounter with I-330, an experience he refigures as a mathematical problem: "$L = f(D)$, love is the function of death" (8) and should therefore be avoided. In his mind I-330 is associated with the childhood terror: "The woman had a disagreeable effect upon me, like an irrational component of an equation which you cannot eliminate" (10). She turns out to be a member of a secret society, the Mefi, which is planning a revolution, and under her guidance D-503 visits such subversive places as the Ancient House, with its disturbing, nongeometrical shapes, and the Green Wall, beyond which lies the jungle world of the uncivilized, the irrational realm of emotions and the subconscious. Although D-503 is temporarily attracted to the Mefi, he suspects that they are using him to gain control of the *Integral*; moreover, he is unable to repudiate the safe and conditioned mathematical framework of his life. The Mefi revolution is crushed by the state, which executes most of the insurgents but graciously permits D-503 to undergo a frontal lobotomy instead, to remove any other incipient fancies. This reeducation program is so successful that D-503, who now feels as if "a splinter has been taken out of my head" (217), is able to watch with equanimity while I-330 is tortured to death.

The indictment of scientism in *We* is subtle but relentless. By the recurrent use of scientific logic and particularly mathematical language (the "numbers" of the state are designated as plus or minus according to their usefulness), Zamyatin emphasizes his central point, namely, that the basis of dystopia is total capitulation to safe, rational, mechanical (and hence inhuman) patterns of thought and behavior, to the exclusion of an emotional and creative life.

In Aldous Huxley's *Brave New World* (1932) the external uniformity of Zamyatin's *We* is extended to the realm of psychology, producing identical components of the social machine, conditioned as fetuses not only to perform the tasks allotted to them but to enjoy doing so. Like Zamyatin, Huxley depicts scientific rationalism as closely associated with a totalitarian regime, since philosophically both require the subjection of the individual to the system, and both writers portray scientists as the dehumanized agents of the totalitarian state. In the postwar United States, however, the "system" that subdues individuals is more frequently identified by writers with the military-industrial complex. An early example of this conflation is Algis Budrys's novel *Who?* (1958), in which the symbol of the nonhuman scientist is rendered almost literally as a new form of robot. When Lucas Martino, a brilliant physicist working in West Germany on a military device called K-88, is seriously injured by an explosion in his laboratory, situated near the East German border, a Soviet medical team quickly whisks him off to a hospital. In due course a bionically engineered "person," half resembling Martino and half metal, returns across the border. The FBI spends many weeks trying to establish whether the returned man is the original Martino, a Soviet-convert Martino (the product of brainwashing by his captors), or a Soviet impersonator. Ironically, the FBI is never satisfied that it has discovered the truth, for the bionic Martino, having no personal fallibilities to be manipulated, now represents not only the ideal spy but also the least accessible secret agent. The FBI plays it safe. In order to preclude any possibility of the secret of K-88 being passed to the Soviets, Martino is removed from the project, even though this means the end of K-88 altogether. Ironically, Martino, inside his metal case, remains loyal, having resisted the Soviet attempts to brainwash him. He thus represents, among his other allegorical aspects, the impotence enforced on science by the very military security systems intended to safeguard it.

At least as interesting as the political intrigue is the characterization of Martino himself. We find that, as a student, Martino had believed that the universe was "constructed of perfectly fitted parts,"[7] and in a series of flashbacks we see that he was never able to form meaningful human relationships, because physics always came first for him. After graduating, he progressively lost what humanity he had, becoming the property of a succession of virtually identical military research establishments. Half inanimate, literally as well as symbolically, this image of the scientist represents a risk to society, partly because of his superior knowledge and power in some areas but also because, as a result of his "otherness," he remains permanently isolated from the human fulfillment he seeks. When Martino tries to renew former acquaintanceships, people flee in terror from his well-intentioned efforts as though from Frankenstein's monster. Martino is thus a symbol of the depersonalization of the scientist resulting both from his own inclination and from manipulation by others.

The American novelist Thomas Pynchon was a technical writer for Boeing at the time when that company secured the prime contract for the US Apollo Saturn V rocket. A year later he left Boeing to write his first novel, *V* (1963), about a fictional company, Yoyodyne, Inc., which grows into a postwar aerospace giant, making gyroscopes for aircraft and missiles and becoming intimately involved in military-industrial collaborations. Three years later Pynchon published *The Crying of Lot 49* (1966), in which the same firm's Galactronics Division extends its operations to the missile and spacecraft business. Pynchon relates the figure of the inhuman scientist to a vast social and philosophical system that, by its very nature, defies coherent description and analysis. His classic novel *Gravity's Rainbow* (1973) seeks to account for the meteoric rise of the American Rocket State, in which unimaginable sums of money were heaped on the National Aeronautics and Space Administration (NASA) in its dual role as the flagship of US prestige in space research and supplier of weapons for the West in the Cold War.

Gravity's Rainbow explores a world where the previous historical "explanations"—Newtonian mechanism and statistical physics—which offered a measure of certainty about the theoretical rationality of the universe, have been replaced by the irreducible uncertainty of quantum physics. Pynchon shows how the various systems of science and technology have all created their own servants with a mechanical, reductionist explanation of the world, a statistical perception, or a fragmented picture of "reality." Set in Europe near the end of the Second World War, *Gravity's Rainbow* uses the contemporary sociopolitical breakdown as a symbol of the fragmentation that characterizes twentieth-century postmodernism. The emblem of the new order is the V-2 rocket, which was fired on London and, traveling faster than sound, crashed *before* the sound of its approach could be heard, a violation of the classical system of causality.

Pynchon's characters represent the various traditional systems of scientific thought. Pavlov's experiments with conditioned responses in dogs were essentially an attempt to extend Newton's principles into the biological area, and Pointsman, Pynchon's Pavlovian character, still espouses a clockwork view of the animate as well as the inanimate world. Seeing man himself as a machine, he assumes that human behavior can be predicted in the same way that the engineers plot the path of the rocket using the principles of Newtonian physics. The German rocket engineer Franz Pökler, whose career closely parallels that of Wernher von Braun, is another Newtonian figure; Marcel, the mechanical chess player, a reference to Poe's story "Maelzel's Chess Player" (see chapter 10), is even more symbolic of this reductionist view of the universe, since he becomes indistinguishable from a human being. Yet another character, Roger Mexico, works with statistical models to try to overcome the small perturbations and unforeseen forces. By contrast, the engineer Beláustegui is a "prophet of science," who accepts that every moment has its own value, unconnected with any other, and that chance is the only law. Anti-paranoia, "where nothing is connected to anything, a condition not many of us can bear for long,"[8] is the last desperate response in this bleak contemporary world, which represents the antithesis of classical mechanics. Thus the GI Tyrone Slothrop, engaged in trying to find out how the rocket is assembled, himself becomes disassembled. Pynchon does not develop his characters in depth as individuals; rather they contribute to a panoramic picture of the major thought systems whereby scientists from the seventeenth to the twentieth centuries have attempted to make sense of the universe. All are essentially impersonal in their basic assumptions; all regard man as something less than human—as a Cartesian machine, a set of statistical predictions, or a random array of chance molecules. In Pynchon's view, both extremes—the classical mechanics world of order and determinism on the one hand and the quantum physics world of disorder and unpredictability on the other—are equally sterile and destructive of humanity.

Dehumanized Individual Scientists

Whereas the characters of the preceding section were presented as inevitable consequences of their society's preoccupation with scientific materialism, many of the more individualized scientist characters of twentieth-century fiction have been depicted as active agents in the depersonalization of Western society. Rather than merely gravitating toward a career that rewards their particular strengths, these characters deliberately exploit their own personality deficiencies in a discipline that empowers them because of their technical qualifications.

As well as his dystopia, *Brave New World*, which was discussed in the previous chapter, Aldous Huxley, who had ample opportunity through his family connections to observe real-life scientists, also launched some of the most vituperative attacks on the evils perpetrated by the impersonal, unemotional scientist. His characters, mostly biologists, are shown to be all the more dangerous because they pass for normal in their society, acquiring honors for their intellectual prowess while wreaking misery, degradation, and death on those around them. The personal failure of these characters is located within a relatively narrow causal range, but the cumulative impact in successive books became increasingly vicious as Huxley explored the far-reaching effects on others of the socially accepted peccadilloes permitted to scientists.

Two particular aspects are prominent in his attack. The first is the effect that reductionist methodology has on a scientist's relations with others. Thus Shearwater, the physiologist of *Antic Hay* (1923), modeled on Huxley's contemporary, the eminent biologist J. B. S. Haldane, is so involved in his experiments with one part of the human body, the kidneys, that (like Shadwell's Sir Nicholas Gimcrack, and for similar reasons) he is quite oblivious of his wife's infidelity. A more sinister example of such reductionism is provided by Dr. Sigmund Obispo, the biologist of *After Many a Summer Dies the Swan* (1939), whose particular form of sensuality, considered by himself as intellectually refined, involves tormenting the object of his lust in order to study her reactions, as though she were one of his laboratory animals. Obispo's first name is an obvious reference to Freud, whose reductionist premises Huxley considered demeaning of human beings. Significantly, Obispo's research into artificially increasing the human life span yields results very different from his glorious expectations: longevity is found to be inextricably associated with regression to the savagery of nonhuman ancestors.[9]

Second, Huxley's scientists are deficient in all areas other than the intellect. Lord Edward Tantamount of *Point Counter Point* (1928), a fictional portrait of J. B. S. Haldane's father, J. S. Haldane, professor of physiology at Oxford, is described as being "in all but intellect a kind of child."[10] He immerses himself in his laboratory because he is unable to cope with the world of living people and relationships. "In the laboratory, at his desk, he was as old as science itself. But his feelings, his intuitions, his instincts were those of a little boy. Unexercised, the greater part of his spiritual being had never developed" (26).[11]

Huxley's most detailed depiction of an intellectually brilliant but emotionally underdeveloped scientist is Henry Maartens, of *The Genius and the Goddess* (1955). In the world of science he is preeminent and inspiring—"working with him was like having your own intelligence raised to a higher power"—but outside his laboratory he is "a kind of high-class monster," says Rivers, the narrator and physicist who works with him.[12] In personal relationships he is a child, selfishly dependent on his mother-figure wife, Katy. His chronic hypochondria becomes a weapon to enforce total devotion from her, for Henry is adept at inducing a strategically timed bout of pneumonia with the most extreme symptoms, ensuring her return from her dying mother's bedside. Totally immersed in himself, Henry is unable to communicate with others, even his wife and children. He lives "in a state of the most profound voluntary ignorance . . . abounding in preconceived opinions about everything" (70). He speaks with authority on education, and indeed he is knowledgeable about all the theories on the subject, but he can never see their relevance, because he has no interest in actual people. "He had read Piaget, he had read Dewey, he had read Montessori, he had read the psycho-analysts. It was all there in his cerebral filing cabinet, classified, categorised, instantly available. But when it came to doing something for Ruth and Timmy, he was either hopelessly incompetent or, more often, he just faded out of the picture. For of course they bored him. All children bored him. So did the overwhelming majority of adults. How could it be otherwise? . . . Humanity was something in which poor Henry was incapable, congenitally, of taking an interest" (70–71). Unlike his comic predecessor Sir Nicholas Gimcrack, however, Henry represents an insidious evil in society, first, through his emotional blackmail of others and, second, because his complete divorce of theory and practice, and his lack of interest in the latter, mean that he allows his research to be used for destructive ends. Huxley makes brief but specific reference to the scientists who worked on the atomic bomb during the war when, later in his life, Henry makes a series of tape recordings reminiscing "about those exciting war years when he was working on the A-Bomb! Of his gaily apocalyptic speculations about the bigger and better Infernal Machines of the future!" (127).

Yet Huxley also presents Henry as a pathetic, less-than-human figure, an empty shell as bereft of contact with himself as with others. "Between the worlds of quantum theory and epistemology at one end of the spectrum and of sex and pain at the other, there was a kind of limbo peopled only by ghosts. And among those ghosts was about seventy-five per cent of himself" (71). Of the tapes, Rivers remarks: "You could have sworn that it was a real human being who was talking. But gradually, as you went on listening, you began to realise that there was nobody at home. The tapes were being reeled off automatically, it was the *vox et praeterea nihil*—the voice of Henry Maartens without his presence" (127).

If Huxley had a rich field of observation to draw upon in his depiction of scientists, Charles Percy Snow, a physical chemist by training, had an even more intimate knowledge of the procedures of the research world, and his depictions of scientists in the laboratory carry particular weight for their realism and revelation. Even in his first novel, *Death under Sail* (1932), Snow suggests that the single-minded pursuit of science betokens a failure to grow past the childhood desire for the secure, the familiar. One of the characters summarizes the limitations of another, William Garnett, who, though practical and efficient in matters mechanical and set for a brilliant future in science, is "emotionally underdeveloped in lots of ways.... Like a good many scientists, he's still aged fifteen except in just the things that his science encourages.... I mean he's at home with *things*, at home with anything where he can use his scientific mind—but very frightened of emotions and people, utterly lost in all the sides of life which seem worthwhile to most of us."[13]

The characters of Snow's novel *The New Men* (1954), working at the fictional equivalent of The Atomic Energy Research Establishment at Harwell on the development of the atomic bomb, will be discussed as examples of the amoral scientist in chapter 14, but his most authentic and partly autobiographical study of an individual scientist lacking emotional qualities is the character Arthur Miles, whose career is traced in *The Search* (1934). As a child, Miles was first introduced to the world of science by means of a telescope, often used in literature to symbolize being distanced from one's immediate surroundings; later he is encouraged to make a career in physics, focusing all his energies on research. When he first finds himself attracted to a girl, he thinks: "The only thing outside myself has been my work. Is this girl going to upset *that*?"[14] It soon becomes evident that the answer to this question is no. As he doggedly pursues a research career at Cambridge during the years of Rutherford's work on the disintegration of the atom, the woman finally rejects his proposal on the grounds that "it'd be as bad as marrying a man with a faith. It *would* be marrying a man with a faith. . . . You oughtn't to marry. Perhaps you ought to be celibate" (120). Thus far Snow's characterization follows the conventional Romantic pattern of the scientist's rejection of relationships in favor of research, but he then introduces an interesting variation on the theme, for Miles, after some reasonable successes and minor setbacks in his laboratory work, loses enthusiasm for research and instead seeks administrative power as director of a prestigious new scientific institute. Ironically, although he had earlier resisted the temptation to further his career by selectively removing a damaging piece of counterevidence from an otherwise convincing set of results, he now misses out on being appointed director of the Institute of Biophysical Research because he has failed to check one of his assistants' experimental results before including it in an important paper he published in *Nature* and is held responsible for scientific fraud. Miles accepts the justice of his rejection,[15] but is unwilling to rehabilitate himself in the scientific community through years of hard work in a second-rate post, and like H. G. Wells and Snow himself, he turns instead to literature, writing novels on the sociology of science.[16]

In his characterization of Miles, then, Snow reverses the normal causal sequence. Most of the characters discussed in this chapter are shown to be emotionally deficient because of their exclusive commitment to science, but Snow suggests a two-way causality: Miles fails to become passionate about science precisely because he is deficient in emotions. In retrospect, then, it is not surprising that Miles becomes more interested in manipulating committees and administrative posts than in the hours of slog in the laboratory. The former are concerned with power, the latter with dedication.

If lack of feeling is a limitation in a biochemist, it must inevitably appear more so in a medical scientist, whose research impinges directly on his patients. This is the major theme of Irving Fineman's novel *Doctor Addams* (1939), which focuses on a brilliant medical biophysicist whose research into electrophoresis has been applied successfully in a wide range of medical and industrial problems. Addams's sinister qualities are at first masked by the immediate benefits of his work to his grateful patients, who accept his dismissive attitude toward them. Fineman explicitly subverts the stereotype of the dedicated doctor by insisting that Addams has no interest in medical practice or even the small but necessary clinical component of his research. In his clinic the patients are selected, not on the basis of their need for treatment, but according to whether their complaints are useful to his research. *Doctor Addams* was one of the first novels to raise the specter of artificial interference in the process of conception, and Fineman indicates that, in his view, this is the mark of an impersonal scientist, who regards his patients as so many objects. When Addams's research proves to have application in both contraception and uterine contractions, he, like Frankenstein, clearly sees himself in the role of the creator, far superior to those to whom he dispenses joy or sorrow. Indeed, unlike his father, a traditional, kindly general practitioner, Addams despises humanity and sees no future for it. He writes to his father: "In this pursuit of mine ... I must admit, frankly, to no concern for such visions, optimistic or otherwise, which you insist that men of good will, and especially scientists, should have. . . . To that tradition [of science] I am indeed so devoted that the idea of any further ameliorative service I might perform seems inconsequential."[17]

For Addams, research "had to do with the persistent breakdown of an infinite and timeless barrier of darkness, against which the daily disorders and confusion of men became insignificant and negligible" (292–93). The clearest condemnation of Addams comes from one of his students, who in the middle of Addams's lengthy discourse on the movement of blood cells in an electric field recalls him to the fact that the patient before him is dying: "It was startling to hear the word 'patient'; he had been thinking of tissues and electric potentials, of a process, not of a human being" (48).

For the reasons outlined at the beginning of this chapter, the unfeeling scientist in twentieth-century literature was very likely to be a mathematician or computer scientist, obsessed with the desire for mathematical clarity and order in all areas of life. Prior to the proposal of chaos theory, which sought to mathematize the anomalies as well as those phenomena that conformed to the rules, events that challenged a mathematical thesis were discarded as experimental error. A further reason, and one that recurs in twentieth-century literature, is the suggestion that pure mathematics, in particular, offers an escape from the complexity of emotional entanglements.

Max Frisch's drama *Don Juan oder die Liebe zur Geometrie: Eine Komödie in fünf Akten* (1953) presents a mathematician who wishes to escape totally from the changes and chances of this uncertain life in order to study geometry as a pure philosophical abstraction. Unlike the authors considered previously in this chapter, Frisch is not interested in condemning his protagonist, only in explaining him. His choice of the unlikely figure of Don Juan, the notorious libertine, to fill such a role may be intended to indicate some skepticism about the Don's alleged desire to escape from relationships in order to embrace the static absolute of mathematical laws as ultimate perfection. Equally, it may be Frisch's attempt to locate the psychological basis for the legend in a restless dissatisfaction with the emotional complexities of amorous relationships. Fleeing from his fiancée, Juan tells his friend Roderigo: "I feel freer than I have ever felt before, Roderigo, empty and alert and filled with the masculine need for geometry.... Have you never experienced the feeling of sober amazement at a science that is correct? ... at the nature of a circle, at the purity of a geometrical locus? I long for the pure, my friend, for the sober, the exact. I have a horror of the morass of our emotions.... Do you know what a triangle is? It's as inexorable as destiny.... No deception and no changing mood affects the issue."[18]

Secure in the absolute certainty of two- and three-dimensional geometry, Juan becomes agitated when the bishop, representing a metaphysical view, introduces the subject of four-dimensional geometry, for this imports complexities and uncertainties, the mathematical counterpart of the emotional entanglements from which he is fleeing. In Frisch's treatment, Don Juan's dissatisfaction with women arises because they remind him that he, unlike a geometrical figure, is incomplete in himself: "My dislike of Creation, which has divided us into men and women, is more intense than ever.... What a monstrous mistake that the individual alone is not a whole!" (150). In his postscript to the play Frisch adds: "If he lived in our own day, Don Juan (as I see him) would probably concern himself with atomic physics: in search of ultimate truth.... As an atomic physicist too he would sooner or later be faced with the choice: death or capitulation—capitulation of that masculine spirit which obviously, if it remains autocratic, is going to blow up Creation as soon as it possesses the technical ability to do so" (158–59). Frisch might, alternatively, have made Don Juan a computer scientist. Artificial intelligence epitomizes a simplified world where everything is reduced to 1 or 0, where reason is undiluted by physical and emotional experiences.

Almost without exception, the computer scientist is depicted as wholly absorbed in the computer as an end in itself rather than as a means to achieving other goals. The resultant irony, explored by several writers, is that such scientists eventually end up being controlled by the computer, sacrificing their own human traits in order to become more "acceptable" to it.

In his short story "Die Schalttafel" (The switchboard) (1956) Erich Nossack describes a chemistry student who has mapped out his plan for life on a switchboard, so that at each moment he can calculate the possibilities and probable consequences of any proposed action.[19] His purpose is to choose always the most conformist course. That is, he has effectively programmed himself to become as much like a computer as possible, eliminating freedom and spontaneity.

A similar critique of pure reason underlies Heinz von Cramer's modern parable "Aufzeichnungen eines ordentlichen Menschen" (Notes of an orderly person) (1964), about a mathematician who has constructed for himself a completely regulated life as insulation from the outside world. For him, order has not only a pragmatic value but also the quality of metaphysical truth: "I love order. I am a scientific man, not merely a scientist. That is, my profession has taken possession of me, body and soul. However, order has, for me, also a metaphysical meaning; a threshold on the other side of which it becomes apparent as a universal principle, a principle of creation."[20] The logical extension of such obsession with order is his progressive identification with a supercomputer, and in von Cramer's story this is symbolically realized. The mathematician comes to consider his computer as a living organism, to which he ascribes emotions and personal needs. When the computer performs operations for which he has not programmed it, he feels subordinated by it and makes an unsuccessful and increasingly frantic attempt to establish that he is really its master. Unable to communicate with the computer through his own language, he resorts to its limited binary language—0/1—thereby effectively reducing himself to its level. He then enters into a personal contest with the computer, progressively seeing himself as a modern Frankenstein at the mercy of his own creation. To his frenzied and hallucinating mind the machine appears to transmit to *him* the commands with which he has programmed *it*, to observe him, analyze him, and rule his life. His last, irrational diary entries indicate his Kafka-like metamorphosis into an inferior computer. Thus von Cramer locates the causes of computer dominance in the minds of the programmers, who endow their inanimate machines with human qualities while relinquishing their own humanity.

The seductive aspects of mathematical order as a refuge from the frightening complexities of life have also been explored by the Austrian writer Hermann Broch, who had himself studied mathematics and philosophy. Richard Hieck, the protagonist of Broch's *Die unbekannte Größe* (The unknown quantity) (1933), is a mathematician and astronomer, who places all his faith in mathematical certainty and logical positivism. Mathematics seems an island of decency in a world of uncertainty and irrationality, and he expects it to provide him with stability and assurance and, ultimately, to lead to the understanding of the whole of life. His intention is that of Descartes: "To grasp all the appearances of Life, to grasp them mathematically and rationally, because through this prolific 'mathematicising' of the world, he would arrive at the totality of his own life."[21]

In his attempts to exorcise from his personality all irrational elements, including emotion, Hieck comes, paradoxically, to ascribe a religious dimension to mathematics, which he now regards as a deliverance from sin and as the source of moral probity. "Within mathematics, something could be accomplished only by those human beings who remain pure from all that is sinful or whatever else one wants to call it" (91). When Hieck invites his girlfriend, also a mathematician, to visit his observatory, he becomes increasingly conscious of the inadequacy of the Einsteinian system he is expounding to account for the emotion he is experiencing. Yet, despite this incident and the warnings of his elderly professor, Weitprecht, that mathematics is no substitute for life; it takes the death of his brother to bring Hieck to the full realization that scientific knowledge expresses only part of a much broader mystical truth.

Broch related this realization to the theory of relativity itself, which he saw as an affirmation of the integral relationship between the observer and his field of observation. Ironically, he regarded many scientists, preoccupied as they were with the search for objectivity, as slow to accept the full implications of this central precept of modern physics. Hence the paradox of Hieck expounding relativity theory while trying, in effect, to subvert it by remaining objective and impersonal.

Stephen Sewell's play *Welcome the Bright World* (1982) goes further in exploring the effect of such computer identification on the devotee; it suggests a close connection between the allegiance to inanimate conceptualizations of the world, denigration of people, rejection of accountability by the scientists involved, and potential destruction of the world by a nuclear war. Sewell presents us with two nuclear physicists, Max Lewin and Sebastian Ayalti, who have given their lives to trying to understand the nature of the electromagnetic forces that hold the atom together and the radioactive disintegration of nuclear particles. In his notes for the play's premiere, Sewell wrote: "This work is currently moving rapidly ahead. Unfortunately, and tragically, so is the work of those scientists and technologists involved in the development of even more hideous nuclear weapons." Within the play he explores this anomaly by showing how these highly enthusiastic and committed scientists insist on remaining detached in their deteriorating personal relationships and in the worsening political situation of a virtual police state. Early in the play, Sebastian and Max discuss whether the scientist has a duty to explain to the public the dangers of nuclear energy, Max arguing for the view that scientists are beyond accountability:

> MAX: How can you talk about something as complex as nuclear energy in front of ten thousand people?
> SEBAST: People have the right to information.
> MAX: Most people don't count, Sebastian![22]

When Rebekah asks Sebastian what he likes about physics, he answers:

> SEBAST: I always wanted to understand the world. . . . I'm an obsessive.
> REBEK: What are you obsessed with?
> SEBAST: Knowledge and discovery are like drugs. . . . Schopenhauer sought relief from the will in art. . . . Contemplating nature gives me the same relief.
> REBEK: From the will?
> SEBAST: From involvement. (51)

Sebastian later proclaims, "I only have one relationship, and that's with my work. . . . I didn't want to get involved; I didn't want to have any scenes; I only want to do my work. . . . I don't want relationships, I don't want involvements!" (62–63).

Mathematics and the particular dehumanizing effect it has on its devotees also interest the Austrian writer Robert Musil. In his early work *Die Verwirrungen des Zöglings Törleß* (The confusion of young Törless) (1906), his adolescent schoolboy protagonist is obsessed by a desire to understand life "with the eye of reason," that is, in the simple, objective terms of mathematics, as a means of escaping from the frightening depths of emotion and mysticism that he dimly perceives are also part of life. Impatient with simplistic analogies and the master's patronizing assurances that he will be able to understand it when he is older and has mastered Kant, Törless craves the certainty that mathematics alone seems to offer. Yet even here, the inexplicable, the mathematical correlative of the mystical, haunts him (as it had D-503 in *We*) in the form of infinity and imaginary numbers such as the square root of –1. Simultaneously, Törless is drawn into an intensely emotional situation of brutal cruelty and exploitation, from which he escapes only after a virtual nervous breakdown. Like Broch's mathematician Hieck, Törless learns the lesson, foreshadowed in both the *Verwirrungen* (confusion) of the title and the introductory epigraph from Maeterlinck, which proposes that for everything there are always two aspects—reason and mysticism—that cannot be reconciled.

In Musil's later story "Tonka" (1924), in the *Drei Frauen* volume, these two perspectives are embodied in different characters. The young scientist, who, significantly, remains nameless, represents the rational component developed at the expense of other aspects of life. Tonka, the servant girl, with whom he enters into an affair, embodies the mysterious, irrational forces of nature. The sex roles ascribed to these opposites are certainly intentional. For Musil, it is the male mind that reaches its fullest intellectual development in the conscious activities of logic, science, and verbal articulateness, so his scientist gains strength and confidence from his work on a new invention. Tonka's needs, on the other hand, are emotional rather than rational; like nature, she seems merely to *be*. The ensuing conflict between reason and mysticism is tragic for Tonka. When she becomes pregnant, and there is some doubt about the precise time of conception, the scientist demands irrefutable empirical evidence for or against her possible infidelity; he cannot accept a truth based on trust in the girl. He therefore abandons her, even though he is equally unsure about her infidelity. The completion of his invention on the day Tonka dies would seem to represent the triumph of reason over emotions; yet, paradoxically, her death also marks the permanent impossibility of any final solution to his uncertainty. Although it is easy to read the story as a vindication of Tonka and the intuitive, emotional perspective she represents, Musil's own evaluation is less clear cut. Here, as in *Törless*, he remains ambivalent, for there is also an implicit Nietzschean justification of the strong, intellectual male vanquishing the weak emotional female.

Musil seemed unable to leave this topic. In his epic novel *Der Mann ohne Eigenschaften* (The Man without qualities) (1930–43), the protagonist, Ulrich, like Musil himself, has tried several careers—the army, mathematical research, engineering, public service—but retains throughout the noninvolved approach to life of the pure mathematician. He analyzes every thought and feeling with disinterested objectivity, ceaselessly dissecting, questioning, and measuring each experience, each person in his life, as an abstraction until, as the title suggests, he loses any identity and cohesion as an individual. Ulrich's obsession with order arises from his quest for meaning. At the start of the novel, he has given himself one year to discover meaning in life; otherwise he will commit suicide. Thus, like Törless and the scientist of "Tonka," Ulrich sees the world of abstract mathematical thought as a bulwark against the unpredictable dangers of the emotions and meaninglessness.

Embarking on his "hypothetical life" as on a mathematical experiment, Ulrich remains detached, skeptical, a relativist. While to the other characters in the novel, this stance seems intolerable, impersonal, and destructive of human relationships, Musil suggests that it has its positive side too—Ulrich's lack of prejudice and his openness to the future. He regards himself as a man of possibilities, a *Möglichkeitsmensch*, continually speculating on the options available but never committing himself to any one of them. As long as they are merely conjectures, they may be discarded at any time without regret. "He sensed that this order was not so firm as it pretended to be; no thing, no self, no form, no principle was certain. Everything is caught up in invisible but never-resting metamorphosis; . . . and the present is nothing but a hypothesis which one has not yet gotten beyond."[23]

Ulrich undertakes two experiments: a "rational experiment" designed to find an intellectual basis for understanding life and a "mystical experiment." Like Musil, he believes that these alternative, even opposite, roads to truth must be combined into a single creative road and that this is the greatest challenge awaiting modern man. But it is the first experiment that concerns us here. In the first phase of his life, Ulrich commits himself to scientific method and the principles of mathematics with the utmost rigor. His friends, predictably, find him cold, hypercritical, lacking in compassion, and, withal, strangely impractical, for it is part of Ulrich's commitment to pure science that it should not be required to be useful. Musil calls modern science "the miracle of the Anti-Christ" (360), and he notes Ulrich's "tendency towards malevolence and hardness of heart" (359).

However, Musil's depiction of Ulrich remains ambiguous. On the one hand, he towers over the other characters, with their limited, egocentric grasp on life; on the other hand, he fails to attain even the partial understanding of life he had expected. Yet Musil, writing about the futility and violence of Austrian society in the last days of the Hapsburg monarchy, does not imply that Ulrich's failure to discover a structure and order among the disorder and pluralism of his society is evidence of his own deficiency. On the contrary, the novel suggests that although belief in a firm, universal order is a delusion, disinterested objectivity may be the only appropriate stance in a time of ideological breakdown and anomie.

A particularly provocative feminist analysis of the inhuman scientist is to be found in Christa Wolf's short story "Selbstversuch. Traktat zu einem Protokoll" ("Self-Experiment: Appendix to a report") (1973). Wolf's story concerns a female scientist, a physiopsychologist specializing in sex-change research and condescendingly regarded by her professor as "the equal of any male scientist."[24] By means of a drug developed by her professor, she has voluntarily undergone a sex change, which is hailed as highly successful. The professor's experiment to "create sex" at will links him with the characters we have encountered in earlier chapters who were focused on creating or modifying life and also indicates that, even in a culture (the DDR, East Germany) that purports to regard men and women equally, he is fixated on sexual differences. Yet before a month of her life as the male Anders (meaning Other) has passed, she rejects her new persona and undergoes a reversal of the process. As in Musil's writing, the sex roles are identified with two different modes of perception. In assuming a male role, the woman of the story is reinforcing the intellectual, scientific aspect of herself and suppressing her feminine, intuitive self. Her determination to be restored to her female role results from her experience of the limitations of a purely rational perception. In her male identity as Anders s/he finds it impossible to speak or write of how she feels. Instead she is aware of "the silence which reigned inside me. Do you know what person means? Mask. Role. Real Self. A prerequisite for language ... must be the existence of at least one of these three conditions. The fact that all of them were lost to me had to mean virtually total silence. You can't write down anything about nobody" (127). Knowing, as the mythical Tiresias did, both sex roles, she opts unhesitatingly for the more complex, nonscientific awareness that Wolf identifies with the female consciousness. The professor, who epitomizes the scientific assumption that he has caught reality "fast in a net of numbers, diagrams, and calculations," is exposed as a figure of ultimate weakness. The woman tells him: "Your ingeniously constructed system of rules, your hopeless addiction to work, all your withdrawal maneuvers were only an attempt to protect yourself from this discovery: that you cannot love, and know it" (130). The narrator, on the other hand, freed from the need to enact a role "as in the cinema," can embark on a new experiment as a woman: "Now my experiment lies ahead: the attempt to love. Which incidentally can lead to fantastic inventions—to the creation of the person one can love" (131). Friederike Eigler comments that this new experiment "can be read as a trope for what Sandra Harding calls the shift from the 'Woman Question' in science to the 'Science Question' in feminism."[25]

In her short novel *Störfall* (Breakdown) (1987), written in response to the Chernobyl nuclear reactor accident of 1986, Wolf further indicts what she sees as the obsessive fascination of male scientists with technical problems at the expense of personal relationships and social responsibility, linking this with their isolation from their emotional sources. Specifically targeting the physicists working on the Star Wars project at the US Livermore Laboratories, Wolf compares their situation to that of boys in a boarding school, deprived of relationships with women or families, and asks whether this is the cause or effect of their intense relationship with their computer. "These were men in an isolation station, without women, without children or friends, without any pleasures other than their work, subject to the most rigorous controls of security and secrecy. . . . I have read that they know neither father nor mother, neither brother nor sister, neither wife nor child (there are no women there . . . ! Is this oppressive fact the reason for the computer-love of the young people or its result?)."[26]

Wolf also focuses on what she has read about the girlfriend of one of these young Star Warriors. Josephine Stein, the girlfriend of Peter Hagelstein, is a pacifist and protests publicly against Hagelstein's work in developing the

X-ray laser for the Star Wars project. Eventually she leaves him, and Wolf casts her, not as Faust's passive Gretchen, but as a new Lysistrata. This is consistent with Wolf's feminist view that women, being marginalized by the patriarchal hegemony, are better able to perceive its defects and to protest against its inhumane principles. So, too, the female narrator of *Störfall*, whose preoccupations are with growing plants in her garden, nurturing children and friends, and meditating critically on the implications for humanity of the Chernobyl disaster, is contrasted throughout with her brother, a computer scientist currently undergoing surgery for the removal of a brain tumor. His condition symbolizes Wolf's belief that a brain that is overdeveloped in relation to the rest of the person is, in fact, diseased. This juxtaposition of personal and public breakdowns, as indicated by the title, reiterates the cause-and-effect question quoted above: Is the uncontrolled technology apparent in the breakdown of the Chernobyl reactor a cause or a result of the disproportionately evolved brains of scientists and technologists, who display a correspondingly stunted emotional development?

Wolf's affirmation of complementarity provides an interesting metaphorical parallel with Heisenberg's uncertainty principle and Bohr's complementarity principle. Increased knowledge in one area occurs at the expense of decreased certainty in another. We cannot know accurately both the position and the momentum of a given particle simultaneously; we cannot overly develop our technologically focused rationality without sacrificing our emotional well-being, and with it our social and environmental responsibilities. Viewed in this light, the preoccupation of the scientist characters discussed in this chapter with objectivity, measurement, and accuracy of definition is marked by an extra level of irony.

As treated by most of these authors (Asimov is, of course, the notable exception), there is at least a suggested link between the impersonal scientist and the amoral one. This in turn carries the implication that morality is associated not with objectivity and rationality but rather with subjectivity and empathy, with sensibilities nurtured by emotional experiences.

This connection is made explicit in Carl Zuckmayer's play *Das kalte Licht* (The cold light) (1955). Factually, the play is based on the story of Emil Julius Klaus Fuchs, a naturalized British subject of German origin and communist sympathies who worked in atomic weapons research establishments in both Britain and America and passed on secret information to the Soviet Union.[27] However, Zuckmayer's interest lies not so much in the moral dilemma of the scientist working on weapons of mass destruction—the subject of the works examined in the preceding section—as in the more general problem of alienated modern man, bereft of any system of values. The moral rootlessness of his atom spy, Christof Wolters, is symbolized by his involuntary movements from country to country. An emigrant from Nazi Germany because of his communist sympathies, he has applied for British citizenship, but his application is delayed by the outbreak of war, and he is deported to Canada as a civilian detainee. When his background in physics becomes known, he is quickly reassessed as an asset and dispatched to London to help in atomic weapons research. Subsequently he is sent to work with the American scientists at Los Alamos, then returns to Britain, where he adopts the religion of science: "In mathematics there is something—idea and reality. It can only be proved hypothetically, of course—yet it's the basis of our whole existence."[28]

Resenting the restrictions and secrecy imposed by the war and believing that science should be international property, Wolters agrees to pass on scientific results to the Russians. His perfunctory stipulations that the information must not be abused indicate a troubled conscience, but Wolters believes there is no going back, either in science (where the point of no return is symbolized by the explosion of the atomic bomb at Hiroshima) or in one's private actions. He thus affirms the determinist doctrine that Dürrenmatt and Kipphardt were also to examine. "There's no way out. What's done, can't be undone" (71).

Wolters is depicted by Zuckmayer as a flawed, incomplete individual, preoccupied with abstract thought and having "a conscience that somehow short-circuited and a kind of a fatal split in his ethical code" (113). He is a man lost and split (like the atom). This represents an interesting twentieth-century variation of the Romantic image of the scientist as cold and unfeeling.[29] As the title suggests, colness is a recurrent image of the play. Although the epigraph (allegedly from a textbook on nuclear physics) relates the title to luminescence—"Cold light is a light which cannot ignite, nor does it generate any heat"—the underlying meaning of the phrase is developed in moral terms through contrast with the warm glow of conscience. In what is effectively the pivotal statement of the play Wolters says: "All my life I've been standing in the beam of a cold light—one that comes from outside and turns you to ice. . . . But one single moment can be a great blaze—transforming everything" (141).

In the face of his moral confusion, Wolters comes to affirm the centrality of conscience, the "inner light," as a guide to action, as "an integral part of our bodies—born with us. . . . It's there to help us when we're lost, and remind us that there's a better and finer—a higher order of things, or quite simply to show us the truth" (129). Eventually he chooses to confess his espionage, even though there is insufficient evidence to convict him, discovering inner peace by doing the duty that lies closest to hand: "Good can only be lived, and like charity, it begins at home!" Thus Wolters effectively discards the public world of social responsibility and returns to the pietistic Protestant ethic of individual integrity, the religious background from which he, like Fuchs himself, had originally come.

The next chapter explores the character of the amoral scientist, the contributory causes postulated by writers for the association of science with amorality, and the effects on society of such a stance.

CHAPTER 14

Scientia Gratia Scientiae
The Amoral Scientist

> We scientists have only one God of free research. And this God says *"Fiat scientia pereat mundus"*—let there be knowledge though the world perish!
> —Heinrich Schirmbeck

A distinctive subset of the stereotype of the impersonal scientist is that of the amoral scientist. Whereas the scientists discussed in chapter 13 affected mainly, perhaps only, their immediate family and friends, the amoral scientist carries the stance of noncommitment into the broader ethical sphere and produces major risk factors for society. Since these scientists are often powerful people, eminent in government policy making or acting as advisers to the military-industrial complex, their impact may be pervasive and insidious.

Compared with mad or evil scientists, amoral scientists are less readily identifiable as evil; they do not pursue science for power or wealth but merely for the apparently modest reward of solving an abstract intellectual problem, sometimes with patriotic intention. As with the utopian scientists of chapter 11, there is often a deceptive air of idealism about such dedicated individuals who sacrifice personal comfort or material gain to the pursuit of truth. Compared with the barons of industry, they may even appear figures of wisdom or innocence. Shades of the Curies hover around them, and Bacon and Newton stand benignly in the background. But although the evil is less obvious, it is no less dangerous. By achieving respectability, even fame, these scientists influence others to dismiss ethical concerns as irrelevant, even inimical, to science and hence to the development and prosperity of society. Whether explicitly or implicitly, these characters adopt the principle that doing X because X is possible is a prerequisite of science.

It is only since the 1970s that ethics committees, which monitor research programs for experiments likely to infringe upon human or animal rights, have been widely accepted as the norm within the scientific community; formerly they would have been regarded as a violation of academic freedom, and even now their presence is tolerated only because it is often a precondition for funding.

There are several reasons for the acceptance, even the cultivation, of amorality within the scientific fraternity. The one most frequently invoked is the assumption that science is value-free, that science in itself is neither good nor bad, only its applications. Many scientists believe that they cannot, and should not, have any influence over the use to which their discoveries are put. They see their role as pursuing knowledge for its own sake and regard this as an essentially noble, even aesthetic, enterprise. Enrico Fermi is quoted as having said in relation to his work on the atomic bomb, "Don't bother me with your conscientious scruples. After all the thing is superb physics."[1] This was the view of many scientists working on the Manhattan Project. Shiv Visvanathan cites Robert Jungk's encounter at Los Alamos with a mathematician whose face was "wreathed in smiles of almost angelic beauty. He looked as if his gaze was fixed upon the world of harmonies. But in fact he told me later that he was thinking about a mathematical problem whose solution was essential to the construction of a new type of H-Bomb."[2]

This aspect has been emphasized in several films about the Manhattan Project, including the documentary *The Day after Trinity* (1980), the story of J. Robert Oppenheimer's involvement in building the first atomic bomb. As we have seen in chapter 12, in this film physicist Freeman Dyson acknowledges the fascination that such power exerts "to release this energy that fuels the stars—to lift a million tons of rock into the sky."[3] However brilliant as physicists, in humanitarian terms these figures are depicted in fiction and film as at least immoral, perhaps even mad: their calculations include human victims as mere bargaining chips in their war games. If challenged, they frequently take refuge in appeals to the inevitability of events, which have proved to be beyond their control.

This stance was bolstered by the widespread fatalistic belief of many scientists that discoveries will emerge inevitably when the intellectual climate is right. Although at the time of their discovery important breakthroughs in research may seem to be the work of particular brilliant scientists, in hindsight it often appears that they would necessarily have arisen within a short time. The fact that many major projects are worked on simultaneously by several research groups in competition lends support for this view. This assumption of the inevitability of scientific discovery was a particularly significant factor during the Second World War, when research on the atom bomb was fueled largely by the argument that the "other side" was close to making the breakthrough first. It was Albert Einstein's letter of 2 August 1939 to President Roosevelt suggesting that Germany might already have begun such a program that initiated the US project to develop the atomic bomb.[4] The belief that the enemy might "get there first" was, of course, exploited both by the military as war propaganda and by the scientists themselves in order to obtain funding for their research. A similar argument was later used throughout the Cold War for the same purposes. The argument for multiple possibilities of discovery appears to absolve scientists from specific responsibility or, ultimately, any responsibility at all, for their work. They can claim that they are at most midwives assisting at a birth that would have occurred without their particular intervention and that to refuse to work on a particular project allows another research group to reap the honors and rewards.

A related factor in peer-group acceptance of the amoral scientist has been the growth during the twentieth century of the research team. As science has become increasingly complex and expensive, it has become virtually impossible to undertake or finance it other than as a group project. While there may officially be a team leader, who functions as the figurehead (and who may well receive the public honors for the team's success), the other members usually regard themselves as being equally important in achieving the results. It therefore becomes easy for individual scientists to disclaim responsibility. Each is only one member of the team.

In literature, however, amorality has rarely been treated favorably, and where it is associated with science the authorial voice has, in most cases, been highly critical of the pursuit of truth for its own sake. To most writers from a humanities background these teams of scientists, immured in their vast and often secret research institutions, are faceless and irresponsible, using the impersonal focus of their work to justify a position of ethical minimalism. Their disregard of the human factor in their research is interpreted as evidence of their alienation from their own humanity; thus many of the characters examined here are depicted as emotionally retarded, like the impersonal scientists of chapter 13.

This amoral and pragmatic stance is seen by many writers as symptomatic of Western technological society, bent only on extending the boundaries of knowledge and dominating both the natural and the human world. Since the 1970s feminist critics have related this amoral, rationalist hegemony to the emotionally deficient conditioning of males in Western society. The former atomic physicist Brian Easlea has explored this hypothesis in relation to the cult of nuclear weapons research as a masculine alternative to childbirth.[5]

Symptomatic of the paradigm of amorality is the realization in the twentieth century of two of the alchemists' goals: the artificial creation of human beings (see fig. 21) and the discovery of almost limitless power. Both are important symbols in the analysis of the technological society, and both feature largely in the presentation of the scientists discussed in this chapter.

Because of their involvement in the development of atomic weapons physicists have been a prime target for charges of amorality. Writers have felt compelled to explore the motives that could lead brilliant scientists, many of whom had at some stage embraced a creed of internationalism, to collaborate in producing weapons of unprecedented mass destruction. In *Night Thoughts of a Classical Physicist* (1982) Russell McCormmach suggests how easily a course of action fundamentally inimical to both the moral stance of the individual and the principles of science as laid down by Francis Bacon and reaffirmed in the scientific norms (CUDOS) of sociologist Robert Merton[6] can be justified in terms of high-sounding principles. His protagonist, Jakob, an aging German physics professor, reflects during the First World War: "It is no betrayal of our scientific creed for us to yield to the dominant pull of

Figure 21. *The Test Tube Baby,* by Albert Robida. From *Vingtième Siècle*, 1883.

patriotism, despite what scientists abroad may say about us. To work for the fatherland, we temporarily sacrifice what is essential to our work as scientists, but we do not for a moment renounce our desire for a peaceful world that knows no national scientific barriers. . . . In science and only in science, [people] work seriously together, in harmony instead of in conflict, in selfless co-ordination instead of in egoistic isolation."[7]

Questions raised concerning the social responsibility of scientists revolve around the same recurrent issues: ethical considerations of particular research projects and how they mesh with the traditional premise that science is value-free; the degree to which scientists can be held responsible for the use to which their discoveries are put by others; the control they can hope to retain; and the accuracy with which potentially harmful consequences can be predicted.

Because the power and status of scientists, albeit indirect, far exceeds that hitherto exerted by any previous persons or groups in human history, the number of novels, plays, and films featuring amoral scientists has increased dramatically, especially since 1945. Those chosen for discussion here by no means exhaust the subject but have been selected as representative.

The Bomb Makers

It is one of the paradoxes of the twentieth century that although, prior to the 1940s, fictional characters who invented and proposed to explode bombs were considered, without exception, to be both mad and evil or, at the very least, seriously misguided, the actual scientists involved in the construction, testing, and use of weapons of mass destruction were, for the most part, regarded by their contemporaries as highly intelligent individuals, their research being funded with an ever-increasing slice of their nation's wealth. Only in literature was their humanity or their sanity systematically questioned, as writers explored the motivation of seemingly idealistic scientists to collaborate with governments and the military-industrial complex.

These characters frequently allege the inevitability of events beyond their control. Albert Einstein's own explanation for the involvement of physicists in atomic weapons research represents a prototype of this "argument": "Scientists have been driven by the intellectual fascination of complex questions, and on occasion they have found added justification for their work in the hope that their activities might transform the world. They have often offered policy makers undreamed-of opportunities or have prodded the policy makers to set limits to their ambitions. But ultimately the scientists have worked as the servants of policy. Those who would praise or blame the weapons scientists for the consequences brought about by the products of their work should first direct their attention to the authors of policy."[8] Deconstructing this statement, we can see the image of the noble scientist presented in its most attractive form—all of the nobility and none of the responsibility. This was undoubtedly the way most weapons scientists preferred to see themselves, and it is the self-image of the characters discussed here. The authors, however, are less permissive; they have inclined more toward the view put forward by the physicist Freeman Dyson, who was, himself, involved in research on the atomic bomb:

> Up to the year 1982, six countries had overtly acquired nuclear status: the United States, the Soviet Union, Britain, France, China and India. It was certainly true in the first four cases, and possibly true in all six, that scientists rather than generals took the initiative in getting nuclear weapons programs started. In each case of which the world has knowledge, scientists were motivated to build weapons by feelings of professional pride as well as of patriotic duty. The construction of the bomb was a technical challenge that stirred their fiercest competitive instincts. In each case the scientists felt themselves to be in competition with the scientists of some other country.... It is no great exaggeration to say that the British and French programs were driven by professional pique of scientists rather than by a careful consideration of strategic needs. The same thing might be said of the American hydrogen-bomb program.[9]

The force of this statement is accentuated when we take into account the fact that it was already evident to American physicists toward the end of 1944 that Germany had abandoned its atomic bomb project and hence that the alleged rationale of needing to perfect the bomb before Hitler could do so was no longer valid. Joseph Rotblat, the only physicist to leave the Manhattan Project for reasons of conscience, in discussing why his colleagues remained, asserted that "the most frequent reason given was pure and simple scientific curiosity—the strong urge to find out whether the theoretical calculations and predictions would come true. These scientists felt that only after the test at Alamogordo should they enter into the debate about the use of the bomb."[10] By then, of course, the project had attained its own political momentum. There would always be a new pretext to justify its continuation—if not Hitler, then the Soviets. The seeds of the Cold War were already germinating in the heat of Los Alamos before the Second World War ended. Rotblat's statement about the "urge to find out whether the theoretical calculations and predictions would come true" when the consequences would be the death of thousands of innocent people and the long-term irradiation of the environment, may well seem incredible lunacy or the most inhuman malevolence, yet it encapsulates the values of the amoral scientist in literature and of many highly respected atomic physicists in real life.

Some satirical treatments of this theme dwelt on a symbolic, representative character, one of the earliest being Carl Sandburg's black-comic poem (dated, significantly, August 1945) about a seemingly harmless, but sinister atomic physicist, "Mr. Attila."

> They made a myth of you, professor,   They didn't think it, eh professor?
>   you of the gentle voice,   On account of your're so absent-minded,
>   the books, the specs,   you bumping into the tree and saying,
>   the furtive rabbit manners   "Excuse me, I thought you were a tree,"
>   in the mortar-board cap   passing on again blank and absent-minded.
>   and the medieval gown.
>   Now it's "Mr. Attila, how do you do?"
>   Do you pack wallops of wholesale death?
>   Are you the practical dynamite son-of-a-gun?
>   Have you come through with a few abstractions?
>   Is it you Mr. Attila we hear saying,
>   "I beg your pardon but we believe we have made some
>   degree of progress on the residual qualities of the atom"?[11]

Despite his mild manners and timidity, Mr. Attila has committed mass murder as certainly as did his barbaric namesake. A similar indictment of the superficially comic scientist who perpetrates near disasters underlies several of the films featuring absentminded professors, for example Doc Brown in *Back to the Future* (1985) and Wayne Szalinski in *Honey, I Shrunk the Kids* (1989).

The first treatment of physicists working on the bomb was Bertolt Brecht's second version (1945–47) of *Leben des Galilei* (Life of Galileo). This modification of the original version (1938–39) was written in the United States after the bombing of Hiroshima and Nagasaki. Brecht's plays were conceived in the tradition of the epic theatre, intended to produce a critical response to political ideas, and he wrote in his introduction: "Overnight the biography of the founder of the new system of physics reads differently. The infernal effect of the Great Bomb placed the conflict between Galileo and the authorities of his day in a new, sharper light."[12] Unlike his earlier protagonist, a courageous rebel against repressive authority, this revised Galileo is the prototype of the amoral scientist pursuing research without concern for its consequences. He commented: "The scientific work does not outweigh the social failure," adding later, "In the end, Galileo is a promoter of science and a social criminal."[13] In his notes on the play Brecht amplified this: "Galileo's crime can be regarded as the 'Original sin' of modern natural sciences. From the new astronomy, which deeply interested a new class—the bourgeoisie—since it gave an impetus to the revolutionary social current of the time, he made a sharply defined special science which . . . was able to develop comparatively undisturbed. The atom bomb is, both as a technical and as a social phenomenon, the classical end-product of his contribution to science and his failure to society."[14]

Brecht's third version of *Galileo* (1953–55) is an even more rigorous criticism of the divorce of scientists from society and their cult of *scientia gratia scientiae*. Galileo becomes the central focus of Brecht's condemnation of the inhumanity of contemporary science and the irresponsibility of scientists.

In charging Galileo with social treason, Brecht set him in an anachronistic framework, one that did not exist in the seventeenth century. At worst, the actual Galileo's "sin" was one of omission: he turned his back on the dawning enthusiasm of the bourgeoisie for practical knowledge and isolated himself in the pursuit of pure science. Compared with the active sins of commission of those who developed nuclear weapons, his "treason" appears little more than a minor ethical failing. Yet in another sense Brecht was correct in placing Galileo at a critical moment in the development of science, the moment when, according to Arthur Koestler, the confrontation between Galileo and the church resulted in a progressive divergence between scientific and technological progress on the one hand and the affirmation of a moral and spiritual dimension on the other.[15] Viewed in this light, it is arguable that Galileo's withdrawal from society, as depicted by Brecht, was indeed a precursor of the behavior of Oppenheimer and his colleagues working on the Manhattan Project in careless isolation.

Brecht's treatment is deliberately stylized and universalized, consistent with his technique of *Verfremdungseffekt* (distancing effect).[16] When the bomb makers appear in person as differentiated characters in other twentieth-century plays or novels, the presentation, often based on carefully collected biographical detail, is still critical but varies markedly in the degree of condemnation assigned by the authors to individuals. Pearl Buck's purpose in *Command the Morning* (1959) is to explore how otherwise well-intentioned individuals can engage in military research. Dürrenmatt's play *The Physicists*

(1962), on the other hand, poses the question of how a brilliant physicist can avoid having his ideas misused. In other treatments the focus is satirical, as in Vonnegut's Felix Hoenikker, the physicist of *Cat's Cradle* (1963), or overtly condemnatory, as in Schirmbeck's Richard Tzessar in *Ärgert dich dein rechtes Auge* (1957).

Often these characters are accounted for in terms of moral schizophrenia. In other areas of life they may be sensitive, humane individuals, but they are unable to see any anomaly in working on a project designed to destroy thousands of civilians. Some of the most trenchant criticism of the atomic physicists in this regard comes from German writers of the two decades following the war. The inventor protagonist of Wolfgang Borchert's *Lesebuchgeschichten* (1946) begins by making home appliances but subsequently, without any sense of immorality, moves on to making bombs. His mind is so rigidly compartmentalized that he can agonize over a wilting flower while regarding society as merely a market for his products—whether appliances or bombs. Borchert reworks the often-remarked phenomenon of Nazi concentration camp supervisors enjoying classical music while perpetrating horrors against humanity.

One of the first realistic portrayals of research scientists involved in a political dilemma was C. P. Snow's novel *The New Men* (1954). Snow, a physical chemist by training, worked during the Second World War in the British civil service, selecting scientific personnel. He was therefore uniquely equipped with inside knowledge about a group of physicists engaged in nuclear fission research. Lewis Eliot, Snow's nonscientist narrator, explicitly comments on the absence of ethical anguish among these "new men" at Barford (the fictional equivalent of Harwell, the British Atomic Energy Research Establishment), their naive acceptance of the view that they must race the Nazis to the production of the atomic bomb—"For those who had a qualm of doubt, that was a complete ethical solvent"[17]—and their equally naive belief that mere possession of the bomb would be sufficient to end the war (and all wars).

Given Snow's own background in both science and the public service, it is significant that in his novels politicians by no means always play villain to the innocent scientists. Bevill, the government minister supervising the project comments: "I used to think scientists were supermen. But they're not supermen, are they? Some of them are brilliant, I grant you that. But . . . a good many of them are like garage hands. Those are the chaps who are going to blow us all up" (69-70). Bevill becomes the spokesperson through whom Snow subverts the scientists' complacent belief that the bomb, once developed, would not be used: "Has there ever been a weapon that someone did not want to let off?" (71). Snow also pinpoints the scientists' traditional allegiance to internationalism as one of the major sources of conflict between them and their government during wartime. Even though "many scientists congratulated themselves on their professional ethic and acted otherwise" they still believed in "free science, without secrets, without much national feeling. . . . The truth was the truth and, in a sensible world, should not be withheld; science belonged to mankind" (130–31). Such universalism, while a lynchpin of Merton's scientific norms, runs starkly counter to the rationale of military and political hegemonies, grounded in the belief that secrecy is a precondition of power.

Insofar as Snow's scientists have a claim to morality, it rests on their idealism concerning the free, international community of scientists and their revulsion at the idea that the atomic bomb should be used against Japanese civilians. Yet Snow also portrays this idealism as coexisting with a quite amoral sense of elation at the success of the project: "Several men of good will felt above all excitement and wonder. . . . I thought the emotion was awe; a not unpleasurable, a self-congratulatory awe" (166). The real-life model for this ambivalence was almost certainly the humane and cultured J. Robert Oppenheimer, whose private fears about the increasing hold by the military over the products of research in atomic physics did not prevent his much-quoted expression of delight in the H-bomb as "technically so sweet that you could not argue about that. It was purely the military, the political and the humane problem of what you were going to do about it once you had it."[18] As Philip Stern remarked, "If scientists as sensitive as Oppenheimer can indeed wall off their moral sensibilities so completely and successfully, then technology is an even more fearsome monster than most of us realise."[19]

Snow presents a broad spectrum of attitudes and motives among his scientists. Two, Martin and Mounteney, regard science as a private satisfaction; content with research as an end in itself, they have no interest in the social relevance of their work. Walter Luke, on the other hand, is convinced that the justification of science is the power it offers over nature, a power that he, like Bacon, never doubts will be used for the betterment of mankind. Other characters are less accepting of the establishment's position. Determined to protest against the use of the bomb before the decision to use it is irrevocable, two members of the Barford group travel to Washington to put their case. Francis Getliffe, Snow's most intelligent and idealized scientist character, voices the pragmatic argument that, quite apart from moral considerations, it would be insane to use the bomb, because it would lead to a proliferation of such weapons in other countries: "Non-scientists never understood, he said, for how short a time you could keep a technical lead. Within five years any major country could make these bombs for itself" (174). Here Snow is almost certainly voicing his personal critique of the Baruch Plan of 1948,[20] which was based on the premise that, because only the United States at that time possessed the knowledge necessary to construct the atomic bomb, it would retain honorable control over it and prevent any other nations from duplicating the discovery. Subsequent events have vindicated Snow's skepticism on a scale unimagined by Snow himself or, indeed, anyone else at the time. The majority of Snow's fictional physicists, however, like their real-life counterparts, refuse to believe that the bomb will be used and incline to the view that "the fission bomb was the final product of scientific civilisation; if it were used at once to destroy, neither science nor the civilisation of which science was bone and fibre, would be free from guilt again" (178).

When the second bomb is dropped on Nagasaki, even those of Snow's scientists who had given qualified assent to the first are outraged and cynical: "They all assumed . . . that the plutonium bomb was dropped as an experiment, to measure its 'effectiveness' against the other. 'It had to be dropped in a hurry,' said someone, 'because the war will be over and there won't be another chance'" (201). It becomes clear that the scientists feel that the "perfection" of the atomic bomb has changed the rules of the game; "the relics of liberal humanism had no place there" (300). Ironically, therefore, those who have been most opposed to the dropping of the bomb now see the only hope for the future in making more bombs in Britain. Luke speaks for most of them when he voices the conventional argument for the bargaining power of deterrence: "All I can say is that, if we're going to get any decency back, then first this country must have a bit of power" (203).

One of the many interesting aspects arising from Snow's inside knowledge of a research establishment relates to the contrast he draws between physicists and engineers, since it qualifies many of the stereotyped attributes traditionally imputed to scientists in general.

> The engineers . . . the people who made the hardware, who used existing knowledge to make something go, were in nine cases out of ten, conservatives in politics, acceptant of any regime in which they found themselves, interested in making their machine work, indifferent to long-term social guesses.

Whereas the physicists, whose whole intellectual life was spent in seeking new truths, found it uncongenial to stop seeking when they had a look at society. They were rebellious, questioning, protestant, curious for the future and unable to resist reshaping it. The engineers bucked to their jobs and gave no trouble, in America, in Russia, in Germany; it was not from them, but from the scientists, that came heretics, forerunners, martyrs, traitors. (176)

For most of Snow's scientists the great temptation is to capitulate to the doctrine that events are too big for individuals to change. Luke affirms, "It may be so. *But we've got to act as though they're not*" (235). Snow identifies a nexus between the fatalistic belief that events proceed inexorably and the pursuit of scientific research in isolation, and his intimate knowledge of a weapons research establishment must lend considerable weight to his analysis of the factors motivating such scientists.

Whereas Snow strove for apparent objectivity of presentation, Heinrich Schirmbeck's monolithic novel of some six hundred pages, *Ärgert dich dein rechtes Auge* (1957), translated into English as *The Blinding Light*, is strongly judgmental. Despite the weight of realistic detail, Schirmbeck's interest, like Brecht's, lies in the universal rather than the particular; where Brecht used Galileo as an archetype, Schirmbeck employs the mathematician Pascal to suggest the demonic nature of all knowledge. Early in the novel, Schirmbeck's narrator, the young physicist Thomas Grey, imagines Pascal's thoughts upon constructing the first calculator. At first Pascal claims that his motive was solely to free mankind from the burden of mental calculation, but he later confesses that he was moved by his own desire to transcend the limitations of knowledge.

Among Schirmbeck's large cast of scientists, the two most striking representatives of the amoral pursuit of science are the atomic physicist Richard Tzessar and Colonel Elliot, head of a cybernetics establishment. Tzessar, clearly modeled on the American physicist Edward Teller, is engaged in research on the hydrogen bomb, which he sees as bringing the solar process down to earth. Introduced under the chapter heading "Lucifer" and described as "the new Faust," Tzessar not only embodies the pride of Lucifer but rejects the call for moral responsibility as "Callow sentimentality!" Through an interchange between Tzessar and the still idealistic Grey, Schirmbeck explores the dialectical positions adopted by scientists at the time. Predictably, Tzessar adopts the pragmatic doctrine that science is inherently neither good nor bad but value-free. "We [scientists] serve the God of free research, the God who says 'Fiat scientia pereat mundus'—let there be knowledge though the world perish! . . . We have no power to prevent it." Grey, the authorial voice, counters: "You're still a child of the eighteenth century, Professor—the Age of Enlightenment. . . . But those voices of scientific idealism are not the voices we need now. Any argument based on them can be nothing but an anachronism, a threadbare travesty with the devil grinning behind the scenes." He compares Tzessar's attitude to that of the firms that during the war took pride in manufacturing the highest quality gas-chambers and corpse-ovens, "also in the name of science."[21]

Tzessar also invokes the argument, voiced by Snow's scientists, of inevitability and the helplessness of the individual to change the course of events: "Scientific research is governed by its own laws; it follows an ineluctable course that cannot be affected by the actions of individuals" (342). He even resorts to an implicit categorical imperative for his own purposes: "Anyone who can't bury himself heart and soul in scientific work, letting himself be possessed by it . . . has no right to call himself a scientist. . . . Only when he commands *every* possibility, including that of self-destruction, will he be truly man, the image of God and the master of his destiny. . . . Our job as scientists is to put him in the saddle. It is for him to ride the horse" (343). Thus, despite his impressive intellectual power, Tzessar is depicted by Schirmbeck as morally retarded. As one of the other characters remarks, "He's an excellent example of the schizoid nature of the scientist. Science grows in them like a tumour. Their organic balance has been upset" (263).

Nor is Tzessar alone in this debility. As Mary Shelley gave added force to her characterization of Frankenstein by creating the parallel figure of the explorer Walton, Schirmbeck explores the same amoral philosophy in another scientific discipline, cybernetics, implying a universal judgment on the state of contemporary science. Colonel Elliot, head of a cybernetics research institute, is one of the first representatives of this discipline in literature. The institute is located in a hexagonal building, suggesting that its inhabitants, like bees in their hexagonal cells, are committed to the preservation of the hive. Significantly it is under the control of the government's Department for Strategic Information and its head is both scientist and military official. Schirmbeck implies that although the cyberneticists believe they are equal partners in the collaboration and are free to do their own research, they are in fact being subtly manipulated by the military. Elliot, who embodies the worst faults of earlier fictional psychologists and mathematicians, regards cybernetics as an intellectual game that represents the supreme alienation of the scientist from the world of experience and that, at a more sinister level, can be used to control society by rigorous conditioning and propaganda. His position is the sociological equivalent of Tzessar's in physics: while professing objectivity, he recognizes only power. In a speech recalling B. F. Skinner, the behaviorist psychologist, he explains enthusiastically: "By feeding him [man] with selected data ... we can direct his thoughts and emotions along specific channels. That is the art of propaganda ... as social cybernetics. ... It comes down to the problem of scientifically turning bad citizens into good ones" (381). Symbolically flanked by his assistants, the "technical eunuch" Norf and a former military officer, Moras, Elliot does not define what he means by "good" or "bad" citizens, but it is clear that for Schirmbeck cybernetics, by its very nature, subordinates all human and moral considerations to the criteria of efficiency and control, such as Huxley had portrayed in *Brave New World*.

Schirmbeck provides explicit commentary on the evils of the amoral stance of Tzessar and Elliot through the heroic character of Prince De Bary, modeled on the French physicist Louis de Broglie. De Bary tells Grey that as a younger scientist he, too, had been seduced into a life of contemplation focused on the pure beauty of mathematics and theoretical physics. Later he rejected this course as dangerous and immoral insofar as it omits the human factor, the unpredictable, the indescribable, and hence can discover only a part of the truth. Whatever its merits as literature, Schirmbeck's novel remains a tour de force in its formulation of contemporary arguments concerning the involvement or disengagement of scientists in relation to the needs of the emerging military-industrial complexes of the 1950s and puts forward the uncomfortable view, perennially relevant in relation to politics and corporations, that the morality of a project may be gauged by reference to its funding body.

Another vigorous indictment of atomic physicists for their determination to test lethal weapons regardless of the consequences is provided by Heinz von Cramer's novel *Die Konzessionen des Himmels* (The franchise of the heavens) (1961), set in post–World War II Australia and intended as a comment on the British nuclear tests in South Australia during the 1950s.[22] Von Cramer's scientists demand freedom to continue their experiments without being answerable to anyone, and in the climate of the Cold War, political, military, and even religious bodies condone the bomb tests in the name of expediency, even though the Aboriginal population of the area has already suffered terrible injuries attributable to radiation and will continue to do so.[23]

In contrast to these sophisticated and mostly damning depictions of the bomb makers by British and German writers, criticism of scientists working on nuclear weapons emerged relatively slowly in the United States. Immediately after the war science was, as American journalist Daniel Lang records, "synonymous with victory, and its practitioners, notably physicists, were regarded as celebrities. . . . They were not only looked up to as the inventors of a successful secret weapon, but they themselves seemed secret weapons—a hitherto undiscovered national resource, a fraternity of geniuses who could bail the country out of any crisis. . . . If they could bring off so formidable a feat as ending a war, then what problem could possibly stump them in the era of peace that was now at hand?"[24]

The explanation for this disparity between attitudes in America and Europe lies partly in the fact that in the United States the atomic bomb was directly associated with the ending of the war in the Pacific and hence with the saving of American soldiers' lives. The immense loss of Japanese civilian lives in Hiroshima and Nagasaki was off focus and seen as a regrettable necessity, what later came to be designated "collateral damage." Thus the nexus between science and war was viewed by most Americans at this time as a benign cooperation. Moreover, as long as America alone possessed the secret of the bomb (refusing to share it even with its allies Britain and Canada, who had contributed much to America's "success" in the field), nationalist feeling assumed that such knowledge was in good hands (several instances of this assumption appeared in the literature featuring scientists as heroes discussed in chapter 11) and would be used only in the defense of democracy.

Two critical factors led to the reversal of this estimate. First, several leading scientists, including Einstein and Oppenheimer, appalled at the power and destruction that had resulted from their work, embarked on a crusade to educate the public about the dangers of unchecked research. Their aim was to ensure not only that the bomb would never be used again but that there would be international control of atomic energy. To the popular mind this seemed either naive or, in the face of growing fears about the spread of communism, irresponsible and dangerous. When, in 1949, the Soviet Union exploded its first nuclear weapon, the American reaction was one of panic. Having ignored the scientists' warnings that the knowledge gap was not fixed but an ever-diminishing variable, the public immediately assumed that scientists were not to be trusted with their own discovery, since they might be harboring traitorous, internationalist thoughts. The fear and suspicion generated by the Oppenheimer investigation in 1953, which culminated in a witch hunt for atom spies, left, in the public mind, a taint on all scientists. If the former leader of America's science team could fall under suspicion as a security risk, what degree of perfidy might be incubating among his subordinates?

Second, and more spectacularly, the successful launching of the Russian spacecraft Sputnik on 4 October 1957 sent a shock wave of incredulity through the United States, as it was immediately clear that the Soviet Union was now the leader in space research. US scientists were berated for losing the knowledge advantage that had existed at the end of the war and were reassessed as a doubtful assett: useful if kept under strict control, otherwise a dangerous resource not to be trusted.

While this suspicion of scientists was being aired in the media, literary disapproval focused less on their alleged political disaffection than on the ethical issue of their devotion to research regardless of its consequences. Where the popular press pointed to possible un-American activities on the part of scientists at the onset of the Cold War, novelists and playwrights emphasized the inhumanity of the "patriotic" scientists, playing intellectual games while humanity as a whole staggered ever closer to the brink of destruction. This aspect was emphasized in several films based on the Manhattan Project, including *The Day after Trinity* (1981). It was in this ambivalent climate that Pearl Buck's novel *Command the Morning* (1959) was written, reviewing the motivation of the scientists who worked on the Los Alamos project.[25] After spending many days studying the site and interviewing the scientists and their families, Buck integrated real and fictitious characters in her novel, the American counterpart of Snow's *The New Men*. Enrico Fermi features in the cast as the archetype of the scientist who refuses to acknowledge any social responsibility for his research.

Although many of Buck's other scientists begin with scruples on various accounts, these are dissolved by the Japanese attack on Pearl Harbor, after which the decision to make the bomb is for them no longer an abstract question but one involving American lives. This, combined with the belief that the Germans have almost perfected an atomic bomb, produces that sense of inevitability about the project that had been depicted by both Schirmbeck and Snow; it no longer seems possible to stop the research. In Buck's novel the American scientists are portrayed as avoiding individual responsibility for the decisions, resorting to opinion polls among the scientific fraternity, seeking safety in numbers, and ultimately assigning to the military the decision about whether or not to use the bomb. Buck indicates that such avoidance of responsibility is not a position of moral neutrality but a definite evil and implies that the traditional claims for a value-free science will inevitably issue in scientific arrogance in the next generation: the younger scientists, having tasted power, no longer harbor any scruples. The novel ends with the ambivalent attitude of the project leader, Burton Hall. In his description of this younger generation there is both pride and despair: "These kids! We found the divine fire for them and they grabbed it away from us. They're riding into space on the wings of power. . . . I have a hunch that when the first one goes soaring off into space and driving for the moon . . . he'll yell back to the rest of us here . . . 'Yeah, man, I do command the morning!'" (316–17).[26] In this symbol of the scientist challenging, if not playing, God, claiming exemption from the human condition of responsibility, Buck suggests a form of evil all the more insidious for being socially esteemed.

Similar unmistakable criticism of the atomic physicists working on the bomb is found in Dexter Masters's novel *The Accident* (1955), which focuses on the group at Los Alamos, but after the war. Like Buck's novel, it was based on actual incidents and people and thereby signaled to readers the urgency of becoming informed about contemporary issues that might otherwise be dismissed as fiction. In 1946 Louis Slotin, one of the nuclear physicists involved in the development of the atomic bomb, fell victim to an accidentally triggered burst of radiation during a test exercise on the critical mass of uranium 238. Nine days later he died, having suffered terrible physical disintegration, even though the radiation levels received were nowhere near as high as those suffered by the victims of the Hiroshima and Nagasaki explosions. In Masters's novel the life of Louis Saxl, Slotin's fictional counterpart, is reconstructed in detail—the motivation of the scientist, his moral values and scruples, the conflict between the scientific and social conscience, and the battle of both with military directives. Early in the novel Masters draws a parallel between the foreseen possibility of Saxl's "accident," which was considered an "acceptable risk," and the possibility that had been put forward by the Nobel Prize–winning physicist Baillie, that the explosion of the atomic bomb might ignite the whole of the earth's atmosphere (a possibility also discussed in Buck's *Command the Morning*). To the physicists involved, this too was an "acceptable risk," being of the order of only 1 percent, but to the doctor, Pederson, it is incredible that they would "go ahead and explode a bomb that might blow up the world."[27]

In describing the spectrum of the other physicists' responses to Saxl's accident, ranging from the intense pity and horror to anger that this accident was "allowed" to happen, Masters emphasizes the unease of the scientists who had stayed on after the war to work on the Bikini tests and who were then forced to witness the consequences for their colleague of the research they were doing:

> "The three conditions of scientific work are the feeling that one ought not to leave his path, the belief that work is not all intellectual but moral as well, and the feeling of human solidarity." So wrote Szent-Gyorgi, a Nobel Prize winner of 1937, two years before the fission of the atom was discovered. In the spring of 1946 all three of these conditions were lacking at Los Alamos to the scientists who had left their science to take on the building of the bomb, who were having trouble with their beliefs in their work, and whose feelings of human solidarity were the source of the trouble. (46)

Yet ultimately, whatever their reservations, they all return to the development of more bombs. Saxl becomes a useful set of statistics in the ongoing medical research into radiation effects, and the production of the weapons continues unabated.

Although Buck and Masters evolved their new novels from actual events, they had no hesitation in including their own views and interpretation of these events. Heiner Kipphardt, by contrast, used a different, and perhaps ultimately more effective, method. His play *In the Matter of J. Robert Oppenheimer* (1967), based on the tapes of the enquiry into Oppenheimer's security clearance, took documentary realism to new lengths and gave the impression of being solely a transcription of interviews. However, Kipphardt contrives to give his authorial view obliquely by means of rigorous selection and juxtaposition from a much longer transcript. The character of Oppenheimer himself, leader of the Manhattan Project, will be discussed in chapter 15; here the comments of Edward Teller, who appears in person uttering the actual words he spoke at that time before the Personnel Security Board, are pertinent as a measure of how closely the fictional characters discussed above resemble their real-life counterparts. At the enquiry Teller puts forward his faith in military deterrence: "I am convinced that people will learn political common sense only when they are really and truly scared. Only when the bombs are so big that they will destroy everything there is."[28] He thus denies any personal responsibility for the use of the hydrogen bomb he has worked on. When giving his testimony against his former colleague Oppenheimer, Teller is asked if he has ever had any moral scruples about the hydrogen bomb.

TELLER: No.

EVANS: How did you manage to solve that problem for yourself?

TELLER: I never regarded it as my problem.... It is not a matter of indifference to me, but I cannot possibly foresee the consequences, the full range of practical uses, which are part and parcel of an invention.... Discoveries in themselves are neither good nor evil, neither moral nor immoral, but merely factual. They can be used or misused. (83–84)

Here Teller enunciates the prototypical argument that the scientist is not responsible for the uses to which his discoveries may be put, because he is unable to foresee them. Later in the hearing Teller is asked about his reaction when he heard that twenty-three Japanese fishermen in a trawler had drifted into a radioactive blizzard because the wind had suddenly shifted after the Bikini explosion of the hydrogen bomb. Teller replies: "We set up a commission to study all the effects. And we were able to make considerable improvements in the meteorological forecasts for our tests." It is this that prompts Evans to repeat his previous question, "What kind of people are physicists?" (85). Teller, oblivious of the irony, replies: "They need a little bit more imagination and a little bit better brains, for the job. Apart from that they are just like other people" (85). This is the same argument as was put forward by his fictional counterpart, Tzessar, in Schirmbeck's *Blinding Light*.

Martin Cruz Smith's novel *Stallion Gate* (1986) also offers an interesting exploration of the attitudes of Robert Oppenheimer and his group at Los Alamos against the background of the Mexican belief in the sacredness of life. Smith presents Oppenheimer as a relatively limited individual who needs the bomb as evidence of his identity. Anna Weiss, a German Jewish refugee working on the project, remarks, "There wouldn't be an Oppy without the bomb."[29] Just before the Trinity test, Oppenheimer is like a man possessed, his life drained from him, existing only as an intellect. "His entire body seemed to maintain a faint existence only to carry the painfully brooding skull" (240). Other members of the group have been enlisted because of the belief that they need to make the bomb before Hitler's scientists do. When it becomes apparent that there is no such danger, the rationale is transferred to the need to end the war in the Pacific and save American lives. Smith suggests how glibly the manipulation takes place: "Oppy's voice became nearly tender. 'But you tell a mother of a young soldier who died on the beach of Japan ... that you had a bomb that could have ended the war and that you chose not to use it. Tell his wife. Tell his children'" (176).

Anna Weiss retains her sanity throughout the process of modeling the impact of Trinity only by refusing to consider the process as anything other than a mathematical exercise and claims that the other scientists did likewise. "No one looks ahead to after the bomb is used. Or asks whether the bomb should be used or, at least, demonstrated to the Japanese first.... Every day I kill these thousands and thousands of imaginary people. The only way to do it is to be positive they are purely imaginary, simply numbers" (167). In one intense discussion between Oppenheimer and another physicist, Harvey Pillsbury, the latter argues for an end to the project after the German surrender or, at least, a demonstration of the bomb to the Japanese emperor in a relatively uninhabited area rather than an attack on a major city, but Oppenheimer, in words reminiscent of Dr. Strangelove (see chapter 12 above), replies in terms of cost-effectiveness: "A waste. A waste of a bomb and a waste of the soldiers.... There wouldn't be much to see in a camp—not like a city, not like buildings" (175).

In all these representative novels and plays the authors themselves clearly perceive the amoral stance of scientists as a potential risk to society and, more explicitly, as a basic driver in the development of weapons research. The premise that science is neither good nor bad but value-free is taken by scientists to absolve them from responsibility in an area that would otherwise be too terrible to contemplate. Implicit in this reasoning is a suggestion that the scientists need the military (to accept responsibility for the front-line decisions) as much as the military needs the scientists (to produce their weapons). It is precisely this cozy symbiosis that is under attack from these writers who deny scientists the right to play Pilate and wash their hands of guilt.

This point is also made visually in Roland Joffé's 1989 film *Fat Man and Little Boy*[30] (also known as *Shadow Makers*), another reenactment of the Manhattan Project. In analyzing the relationship between General Leslie Groves, the military director of the project, and Oppenheimer, the scientific director, Joffé suggests that Groves acts as Mephistopheles to Oppenheimer's Faust, to their mutual satisfaction. Ultimately, what this Faust desires as much as achieving the "technologically sweet" success of the bomb is the popular acclaim accorded to the great physicist who designed and completed the project, untainted by any blame for the ensuing destruction. For this Oppenheimer needs Groves to take total moral responsibility for the tactical decisions as to how and when the bomb will be deployed.

In his devastating satire on the American rocket state, *Gravity's Rainbow* (1973) (discussed in chapter 13), Thomas Pynchon focuses on another form of amorality associated with weapons research, the claim that scientists are forced to carry out whatever research is funded. By means of his character Franz Pökler, whose career closely parallels that of Wernher von Braun, Pynchon links the German project to build the V-2 rockets during the Second World War with the US weapons research program during the Cold War and exposes the ease with which German physicists such as von Braun, who from 1937 to 1944 had held eminent positions at Peenemünde working to develop military rockets for the Third Reich, were quickly recruited by the United States in a maneuver disarmingly code-named "Operation Paperclip," before they could be "captured" by the USSR or Britain. Any questions about political allegiance were rapidly and conveniently glossed over. Pynchon's character Pökler, after attending the Technische Hochschule in Munich, moves to the Verein für Raumschiffahrt (the German Rocket Society) in Berlin, then to Peenemünde, and finally to Nordhausen, where he constructs the Imipolex, a Schwarzgerät ("black device") for the rocket with serial number 00000. Pökler's politics extend no further than the acquiring of research funding; hence he is quite happy to join the winning side after the war and move to a lucrative US research institute, which is equally ready to accept his services against its new enemy, the Soviet Union.

Although his presentation differs markedly from the other examples in this chapter in being superficially comic, the American-born writer Kurt Vonnegut Jr., who had been a US soldier in Dresden at the time of the firebombing, is also deeply cynical about the trustworthiness of scientists who reject the ethical perspective and regard science as a game. In *Cat's Cradle* (1963) he presents Dr. Felix Hoenikker, another fictional father of the atomic bomb and another emotionally retarded scientist, repeatedly described as a child in his "innocent" irresponsibility for the consequences of his research.[31] For Hoenikker everything is a game: turtles, a piece of string, atomic weapons— to him all are equally interesting and morally indistinguishable. At one level Hoenikker epitomizes the engaging, absentminded professor, so impractical that he is dependent on his daughter to get him dressed in the morning; but Vonnegut uses the stonemason who has made Hoenikker's tombstone to deconstruct this picture for us: "I know all about how harmless and gentle and dreamy he was supposed to be, how he'd never hurt a fly, how he didn't care about money and power and fancy clothes and automobiles and things. . . . How he was so innocent. . . . But how the hell innocent is a man who helps make a thing like the atomic bomb? . . . I never met a man who was less interested in the living" (63).

Besides the doubtful distinction of being a father of the atomic bomb, Hoenikker has also "in his playful way, and *all* his ways were playful," produced a crystalline substance, ice-nine, which, by virtue of having a melting point of 114°F, causes all water in contact with it to crystallize as ice-nine. Hoenikker had originally created this alternative structure of water to assist soldiers on military maneuvers to cross swamps by solidifying the water. He had not foreseen that the reaction would proceed exponentially and would thus potentially be capable of freezing the entire planet. However, instead of instantly suppressing this discovery, Hoenikker has continued to play with it. When he dies, his three children divide up the lethal crystals of ice-nine as their inheritance ("there was no talk of morals" [50]), walking around unconcernedly with their samples of the substance capable of destroying virtually all life on the planet. Ice-nine is thus a symbolic parallel to the nuclear weapons Hoenikker was already working on. As one character comments, "Anything a scientist worked on was sure to wind up as a weapon, one way or another" (27). Nor is Hoenikker unique; Asa Breed, whose name suggests his representative function, the head of the research laboratory where Hoenikker works, evinces much the same values. His platitudinous message to the high school students is characteristic: "The trouble with the world was that people were still superstitious instead of scientific. He said if everybody would study science more, there wouldn't be all the trouble there was" (25). Breed and Hoenikker are contrasted with a convert to the mystic, holistic religion of Bokononism, Dr. von Koenigswald, who says, "I am a very bad scientist. I will do anything to make a human being feel better, even if it's unscientific" (148).[32] Despite the wry humor of his presentation, Vonnegut's condemnation of scientists is fundamentally serious, as, perhaps, is the pseudocomic reason he suggests for their inhumanity: "Beware of the man who works hard to learn something, learns it, and finds himself no wiser than before.... He is full of murderous resentment of people who are ignorant without having come by their ignorance the hard way" (187).

Pragmatic Postwar Scientists

Criticism of the amoral stance of scientists in a military context continued in literary treatments during peacetime. Writers pointed to the resultant dangers to society in a wide variety of hypothetical situations, from space research to materials technology and industrial pollution. In these analyses the nexus between denial of responsibility and desire for power and transcendence continues to be emphasized.

William Broad's *Star Warriors* (1985) is a documentary study of the young weapons scientists at the Lawrence Livermore Laboratory "whose labors had helped to inspire the 'Star Wars' speech of President Reagan and were now aimed at bringing that vision to life." After weeks spent interviewing these scientists, Broad characterizes them as precocious teenagers in all but their technical and mathematical brilliance, yet "at times they seemed to believe that their labors gave them the power to save or destroy the world."[33]

In her novel *Störfall* (1987) Christa Wolf draws specific parallels between the explosion of the atomic reactor in the nuclear power plant at Chernobyl in 1986, occurring as the result of a "peaceful" use of nuclear power, and the militaristic Star Wars program promoted by the Reagan administration as a path to peace and security. She characterizes the young scientists at the Livermore Laboratories, whom she, too, calls "Star Warriors," as modern-day Fausts who, "driven . . . by the hyperactivity of certain centres of their brain, have given themselves not to the devil . . . but to the fascination of a technical problem."[34] As the result of an article she had read about the Star Wars physicists, Wolf was particularly interested in the character of real-life physicist Peter Hagelstein, who had originally wanted to develop an X-ray laser for medical purposes, hoping to win the Nobel Prize for such a humanitarian invention. However, unable to obtain research funding for this project, Hagelstein sold his idealism for the chance to work on an X-ray laser weapon instead. Wolf attributes this subversion of purpose partly to ambition but mainly to the hothouse, rationalist atmosphere of the Livermore Laboratories, where the all-male scientists live isolated from emotional relationships with women, children, or friends. "What they know, these half-children with their highly-trained brains, with the restless left hemisphere of their brains feverishly working day and night—what they know is their machine. Their dearly beloved computer. . . . What they know is the goal of constructing an atom-powered X-ray laser, the core of that fantasy of a totally secure America achieved by transferring future nuclear battles into space" (70–71).

The American Nobel Prize–winning novelist Saul Bellow, whose black humor is not unlike Vonnegut's, also turned his attention to the amoral scientific worldview and its effect on postwar American society. In *Mr. Sammler's Planet* (1970) he extends the discussion of responsibility to include man's proposed excursion into space. Bellow's representative scientist is the Indian physicist Dr. Govinda Lal, who, like H. G. Wells in a previous generation, cherishes an optimistic belief in science as the hope for a utopian future. Discussing with the humanist Mr. Sammler the issue of man's future and responsibility in science, Dr. Lal asserts, albeit with considerable charm, the ethic of "take what you want because it is there": "There is a universe into which we can overflow. Obviously we cannot manage with one single planet. ... If we could soar out and did not, we would condemn ourselves.... There is no duty in biology. There is no sovereign obligation to one's breed. When biological destiny is fulfilled in reproduction the desire is often to die. We please ourselves in extracting ideas of duty from biology. But duty is pain. Duty is hateful—misery oppressive."[35]

Once a disciple of Wells's Open Conspiracy project,[36] Sammler is temporarily attracted to the order and serenity of science, especially in the face of its prevailing opposite, the violence, aggressive individualism, and antirationalist anarchy of modern American society, but he is unable, finally, to accept its "solutions," because they omit too much from human experience. Having observed the atrocities of war firsthand and become more skeptical about human nature, he answers Lal: "As for duty—you are wrong. The pain of duty makes the creature upright, and this uprightness is no negligible thing" (177). Like science itself, Lal (a rare example of an Asian scientist in Western literature at this time) oversimplifies experience in the interests of pragmatism, and it is clear that Bellow aligns with Sammler rather than Lal in insisting on complexity and on a moral frame of reference. The space race, Bellow implies, will not solve man's problems; he will merely take them with him into space.

One very unusual example of the amoral scientist is provided by an American cyberneticist, Sibylle von Koçalski, the protagonist of Jan Hammerström's *Die Abenteur der Sibylle Kyberneta* (1963). This novel is interesting not only because it provides an unusual illustration of a woman scientist aligning herself with the amoral point of view (generally considered a peculiarly male proclivity) but also because the author, far from being critical about the undue influence of cybernetics and computing, appears optimistically to endorse the heroine's program of worldwide control to wrench society into the cybernetic age.

Hammerström was greatly influenced by the American mathematician Norbert Wiener, whose popular books on cybernetic theory had appeared in the 1950s.[37] Inspired by Wiener's example, Sibylle has already devised tables and statistical computer models to calculate and hence predetermine economic and political decisions, and she looks forward to extending this control over other aspects of society. Although Sibylle has to justify her system for world governance to more conservative cyberneticists and economists, her success is interpreted by Hammerström as auguring a turning point in world relations, substituting rationality and planning for imprecision and inefficiency. It might be Wells the utopian speaking, especially when Sibylle expresses the hope that eventually she will establish mathematical functions for determining social morality. This program accords with Wiener's outline of a *machine à gouverner* for the rational conduct of human affairs. Wiener had asked: "Can't one even conceive a State apparatus concerning all systems of political decisions either under a regime of many states distributed over the earth or under the apparently much more simple regime of human government on this planet. . . . We may dream of a time when the *machine à gouverner* may come to supply—whether for good or evil—the present obvious inadequacy of the brain when the latter is concerned with the customary machinery of politics."[38] Hammerström is virtually the only mainstream writer to present this amoral stance favorably or even neutrally. Most portrayals carry a strong element of authorial cynicism, if not overt condemnation, and many link it with an unacknowledged desire for power. In such presentations there is only a thin line between the amoral characters and the dangerous scientists considered in chapter 12.

In this and the two previous chapters we have seen a progression in the depiction of the scientist that is roughly chronological: the power maniac seeking to destroy the world; the scientist with an essentially evil philosophy; the scientist who believes that science per se is of more value than its social consequences or fails to think of the latter at all; and the scientist as exemplar of an amoral perspective, either because of intellectual arrogance, which discounts other facets of human experience, or because of his own personal inadequacies in nonrational pursuits. While this trend may appear to suggest decreasing overt evil on the part of the scientist, it actually involves an increasingly insidious and pervasive danger, because the attitudes become more socially acceptable and are thus permitted greater freedom for maneuver and exercise of power. The end point of such a trajectory would be a society powerless to break loose from the stranglehold that science and technology have upon it, as foreshadowed in E. M. Forster's "The Machine Stops" (see chapter 10 above).

Amoral scientists have frequently been censured for the escalation of events beyond what they themselves predicted. This aspect is most often discussed in relation to the responsibility of scientists, particularly chemists, for environmental pollution. In *Die Sintflut* (The flood) (2007) the German writer Stefan Andres reflected critically on the various arguments used by many real-life scientists to exonerate themselves from moral responsibility for this aspect of their work. His character Lorenz, a chemist who has worked in America, returns to postwar Germany to find the country polluted by a noxious gas. To the assertion that scientists should not be permitted to invent such horrors, Lorenz counters with the standard twentieth-century arguments that the scientists have not invented anything new but merely discovered what was already there, that if one scientist did not publish his discovery, another would very soon do so and thus a scientist owes it to his country to make sure that such discoveries are not preempted by the enemy. These arguments from inevitability, used so frequently about the development of the atomic bomb and about chemical and biological weapons, are found repeatedly in the literature, nearly always in order to be overthrown, at least implicitly, by the authorial voice. Heinrich Böll's *Fürsorgliche Belagerung* (The safety net) (1979), Monika Maron's *Flugasche* (Flight of ashes) (1981), Günter Grass's *Die Rättin* (The rat) (1986), and Christa Wolf's *Störfall* (Breakdown) (1987) all address aspects of environmental pollution and affirm the responsibility of the scientists for such effects, even though the consequences far exceeded their intentions. These twentieth-century eco-novels are the precursors of an avalanche of climate change fiction that, since the 1990s, has become the fastest-growing genre, cli-fi, which will be discussed in detail in a future volume.

CHAPTER 15

Robots, Androids, Cyborgs, and Clones
Who Is in Control?

> He wanted to become a sort of scientific substitute for God. He was a frightful materialist, and that's why he did it all... to prove that God was no longer necessary.
>
> —Karel Čapek

Robots, androids, cyborgs, and clones are all manifestations, real or imagined, of the ambivalence of our society to the increasing immersion of human life in the technosphere, which the sociologist-philosopher Jacques Ellul and more recently Stephen Hawking in the 2016 BBC Reith Lectures equated with a threat to human freedom.[1] In essence this is not a new concern. The impact of technology on society has been an ongoing concern since the Industrial Revolution of the eighteenth century. At that time there was grave concern that mechanization was taking over every aspect of life, replacing natural rhythms with the unceasing requirements of machines tended day and night by factory workers, metonymically referred to as "hands." In the nineteenth century Charles Dickens had devised the striking symbol of "melancholy mad elephants" working endlessly and mindlessly, imposing their rhythms on human laborers.[2] At the same time, English manufacturing supremacy fostered the rise of a new wealthy elite independent of hereditary status and held out the promise of individual advancement on the basis of ability and opportunism.

Robots

The twentieth-century symbol of this ambivalence toward technology was the robot, invented in its modern form[3] by the Czech playwright Karel Čapek. The word "robot" comes from the Czech *robota*, meaning "forced labor," a term that derives in turn from *rab* meaning "slave." Initially this referred to programmed machines that released factory workers from repetitive toil but simultaneously threatened their livelihood, a comment on Henry Ford's moving assembly line, introduced in 1913 for the mass production of automobiles. Subsequently reinvented in increasingly humanlike forms as androids, with autonomous thought and intelligence, robots combined the lure of increased power and freedom for their owners with the threat that, like Frankenstein's creature and countless earlier mechanical creations in fiction, they would rebel against the controls imposed on them and dominate or destroy their masters. As an extension of this filmmakers have evolved a recurrent scenario of the development of a master race of highly intelligent robots, programmed or autonomously motivated to take over and destroy the human race. *The Mechanical Man*, *The Terminator*, *Runaway*, *RoboCop*, the Replicators in *Stargate*, the Cylons in *Battlestar Galactica*, the Cybermen and Daleks in *Doctor Who*, *The Matrix*, *Enthiran*, and *I, Robot* all play on, and exacerbate, such fears.

As presented in fiction and film, robots are essentially hybrids, combining the apparent intelligence of humans with the functionality and programmed obedience of machines. They epitomize efficiency, rationality, absence of emotion, objectivity, and the unswerving pursuit of a predetermined end regardless of the consequences. They became a powerful metaphor for technological culture itself, "the paradoxical status of the human body within the technological framework of modern society"[4] and the fear of technology invading and subsuming human identity. As such, robots and, later, androids have been used in fiction and film to indicate and explore the values and attitudes of their scientist-creators. In most cases they signify their designers' hubris, their obsession with controlling their world absolutely, and their cultivation of efficiency as an ultimate goal. Robotics engineers, it is suggested, have a close affinity to their mechanical creations because they themselves have lost their humanity.

The first in-depth literary study of the impact of robots and the motives of their creators was Karel Čapek's seminal play *R.U.R.* (1921), which concerns the effect on society of robots designed to relieve their masters of toil, leaving them free to enjoy endless leisure. It is this feature—their suggested potential for liberating humanity from the burden of heavy and repetitive tasks—that has contributed to their continuing ambiguous status as objects of both desire and fear, an ambiguity that, as we have seen, has beset science itself from its origins in alchemy. Čapek also explores the motives of the Rossums, uncle and nephew, who have created the firm Rossum's Universal Robots, designated by the play's title. Having accidentally discovered a substance that behaves like protoplasm, the elder Rossum, a physiologist, proceeds to experiment with making artificial beings. He thus exemplifies the hubris that leads scientists to believe that reason (science) is sufficient to create anything required. He is eventually killed by one of his creatures, but not before he has produced a manlike being, in modern terms, an android.[5] Like Frankenstein and Moreau, Rossum, whose name derives from the Czech word *rozum* (reason), was driven by a desire to usurp the role of God, an intention about which he was more explicit than his predecessors. He "wanted to become a sort of scientific substitute for God. He was a frightful materialist, and that's why he did it all . . . to prove that God was no longer necessary."[6]

The younger Rossum, eager to exploit this discovery commercially, and impatient with his uncle's inefficient methods, has simplified the design, omitting as superfluous any aesthetic and emotional elements. Domin, the manager of the firm, explains: "The Robots are not people. Mechanically they are *more perfect* than we are, they have an enormously developed intelligence, but they have no soul" (17, my italics). Domin himself cherishes a utopian vision of a world without work, but it is apparent that this ideal is also strongly tinged with hubris: "Man shall have no other aim, no other labour, no other care than to perfect himself. . . . He will be Lord of creation" (51–52). Čapek is also concerned to show how this philosophy of scientific materialism affects those who espouse it. The scientists at the Rossum factory are almost indistinguishable from each other and from the robots, for they too have become standardized, as though fresh from the assembly line. They are among the first and most stylized symbols of the anonymity and disappearance of individuality in a scientific, industrialized society, where people are interchangeable, defined only by the tasks assigned to them. Eventually the robots revolt against their exploitative, ineffectual masters and kill every human being except one, who is spared only on condition that he work to rediscover the lost formula for creating the robots. The parallel with Frankenstein, begged by the monster to create a mate for him, becomes clear as the play evolves to show that the reckless pursuit of science as an end in itself leads inevitably to the actual destruction of humanity, a humanity that, Čapek suggests, has been effectively sterile and symbolically dead for years.

R.U.R. became the prototype for a succession of robot stories and films, in most of which the scientist-inventors, like those in Čapek's play, lose control over their creations. The first, appearing in the same year as R.U.R., was the

Italian science fiction film *L'uomo meccanico* (The mechanical man) (1921), in which a scientist creates a humanoid robot (really an android) with superhuman speed and strength, controlled remotely by a machine. Ironically the scientist himself loses control of his robot to a criminal gang that employs the mechanical man to facilitate crimes before their leader, like the inventors of mechanized automata in chapter 10, is electrocuted by a short circuit in the controls.

The first film to associate robots with the mad, evil scientist stereotype was Fritz Lang's *Metropolis* (1927), based on the novel of the same name by Thea von Harbou.[7] Rotwang, the deranged and obsessive scientist of *Metropolis*, has been called "the most influential scientist in the history of the cinema,"[8] largely because of the film's powerful visual symbolism. *Metropolis* is identified with both an ancient, Gothic house and the machinery of a bleak, urban wasteland depicted as an expressionist, futuristic city of 2026 characterized by skyscrapers where the "managers" live in luxury while the workers toil underground, running the machines and furnaces that power the city.[9] This "Raygun-Gothic" decor was to become the prototype of mad scientists' laboratories in cinema throughout the century, while Rotwang himself, with his disordered mane of white hair, the high forehead associated with the mad inventor, an artificial, clawed right hand, and eyes "smouldering with a hatred close to madness,"[10] is both the descendant of the obsessed alchemist and the ancestor of Dr. Strangelove. In his art deco laboratory, complete with industrial machinery and flashing lights, Rotwang creates not a male homunculus but a female robot, Maria, by imprinting the likeness of a human Maria onto a machine and then bringing her to life as an erotic and destructive woman, reminiscent of the beautiful automaton Olimpia in Hoffman's *Der Sandmann* (1814), also created by a mad and vengeful scientist, Mea of *La femme endormie* (see chapter 10), and Frankenstein's monster (see fig. 22).

Traditionally robots had been creatures of masculine power and strength created by males independent of women, but the robot Maria emphasizes woman as Other, allowing Rotwang to claim power over both women and the process of human reproduction. As Andreas Huyssen remarks, "By creating a female android, Rotwang fulfils the male phantasm of a creation without mother; ... he produces not just any natural life, but woman herself, the epitome of nature."[11] As the evil counterpart of the good human Maria, who tries to quell the rebellion of the workers and who enacts her virgin mother status by collecting up their children to save them from impending destruction, the evil robot Maria is programmed to destroy the workers of Metropolis. When

Figure 22. Rotwang and the Robot Maria, a scene from Fritz Lang's 1927 film *Metropolis.*

she is carried off by the rebels and burned at the stake like a witch (also an uncontrollable female Other), Rotwang, his life and meaning invested in the now-destroyed mechanical Maria, symbolically falls to his death from the cathedral roof. The moral is heavily underlined: evil scientists and their dangerous creations must be destroyed, especially if that creation is a sexually aggressive female who flouts the rules of society as robot Maria does during her erotic striptease in a brothel.

Another and more contentious variation of the amoral robot maker is Zapparoni, the scientist-creator of Ernst Jünger's *Gläserne Bienen* (1957), translated as *The Glass Bees* (1961). This Prospero-like figure creates aesthetically perfect glass bees, tiny, sophisticated automata that, in their functionality and craftsmanship, combine technology and art, prefiguring the evolution during the 1980s of the powerful microchip. Symbolically, Zapparoni creates for himself a complete culture, endowing his robots not only with physical being but also with the ability to create other robots, a modern version of the "philosopher's stone" the narrator thinks, emphasizing the connection between Zapparoni and the magus-alchemist Albertus Magnus, whose works

Frankenstein studied. At first Zapparoni's world appears utopian, a landscape of marvels, both functional and aesthetic, offered by technology; but through his narrator, the retired military man Captain Richard, Jünger expresses reservations about Zapparoni's pose of godlike detachment from his world, living in happy ignorance and refusing to accept accountability for a Brave New World society. In 1995, on his hundredth birthday, Jünger commented: "Ours is the time of cybernetics, when machines wait on the threshold of thought and human beings are treated as components of the machine-world which can be cast aside when they are no longer needed. In such a period, what of ethics?"[12] In Jünger's analysis, Western society's complacent acceptance of the gifts and controls of technology gives a new and sinister meaning to Bacon's dictum "knowledge is power." The serene and amoral Zapparoni may be even more dangerous to humanity than a recognizably evil dictator. Jünger's novel raises questions that were to be explored more explicitly by Donna Haraway in "A Cyborg Manifesto" (1991) regarding the blurring, possibly the erasure, of boundaries between animate and inanimate, between cyborgs and humans.

With few exceptions fictional robots are portrayed as a threat to humanity, raising the question of who controls the robots, but in stark contrast to the many dystopian narratives of such dangers, Eando Binder's science fiction short story *I, Robot* (1939) introduces an interesting twist in that the moral roles are reversed. The robot Adam Link, created by scientist Dr. Charles Link, is educated, self-aware, and desirous of serving a human master. However, when Dr. Link is accidentally killed, Adam is blamed and pursued by armed men intent on destroying him. At first he retaliates but then, having found and read a copy of *Frankenstein*, he understands the fear and revulsion he evokes, rejects revenge, and, after writing his confession, prepares to self-destruct. Binder's story was highly innovative for its time in breaking away from the *Frankenstein* cliché, and its popularity led to a series of Adam Link stories by Binder, which were later adapted for the American TV series *The Outer Limits* (1963–65).

Isaac Asimov acknowledged the influence of Binder's concept on his robot stories: "It certainly caught my attention. Two months after I read it, I began 'Robbie,' about a sympathetic robot, and that was the start of my positronic robot series. Eleven years later, when nine of my robot stories were collected into a book, the publisher named the collection *I, Robot* over my objections. My book is now the more famous, but Otto's story was there first."[13] Asimov's many stories about robots and the reactions of their scientist "minders" provide a striking contrast to the prevailing view in fiction about the effect of "intelligent" machines.[14] As a biochemist and author of a large number of books on popular science, Asimov was scornful of what he called the "Frankenstein complex" and remained fundamentally optimistic about technological progress. Like H. G. Wells he saw it as relieving humanity of "those mental tasks that are dull, repetitive, stultifying and degrading, leaving to human beings themselves the far greater work of creative thought in every field from art and literature to science and ethics."[15]

Asimov conceded the possibility of danger from robots and computers only for those who feared change and had, in effect, already abrogated their autonomy in a technological age, essentially becoming like the machines they attacked. This is most apparent in his story "Profession" (1957), set in a future world where most of the inhabitants, fearful of change, have had their brains wired and programmed to act in a routine fashion so as to avoid the agony of decision making. Like the citizens of E. M. Forster's "The Machine Stops" (1909), they have become voluntary appendages of the machine as a result of their reactionary paranoia.

Asimov's most popular sequence of stories, beginning with "I, Robot" (1950), was based on an exploration of robots as necessarily "benign," because controlled by the Three Laws of Robotics, a system of ethics designed to prevent any takeover by robots such as Čapek had depicted.[16] Ironically, these laws were accepted and propagated by later science fiction writers as though they had inherent validity. In the broader context, however, Asimov's robots are atypical, for they are so humanized that they blur, if they do not actually deny, the issues being explored by the other writers discussed in this chapter.

Contrary to Asimov's intention, and possibly unrealized by him, the stories themselves subvert this comfortable optimism, for not only are the human characters upstaged by the more intelligent robots that become the problem-solving heroes of these stories, but the more moral human characters are, themselves, governed by the Three Laws of Robotics. In "Evidence" (1946) the politician Byerly, the most "ethical" human in the robot stories, is accused of being a humanoid (android) with a robot's brain, and the robopsychologist Susan Calvin concedes, "Actions such as his could come only from a robot, or from a very honorable and decent human being. But you see, you just can't differentiate between a robot and the very best of humans."[17]

Where problems in the robots' functioning occur, it is almost invariably attributed to a failure in perception or logic on the part of the humans. The interest of the stories thus centers, not on human qualities and emotions, but on the laws of logic and intellectual wordplay; that is, the plot interest is determined by the rules of the robots and, in the later stories, the computer.

The approved stereotype emerging from Asimov's stories is a thoroughgoing materialist and pragmatist who, without a qualm, exploits the solar system in the name of efficiency and human imperialism. In *View from a Height* (1963) Asimov discusses the most efficient means of colonizing the other planets, where existing life could provide an immediate source of food for the prospective Terran colonists; the use of space as a garbage dump for radioactive waste; and an ingenious real-estate scheme for selling off planets broken into asteroids. Indeed, Asimov's heroes bear a striking resemblance to C. S. Lewis's archvillain, Weston (see chapter 12 above).

A similar, optimistic view of artificial intelligence informs Frank Herbert's novel *Destination: Void* (1966). His four scientists aboard the spaceship *Earthling*—a psychiatrist, a life-systems engineer, a doctor who specializes in brain chemistry, and a computer scientist—represent the four disciplines most closely allied with the understanding and development of cognitive science. In the critical circumstances that attend their lone journey through space, they come to the realization that their survival depends on developing high-level artificial intelligence. Herbert's view is clearly that machine intelligence in cooperation with human intelligence is our only hope for the future and that scientists are therefore indispensable for the very reasons that led to their vilification by the majority of novelists discussed hitherto.

Asimov's robots have spawned a lucrative progeny of "cute," harmless robot characters, popularized in films, most famously the astromech droid R2-D2 and the protocol droid C-3PO of *Star Wars* (1977). Like E.T., these robots (they are not androids since they do not resemble humans) are essentially novel pets with just enough initiative to make the games interesting but always, in the long run, deferential to their humans. In some ways, the complacency they generate could be regarded as the most sinister response of all. Most of the writers and filmmakers represented in this section, however, are far from complacent. They regard robots as special kinds of computers whose creators evince all the failings of the computer scientists in chapter 13, with the added flaw of trying to simulate human actions and, in the case of more sophisticated androids and cyborgs, create alternative human beings.

In contrast to Asimov's collection of the same name, the dystopian film *I, Robot* (2004) also depicts a scenario where robots have become dangerous to humans because a seemingly benign scientist, Dr. Alfred Lanning, has created an artificial intelligence computer, VIKI (virtual interactive kinetic intelligence). VIKI determines that humans are on a course headed for extinction and, because of the First Law of Robotics, this must be prevented. VIKI therefore creates a Zeroth Law stating that a robot may not harm humanity or, by inaction, allow humanity to come to harm. In effect this means ensuring survival of the race by commanding robots to deprive individual humans of their free will if their proposed course of action threatens survival of the species. Powerless to prevent VIKI's plan, Lanning is an example of the scientist out of control. To alert Calvin to what is happening he contrives his own death by means of a unique robot that has been engineered to bypass the Three Laws. Although Calvin and another scientist eventually destroy VIKI, removing the immediate threat, the fragility of Asimov's Three Laws as a safeguard against robots remains, with consequent risk to humans. Again, the overriding message is that robots cannot be depended upon to act according to human morality because they are programmed with different premises, consonant with the aims and values of their designers.

Androids

Typical of these dystopian predictions is the bleak, postholocaust vision of Philip K. Dick, whose mad scientist Dr. Bluthgeld was considered in chapter 12. Like Asimov, Dick insists on the inevitability of machine intelligence, though he consistently refers to both anthropomorphic electronic constructs and humans who have become machinelike in their behavior as "androids." Dick's two nonfictional statements on this subject, "The Android and the Human" (1972) and "Man, Android, and Machine" (1976), focus on the centrality of the relationship between humans and machines and the question of who is really human. "Our electronic constructs are becoming so complex that to comprehend them we must now reverse the analogizing of cybernetics and try to reason from our own mentation and behavior to theirs."[18]

In "The Android and the Human" Dick defines the characteristics of the android mind that distinguish it from the fully human mind as deficiency in feelings, inflexibility, unswerving obedience, and most of all predictability—the first two qualities being similar to those ascribed to the figure of the scientist by the writers discussed in chapter 13. Dick, however, is less sure that the impersonal qualities can be so readily identified with one group. In *Vulcan's Hammer* (1960) the machines act as destructive humans, while in *The Man in the High Castle* (1962) the humans have become destructive machines. *Do Androids Dream of Electric Sheep?* (1968) is another parable of biological indeterminacy: how to distinguish the "perfect" replica or android from humans, the dilemma in Asimov's story "Evidence" (see above). The test devised to discover who is an android revolves around an empathy test, but such a criterion is already loaded with cultural assumptions and prejudgments. In essence Dick explores the premise that man has two brains in one skull, approximating an intellectual, unfeeling personality and an empathizing, intuitive one. Thus in Dick's reading, the impersonal scientist is only a special example of twentieth-century man for whom the most fitting symbol is the android.

Almost a century after the release of Fritz Lang's *Metropolis* a similar scenario formed the basis of Alex Garland's film *Ex Machina* (2015), in which a scientist, Oscar Isaac, at the request of Nathan, the billionaire CEO of a software company, has engineered a beautiful female android (gynoid), Ava, with artificial intelligence. Caleb, a software programmer in Nathan's company, is ordered to conduct a Turing test[19] on Ava to determine if her intelligence is human. Caleb falls in love with Ava and attempts to free her from Nathan's prison but after a scuffle in which Nathan is killed, Ava escapes, taking skin and clothes from superseded androids, and leaving Caleb imprisoned in the facility. Like *Frankenstein*, *Ex Machina* leaves us with the ongoing threat of an android at large, but the threat is more insidious because, unlike the monster, Ava mimics an attractive female, whose real nature will be undetected in society. Like Alan Turing himself, *Ex Machina* poses questions about artificial intelligence and how, if at all, we could distinguish it from human intelligence. Unlike the subservient female robots in Ira Levin's satirical thriller *The Stepford Wives* (1972, film adaptations 1975 and 2004) and the genetically engineered replicants in the film *Blade Runner* (1982), adapted from Philip K. Dick's novel *Do Androids Dream of Electric Sheep?* (1968), Ava shows the feisty rebelliousness of Rotwang's robot Maria, reviving the threat of the femme fatale now endowed with increased powers and intelligence.

Klaus Benesch argues that

> fear and admiration, euphoric immersion in, and paranoid withdrawal from, the encompassing technosphere, are indeed the two major stances associated with modern technology. Yet in both cases, the way in which we confront technology is determined by much more than just the efficacy of the machine itself. It is evenly determined by the desire to construct human identity as basically different from the realm of the technological, either in the sense that we experience the staggering advancement of technology as a threat to the culture of

humans or that we think of machines as dumb, unintelligent tools which are created and controlled only by the superior mind of the engineer.[20]

Because of the prevalence of this latter perspective the question of ethics in relation to robots has, until recently, focused on whether it was ethical to use robots for particular purposes and on potential malpractice by robots. More recently a converse concern has arisen about the way intelligent robots, which may potentially have consciousness, motivation, and emotions programmed into them,[21] are treated by their creators or owners. The Polish cyberneticist-novelist Stanislaw Lem was possibly the first to suggest that ethically a robot with humanlike intelligence must be regarded as a human,[22] but what was then only a hypothetical question is becoming imminent. Donna Haraway in "A Cyborg Manifesto" (1991) claims that the concept of the cyborg involves a rejection of definite boundaries between "human" and "animal" and between "human" and "machine": "By the late twentieth century, our time, a mythic time, we are all chimeras, theorized and fabricated hybrids of machine and organism; in short we are cyborgs. The cyborg is our ontology; it gives us our politics. . . .The relation between organism and machine has been a border war. The stakes in the border war have been the territories of production, reproduction, and imagination."[23]

The film *Surrogates* (2009) depicts a world where most people have willingly acquired remotely controlled androids called "surrogates," through which they live idealized lives vicariously from within the safety and comfort of their homes. Seemingly just an extension into the physical world of creating a desired avatar as a persona in cyberspace, this retreat from reality leaves the "operators" of the surrogates more vulnerable to electronic viruses or electromagnetic pulses that can destroy the operator as well as the surrogate.

When robots are the norm as domestic servants, pets, toys, carers, and android sexual partners, should we acknowledge that they have the same legal and moral rights to leisure and enjoyment as human servants rather than slaves? Arguments for such rights have been based on analogy with our evolving treatment of animals and with the premise that ethical behavior toward robots would include an extension of Kant's imperative,[24] rather than our current focus on what robots contribute to us as social and emotional beings.[25]

In 2016 there was a draft motion before the European Commission to consider "that at least the most sophisticated autonomous robots could be established as having the status of electronic persons with specific rights and obligations." The VDMA (German Engineering Federation) director believes this is decades away, but already there are concerns that increasing numbers of robots would adversely affect employment and the viability of social security systems.[26]

However, there are, as yet, few examples in fiction or film where such ethical views are supported. In the cases of HAL in the film *2001: A Space Odyssey* (1968), of the domestic servant and child carer robots (called "synths") in the TV series *Humans* (2015–), in the surrogates of the film *Surrogates* (2009), and of Ava in the film *Ex Machina* (2015), humans' faith in the ethical reliability of the robots is shown to be dangerously misguided.

An interesting exception to the widespread condemnatory attitude to cyborgs and androids is the scientist character Seven of Nine in the *Star Trek: Voyager* series. We are informed that Seven of Nine was a human, Annika, before being assimilated by Borgs as a child and becoming a Borg drone. Subsequently she was transported aboard the Federation ship to work with Captain Janeway against another species, but her presence results in conflict on the ship as the crew are suspicious of her, even when most of her Borg implants have been removed. Indeed her loyalty to the Federation is never entirely sure. She remains Other, both visually because of her costume, intended, like the Borg implant over her left eye, to indicate her cyborg nature, and her fluctuating allegiances. The cyborg perspective allows for this uncertainty and, as in Haraway's "A Cyborg Manifesto," mounts a challenge to our assumptions about connections and disconnections between bodies and mental processes.[27]

Clones

The concept of cloning carries a strong emotional and deeply divisive power not only because it dramatizes in the starkest form bioethical concerns arising from many fields of bioengineering, actual or potential (artificial creation of human beings, genetic manipulation, unregulated biotechnology, disempowerment), and because it threatens the Western belief in the uniqueness of the individual, but also because, in popular culture, it suggests that we are *nothing more than* our DNA. "Geneticization" is the term used by geneticist Abby Lippman[28] to indicate this increasingly reductionist view of humanity as a collection of genes, such that not only physiology but medical issues, appearance, psychological states, and behavior are regarded as the direct result of an individual's DNA, which only the wealthy can afford to have modified. The completion of the Human Genome Project in 2003 has been taken as further support for, even proof of, the mechanistic basis of the individual—a recycling in molecular terms of La Mettrie's *l'homme machine* (machine-man).[29] Yet, as Dorothy Nelkin and M. Susan Lindee point out, although the genome may have replaced the soul as the "location of the true self," it has itself assumed in popular thought the status of the sacred, such that tampering with it is regarded as "playing God"[30] and elicits much the same combination of fascination and horror as the creation of a homunculus did in medieval times.

This cultural meaning of DNA, as distinct from its scientific or medical meaning, has been forged largely by the many novels and films that present cloning as a threat to society. They invoke deeply embedded cultural memories of *Frankenstein* and *Brave New World* and link to the succession of fictional biologists, and later geneticists, who tampered with the sanctity of life (see chapter 12). In 1993, Michael Lemonick argued:

> When it comes to dealing with cloning, ethicists and science-fiction writers have almost indistinguishable job descriptions. Both groups propose hypothetical situations in which cloning might happen, then examine the likely implications. The only real difference is that ethicists respect the laws of plausibility and don't waste much time on scenarios that probably won't ever come to pass. Science-fiction writers trash those same laws with creative gusto. The result has been a relentless stream of outrageous books, movies and television shows, beginning with Aldous Huxley's *Brave New World*, published 61 years ago, and continuing through the summer's box-office behemoth, *Jurassic Park*.[31]

Radiation-driven mutations (typified by *Godzilla*, 1998 and 2014, but including dozens of other monsters), eugenics (*Gattaca*, 1997) and cloning (*The Boys from Brazil*, novel 1976, film 1978; *The Cloning of Joanna May*, novel 1989, film 1992; and *Jurassic Park*, novel 1992, film 1993) are routinely presented as the ultimate goals, invariably with collateral damage, of genetic experiments. *The Sixth Day* (2000) also promotes genetic essentialism. In its near-future society cloning of animals such as pets, and of human organs, is permitted but cloning of entire human beings, though possible, is forbidden by the "Sixth Day" laws.[32] Since identity is presented as contained wholly in the DNA, the illegally cloned humans are identical with their prototypes, leading to several Shakespearean-like scenarios of mistaken identity. Dr. Griffin Weir, the scientist who invented the illegal cloning process using protoplasm "blanks," works for Replacement Technologies, and does the bidding of its CEO, is not without his nemesis: his wife, dying of a terrible illness, begs him not to clone her again.

Where cloning is sought as an aim in its own right, it almost invariably functions in fiction and film as a condemnation of the scientific mind behind it. Michel Houellebecq's novel *Les Particules élémentaires* (Elementary particles) (1998) and the 2006 film *Atomised* based on it trace the unhappy lives of two half brothers, one of whom, Michel Djerzinski, becomes an introverted molecular biologist. His discoveries in the area of cloning lead eventually to the elimination of love from the reproductive process. Humans are assumed to be no more than particles, to which they decay and from which they can be assembled. This is Frankenstein's thinking at a molecular level, as he assembles his creature from dead body parts, without recourse to a mate. While Houellebecq makes no overt criticism of this alternative system of reproduction, the loveless lives of the two half brothers suggest a level of cultural pessimism that could only be increased by the elimination of sexual reproduction.

Ira Levin's novel *The Boys from Brazil* (1976, film 1978) leaves no doubt about the author's judgement of cloning. The novel begins as a mystery: ninety-four male civil servants at widely different locations are killed when they turn sixty-five. It is then revealed that the man ordering this bizarre operation is Dr. Josef Mengele, a German SS officer and physician who undertook inhumane medical experimentation on prisoners at Auschwitz, including experiments with identical twins, and who, after the war, escaped to Brazil. His immediate purpose is to kill the adoptive fathers of ninety-four male children aged around thirteen, all cloned from Hitler's DNA and gestated in surrogate mothers some twenty-three years younger than their civil-servant husbands. In Mengele's deranged mind, the boys will become more authentic copies of the Führer and more likely to promote his goal of reviving the Third Reich if they relive events comparable to those of Hitler's own life, specifically the death of his father at age sixty-five, when Hitler when himself was thirteen. The cloned boys are destined to experience a similar loss, the death of their adoptive fathers. Mengele is stalked by an elderly Jewish Nazi hunter, Liebermann (modeled on the real-life equivalents Simon Wiesenthal and Serge Klarsfeld), and finally brought to a violent end, mauled by Dobermans. Mengele is clearly stereotypical of the mad, evil scientist, but Christopher Rose has pointed out that perhaps the most interesting feature of *The Boys from Brazil* is that there is another scientist, the unnamed professor who explains to Liebermann (and hence to the readers) how the Hitler clones could have been created and shows a video clip of nuclear transplantation. He represents the authoritative, unbiased scientist, able to communicate science to the layperson.[33] As in most fiction about cloning there is the insidious suggestion that the consequences continue uncontrolled. Liebermann cannot bring himself to allow the cloned children to be killed and burns the list of their names, believing that the role of nurture will modify genetic inheritance. However, one of the boys pores sadistically over the photographs of Mengele's savaged corpse, suggesting that the cloning experiment may indeed have succeeded.

In Fay Weldon's novel *The Cloning of Joanna May* (1990, film 1992) the mad scientist is replaced by an evil businessman, Carl May, CEO of Britnuc, a company that owns nuclear power stations. Psychologically damaged as a child, he organizes an abortion for his wife, Joanna, on the pretext of not wanting children but actually to obtain her ovum, from which his employed scientist, Dr. Holly, creates four clones with different stereotyped personalities. May's motivation, shared by Holly, springs partly from a misogynist view of all women: Joanna, beautiful and apparently competent and well balanced, "was still a woman, and therefore liable to extreme, hysterical and unhelpful reaction: she was a creature of the emotions, rather than reason. That was the female lot."[34] This judgement is reinforced by their view of the natural world as the product of a blind, meaningless evolutionary process and hence chaotic and full of human misery: "An evolutionary process that caused so much grief could surely be improved upon by man: genetic engineering would hardly add to the sum of human misery . . . and might possibly make things a good deal better" (116). May's and Holly's motives, then, may seem to differ substantially from those of the obsessive researcher of the Frankenstein mold, bent on pursuing knowledge for its own sake, but at another level their hatred of the natural world and in particular the feminine body can be seen as a similar attempt to conquer and control nature by scientific reason.

The long-running Canadian TV series *Orphan Black* (2013–16) also focuses on cloning as a means of directing evolution. A scientific movement called Neolution has created clones, which it keeps under surveillance through monitors for the purpose of controlling the creation of human life. Violently opposed to the Neolutionists is a secret religious group, the Proletheans, who believe clones are abominations and pledge themselves to assassinate all clones. Here clones function as a focus for the struggle between scientific control of human life and prohibitions based on religious determinism, a rerun of the Darwinian controversy 150 years earlier.

The best-known example of cloning is the film version of Michael Crichton's novel *Jurassic Park* (1992), which draws heavily on *Frankenstein* and *The Island of Doctor Moreau*. Crichton's box-office success results from the combination of deeply embedded cultural fears about experimental biology and the enduring fascination with dinosaurs. With its full display of special effects, the film also affirmed the power of genetic engineering to release monsters that had previously been the province of speculative science fiction.

In the early 1990s there had been scientific speculation that paleo-DNA might be recovered from insects imprisoned in amber, and Crichton extended this to create a scenario in which the insects are the repository of dinosaur DNA, which could be used to produce dinosaur eggs.[35] The driver of the Jurassic enterprise is not a scientist but John Hammond, capitalist entrepreneur and CEO of InGen, intent on creating an amusement park inhabited by living dinosaurs cloned from DNA retrieved from fossilized mosquitoes trapped in amber. He seduces the brilliant molecular geneticist Henry Wu by offering to release him from the competitiveness and regulations of his university position and providing him with limitless resources for research, thereby demonstrating the power of money to determine the course of research. Both Crichton's novel and Spielberg's film *Jurassic Park* (1993) confirm the move from the individual mad scientist working alone and for his own ends to the scientist in the pay of corporations, doing their bidding and ignoring the potential for disastrous mistakes. Crichton had begun the novel with a comment about unregulated biotechnology of unprecedented power, and when the film was due for release, he reiterated his warning about the dangers of commercializing molecular biology, "the most stunning ethical event in the history of science."[36] Kim Newman also has noted this shift from scientist to entrepreneur as villain. He may not have a prosthetic hand, but Hammond is disabled in other ways and walks with a stick; he has the amoral, uncaring attitude commonly ascribed to scientists, but his motive is not discovery: it is money. Newman suggests that "we are no longer worried about Frankenstein creating a monster, but we'd be worried about the Frankenstein Corporation mass-marketing franchise monsters on every high street."[37] The scientists have passed their responsibility as to what accidents might (and do) occur to the corporation that employs them. As Spencer Weart has observed, "Many of the fears about science and technology [in the cinema today] are actually not fears about science and technology itself—they are concerns about the social system, expressed by people who feel they do not have control over the decisions being made."[38] Spielberg's film directed the power and immediacy of special effects to the dinosaurs and the dangers they pose when they escape. Despite these confronting visual sequences the film is ultimately less sinister than *The Boys from Brazil*, because the cloned creatures, though of terrifying size, are not humans.

It is often forgotten that the "heroes" who avert the immediate danger of the mass escape of the dinosaurs in *Jurassic Park* are also scientists—Dr. Alan Grant, a paleontologist and paleobotanist, and his graduate student, Dr. Ellie Sattler. They, along with mathematician Dr. Ian Malcolm, show deep concern about the cavalier attitude of Hammond when the outcome is so uncertain. Grant concludes, "We never have complete control," while the mathematician urges restraint: "Don't you see the danger inherent in what you are doing here? Genetic power is the most awesome force the planet's seen but you wield it like a kid that's found its dad's gun!" The scientists are correct. Despite their best efforts two of the dinosaurs are seen escaping from the island, implying that scientific knowledge cannot be controlled.

At the time of publication of *Jurassic Park* and the release of the film, the concept of cloning, whether of children or dinosaurs, was merely science fiction, but in 1997 the Roslin Institute's creation of Dolly the sheep by cloning irreversibly changed the scale of possibilities. Greeted with almost universal public abhorrence, the Roslin announcement evoked visions of breeding armies or, alternatively, the creation of "perfect" individuals against whom others would be measured and found wanting. There was strong public feeling that research into human cloning should be halted.[39] Yet the film *Gattaca*, released in the same year as Dolly, prefigures a scenario where genetic engineering is not imposed on a society but eagerly embraced by its members, whose minds have been colonized by the possibilities of science. Indeed the genetics-based neologism *gattaca*[40] has become shorthand for a society where biotechnology has produced a two-tier society based on discrimination against the "genetically unmodifieds." There is no need for a mad scientist: he has been internalized by the whole society, by every family desiring the best future for its children and the almost universal belief that only the genetically engineered are "valid" and able to perform the important functions of the society; others are dismissed as "in-valids." Usually the intentions of the parents are presented as honorable but misguided. In the German film *Blueprint* (2003), a gifted pianist and composer, learning that she has muscular dystrophy, persuades a reproductive researcher to create a clone child so that her daughter can continue to express her musical gifts. However, on discovering her origins the daughter is devastated and rejects her mother and the music world. As in *Godsend* (2004) and *My Sister's Keeper* (2009), the desire of parents to prolong the life of themselves or a dead or dying child becomes a cautionary tale about unforeseen dangers that split the families the cloning process was intended to save.

Already we have seen the enthusiasm with which genetic engineering to prevent inherited diseases or disabilities is welcomed as a medical breakthrough and embraced by parents who carry genetic risk factors. How big a step is it from substitution of "good" genes for compromising ones to the engineering of multiple designer genes? And do we fully consider the potential dangers in a social welfare sense? These are issues that film can explore ahead of reality. David Kirby has pointed out that *Gattaca* is unusual in its "serious exploration of the bioethical issues surrounding human genetic manipulation" and actually rejects the genetic essentialism promoted by most fiction and films that deal with cloning.[41] Insofar as Vincent Freeman (the only free man in the film), a "faith-birth," is able to transcend his "disability" by "borrowing a superior genetic identity of a 'valid,'" a hazardous undertaking which succeeds only because, through his intelligence, determination, and opportunism, he demonstrates that he is more than the sum of his genes.

Caryl Churchill's play *A Number* (2002) presents a moral inquiry into what cloning a human child might mean within the dynamics of a family. Salter, whose wife committed suicide and who had drinking problems attempted to bring up his two-year-old son Bernard but became increasingly neglectful of the child. As Bernard's behavior deteriorated he sent him away to an institution and paid a scientist to create a clone of the child, so that he could have a second chance as a parent and bring up a perfect child, also called Bernard. At age thirty-five the second Bernard encounters his older brother, then forty, and each son independently accuses his father of failing to value individuality. Salter's attempts at self-justification are unacceptable to the sons, and the elder son murders his younger brother. Thus far the play raises issues about the morality of cloning, of thinking that an individual can be replaced by a cloned sibling. The brothers and father are incensed to learn that "some mad scientist has illegally" produced not one but twenty-one clones, the others being unaware of their parentage. In the last scene, however, Salter arranges to meet one of the other clones, Michael Black, also thirty-five. Black, a happily married teacher of mathematics with three children, takes an entirely different view of cloning: "I think it's funny, I think it's delightful, . . . all these very similar people doing things like each other or a bit different or whatever we're doing, what a thrill for the mad old professor if he'd lived to see it, I do see the joy of it."[42] By way of consolation he tells his father, "We've got ninety-nine per cent the same genes as any other person. We've got ninety per cent the same as a chimpanzee. We've got thirty percent the same as a lettuce. ... I love about the lettuce. It makes me feel I belong" (62). He embodies the view that one's identity, one's essential self, derives less from genetic inheritance than from nurture and environment: one can make one's own life irrespective of one's genes, thereby countering the fear of human cloning.

The creation of clones that preserve their own identity and live independent lives was sufficiently provocative, but the concept of clones created specifically for organ harvesting was even more repugnant. Michael Marshall Smith's macabre noir thriller *Spares* (1996) is set in a future when the wealthy are cloned at conception so as to have replacement organs ("spares") ready to harvest when they need them: the perfect health insurance. The clones are kept like mute animals on a farm tended by droids, their only purpose to provide body parts as needed.[43] In the 2005 film *The Island*, set in 2019, Lincoln Six Echo and Jordan Two Delta live in a community within an isolated compound, run by a scientist, Dr. Merrick. The conditions are considerably better than those endured by the clones of *Spares*, and the residents are induced to believe that, with the exception of one contagion-free island, the world outside the compound is too contaminated to support life. Each week a lottery is conducted, and the highly desired prize is relocation to the island. However, Lincoln discovers that he and all the inhabitants of the compound are clones of wealthy sponsors in the outside world who need organ replacements, and the "prize" of removal to the island is actually a cover for the harvesting of their organs or other body parts.

Kazuo Ishiguro's dystopian science fiction novel *Never Let Me Go* (2005, film 2010) also traces the lives of young people bred as clones to provide vital organs until they "complete," that is, until their cumulative donations kill them. As children the clones live in an experimental boarding school, where they are encouraged to lead healthy lives in order to be "successful" donors. They are aware of, and accept, their future purpose with resignation. Ishiguro indicates that this process is accepted by large sections of the society. It is no longer a question of family donors or even of paying for organs: cloned children are specially bred with the sole purpose of providing replacement organs for those who need them.

These are large-scale versions of the scenario in Jodi Picoult's novel *My Sister's Keeper* (2004, film 2009). In this narrative a child is conceived by selective in vitro fertilization specifically to be a genetic match for her older sister, who has promyelocytic leukemia and requires successive bone marrow and organ transplants, which her sister is expected to donate. Here the mother has effectively taken over the role of the mad scientist, forcing this medical "solution" on both her daughters, "playing God" with what is scientifically possible. Like the children of *Never Let Me Go*, the sisters acquiesce in this process until the donor child finally rebels.

More extreme still is the dystopian scenario in *Oryx and Crake* (2003), the first part of Margaret Atwood's *MaddAddam* series. This postapocalyptic world has reverted to a two-class society dominated by wealthy, multinational corporations whose scientifically trained members live in privileged compounds, while the lower classes live outside in the Pleeblands. Distinct from both groups are the Crakers, who have been genetically engineered by a brilliant scientist named Glenn (nicknamed Crake) to act as models for the genetic manipulation of children. Crake also produces an alleged wonder drug, the BlyssPluss pill, which is widely distributed, causing an intended pandemic that wipes out most of the human race. Crake is in many senses the irresponsible, mad scientist, who believes he is justified in using the whole world as his laboratory. Although he is a flat character, his significance transcends his plausibility because, while the novel may appear to be a futuristic tale, it is more accurately seen as speculative fiction, critiquing modern life by envisioning the society that might potentially result from contemporary scientific and cultural trends of materialism, market-driven technologies, genetic engineering, and pharmaceutical monopolies.

Michael Crichton's *Next* (2006) also extrapolates from the potential of contemporary genetic engineering research and patenting of the human genes to a society in which people no longer have legal rights to their own tissue and transgenic organisms are created and discarded at will. As in *Jurassic Park* the characters are reduced to agents for the production of genetically engineered organisms, and even the rationale for the science is simplistic. BioGen has succeeded InGen, but, as in *Jurassic Park*, the geneticists are in the pay of company interests.

The impact of films such as *Jurassic Park* and *Gattaca* in producing vivid images that reinforce societal responses to perceived scientific issues, whether or not those images have any factual validity, should not be underestimated. They represent the new Frankenstein narrative, new ways to "make a human" with all the possibilities for good or evil that this presents. As genetic essentialism becomes the defining factor for explaining human capabilities, personality, and social capital, it offers to those who can afford to embrace it the allure of predetermined success—children with no genetic "defects" however these are defined, a new "humane" form of eugenics, a new, compressed evolution that will inevitably select against those with the "wrong" DNA.

Ironically, despite the doom-laden scenarios depicting the proliferation of robots, androids, and clones, the presentation of such technological possibilities ultimately paves the way for their normalization in the minds of readers and especially of film audiences. The concrete images we see become what Suchman et al. call "performative artefacts"[44] that rehearse the sociological possibilities of technologies in the making, and even though these may be depicted as frightening or undesirable, they persuade us to cross the mental barrier of disbelief. As David Kirby argues, the characters in fiction or film "treat these technologies as a 'natural' part of their landscape and interact with these prototypes as if they are everyday parts of their world . . . making them socially relevant."[45] Novelist Richard Powers predicts: "It seems to me that germ-line modification is not that many years away and will emerge gradually and inexorably as a by-product of all the different kinds of genomic research currently being pursued. What we can only think of in terms of science fiction is about to become social fact, and none of our institutions are ready for the transformation. Perhaps fiction can provide a way of thinking about the revolution in life that other disciplines are bringing about but are not yet equipped or permitted to evaluate."[46]

When medical researchers announce the successful implementation of genetic engineering in overcoming life-threatening or life-diminishing genetic defects, we are presented with a heroic breakthrough that both builds on our knowledge of the technology we have encountered in fiction and film and eventually overcomes the horror and fear those media originally generated. Our fears are mitigated by ethics committees that arbitrate on whether research projects may go forward, withholding funding if they do not conform to guidelines, and by the voluntary agreement of some scientists to honor a global pause in genome engineering of humans, even though the CRISPR (clustered regularly interspaced short palindromic repeats) technology to do so is available and the medical benefits could be immense. Equally, the financial rewards of commercializing genomic enhancement would be highly tempting.[47] At the personal level, hope overtakes fear.

CHAPTER 16

Pandora's Box

> What was once thought can never be unthought.
> —Friedrich Dürrenmatt

> We, the physicists, find that we have never before been of such consequence and that we have never before been so completely helpless.
> —Heiner Kipphardt

Unlike the characters of the previous three chapters, who were all, to a greater or lesser degree, depicted as morally reprehensible and a major source of risk to society, the scientist figures discussed here neither intend evil to result from their work nor deliberately isolate themselves from their fellow human beings; on the contrary, most begin with high moral intentions. Nevertheless, they have all, for one reason or another, opened a Pandora's box of physical, social, or environmental disasters, over which they have lost control in either the technological or the managerial sense. Either they do not foresee the full consequences of their discoveries and therefore, like their archetypal predecessor, Frankenstein, are at the mercy of a creation that cannot now be "undiscovered," or else they expect to have more authority over its deployment than, in the event, they are permitted.

The theme of the scientist hoist by his own petard has always been popular among those who feel threatened by the power of a knowledge they do not understand; it assumes a pleasing quality of moral justice for those who regard science as exceeding the limits of what it is "proper" for humanity to know. In addition, there is often a perceptible element of schadenfreude on the part of writers who resent the power and prestige that Western society accords its scientists compared with its evaluation of the humanities.

Unforeseen Disasters

In terms of unforeseen consequences, the twentieth century produced an alarming spread of examples, from proliferation of nuclear weapons to accidents in nuclear power stations, robotics, genetic engineering, cybernetics, and many more. In these situations the scientists responsible are usually "invisible" because the process has involved such large numbers of collaborators—scientists, engineers, technicians, and administrative officers—that responsibility cannot be laid at any one door. Instead novelists have created scenarios involving one or a few representative scientists to signify the many. Some of these have been discussed in the preceding three chapters as individual mad, evil scientists, unemotional scientists, or amoral scientists.

With the advent of the Cold War and the paranoia of the McCarthy era in the United States, mutant horrors resulting from radiation mingled with fears of Soviet invasion released on cinema screens a succession of invading aliens and inhuman monsters, most let loose by irresponsible scientists. In *The Thing* (1951), based on the science fiction novella *Who Goes There?* (1938) by John W. Campbell, scientists in Antarctica discover an alien spaceship buried in the ice. They recover the alien pilot, who, unrealized by them, can mimic the cellular form, memory, and personality of any creature it devours. Out of curiosity the researchers bring it in from the permaculture to reside at the base. Thereupon it kills and then impersonates the physicist and proceeds to do the same with other members of the crew. As a consequence (in a reference to contemporary fears of the atomic spies) they can no longer be sure who is a genuine colleague and who has been colonized by "the Thing." The film is intended to incite such fears. Its last warning message is "Keep watching the skies." Like most mad scientist films, it is antiscience and antiscientists. The latter are depicted as foolishly curious, unable to see that their desire for knowledge is endangering the human race in general and the United States in particular.

The Thing was followed by a long sequence of films featuring mutant monsters unleashed or created by the radiation from bomb blasts, roaming the earth and terrorizing cities. *Godzilla, King of Monsters!* (1954), the prototype of these films, featured a dinosaur awoken by the US atomic explosion in Japan. It was quickly followed by other films featuring giant mutants produced by radioactive isotopes—enormous ants in *Them!* (1954), *Gigantis the Fire Monster* (1955), a huge octopus in *It Came from Beneath the Sea* (1955), and *Tarantula* (1955). *Attack of the Crab Monsters* (1956); *The Phantom from*

10,000 Leagues (1956); *Rodan* (1956), in which a mutant pterodactyl destroys cities; *X the Unknown* (1956); *The Incredible Shrinking Man* (1957); *Behemoth the Sea Monster* (1959); *The Black Scorpion* (1959); and *Mothra* (1961) featured some of the many radiation-produced monsters. Joe Kane suggests that these creatures are "instruments of punishment for nuclear misuse,"[1] and the punishment is visited on society for trusting its scientists. Thus the creatures have no persona or depth; they are simply instruments of retribution. These horror films imagine what it would look like if . . . The scenario characteristically revolves around the social chaos resulting from some intrusion (from outer space or from the laboratory) for which scientists are responsible through their pursuit of curiosity.

In the rest of this chapter we look at those who suffer remorse or disempowerment resulting from their research.

Scientists against Authority

A different kind of disempowered scientist, a peculiarly twentieth-century figure, is the one who has lost control over the implementation of his research to some organization, usually the government-military complex or a multinational company. His own moral frustration may even drive him to subvert government authorities, incurring the label of traitor. This theme of what we may call "ethical treason" emerged during the Second World War, when scientists in defense work suddenly found themselves involved in research on atomic weapons. Many of the troubled scientists in the literature written during and after the Second World War struggle for control of their intellectual property against powerful authorities, who demand their allegiance in projects that their conscience opposes. This was the situation confronting many physicists who worked on atomic weapons during the war. They alone realized the full extent of the destructive power of the weapons and the inhumanity of using them against civilian populations, but their governments, arguing only from political expedience, demanded that the project be completed, lest the enemy preempt them. A large number of novels and plays examine the struggles of these ethically superior scientists anguishing over a moral issue that their governments are incapable of understanding or addressing. The real enemy is not the bomb but the government-military establishment that demands it.

The first exploration of this situation emerged in Germany, where, even in the years preceding the declaration of war, there was a perceived conflict between the scientists' search for truth and the requirements of the Nazi military program.² When restrictions on writers made it impossible for them to explore the issue overtly, they resorted to a veiled treatment of the subject through the depiction of historical figures, Kepler and Galileo providing obvious parallels. As in the case of Brecht's *Galileo*, the hegemonic authority in these treatments is overtly the medieval church, but the historical situation clearly points to the contemporary one, where the power demanding absolute allegiance is a totalitarian government.

Hans Rehberg's play *Johannes Kepler: Schauspiel in drei Akten* (1933), set during the Thirty Years' War, shows Kepler as a martyr persecuted by Protestants and Catholics alike.³ Neither of these authorities disputes the truth of Kepler's ideas, but they argue that truth must be subordinate to authority and social order. Kepler the mathematician, on the other hand, admits of no authority higher than the truth, since truth comes from God: "God has directed me and shall I throw away what the everlasting God has given me, for the sake of a duke?"⁴ This is a clear parallel to the conflict between the scientists' conscience and the increasingly totalitarian German state.

Max Brod's *Galilei in Gefangenschaft* (1948) also concentrates on the conflict between the allegiance of the scientist to truth as he knows it and the power of authority that tries to suppress that truth. Galileo is less constant than Kepler, whose spirit never succumbed to authority. In the face of the absolute refusal of the church to tolerate any contradiction of Aristotle, Galileo recants his Copernican beliefs, buying time not only to complete his scientific work but also to evolve a philosophical and ethical reassessment of the nature of truth, which had clear implications for Brod's contemporaries: "A truth for which man is not prepared to die is no truth. . . . A man who is not prepared to die for the truth is no man. The truth demands a sacrifice like every great sacrifice demanded of us. . . . Ironically, I was almost a martyr for the Truth against my will though. . . . Now it is different. Now I want it."⁵

Frank Zwillinger's *Galileo Galilei, Schauspiel* (1953) presents a somewhat different portrait of Galileo, as a man who in old age believes he has failed both science (because he did not accept martyrdom for the truth) and the church (because although he recanted outwardly, he has never recanted in his soul) and has therefore lived a double lie. However, Zwillinger's play is more optimistic than most literary treatments of the Galileo story in that his protagonist's dilemma is resolved, albeit somewhat glibly, even dishonestly, when the archbishop of Sienna assures him that, owing to a technicality, the Inquisition's ban on the teaching of the Copernican system is invalid, and hence such teaching is not heretical.⁶ Zwillinger thus removes the scientist's anguish by means of an external authority that accepts the responsibility, a resolution that was fundamentally irrelevant to the contemporary situation of the Cold War.

In England and the United States it was possible to explore the contemporary political situation more explicitly, although even here an element of self-censorship is suggested by the ambivalence writers display toward their scientist characters involved in a conflict with military and political authorities. These scientists are not portrayed as heroic or even as having any clear idea of alternative courses of action. They are therefore distinct from the idealistic rebel scientists, to be considered in chapter 18.

Two novels of wartime research by the British novelist Nigel Balchin depict scientists as neither great discoverers nor potential saviors of their country but ordinary men caught up in a protracted state of emergency. Far from defying the establishment, they are exploited by the politicians, army personnel, or government administrators, whom they despise both intellectually and morally. Their "heroism" can consist of no more than uncompromising honesty in the face of the secrecy, power seeking, and hypocrisy around them. Despite his physical disability Sammy Rice, the physicist of *The Small Back Room* (1943), almost achieves a moment of glory when he is called upon to defuse a bomb, but at the crucial moment the physical effort to unscrew it proves too great, and he has to ask for help. His self-disgust, apparent in his statement that "the thinking had been all right but when it came to doing anything I'd been gutless as usual,"[7] signifies both his heroism and his inadequacy. Balchin implies that although Rice's determination may be the only possible kind of heroism remaining in the real world, it is significantly limited.

In *A Sort of Traitors* (1949), its title a quotation from Shakespeare's *Richard II*,[8] Balchin explores the various temptations that assail a group of scientists working on antibody control of epidemics during the war. Here the element of external control hinges on government censorship relating to publication of results, lest they be used by the enemy to start an epidemic.[9] Balchin's scientists enact different responses, none of which is morally satisfactory. One of the young and idealistic members of the team, Marriott, is influenced by a speech given to the Junior Research Association urging the scientists to "throw off the habit of silence and retirement which is the scientist's curse, and insist that the scientist's influence in the world shall be in proportion to his contribution."[10] When he is told of a fictitious international open publishing syndicate, he is almost drawn into a network of scientific espionage. Another scientist, Sewell, decides to subvert the publication ban by sending copies of his work to the heads of relevant scientific organizations throughout the world and apprising his government of the fact afterward. But neither of these desperate expedients comes to fruition. Sewell admits that his allegedly high-minded concern for "the interests of humanity" is basically self-interest. Thus Balchin's scientist heroes are essentially ordinary, powerless men, often physically handicapped in some way to emphasize the point. In the world of politics and a military-orientated government, only a qualified heroism, involving moral anguish, is permitted to the scientist.

This image of the scientist as essentially honest and well-meaning but unable to oppose the machinations of power politics continued into the Cold War. Upton Sinclair's topical play *A Giant's Strength* (1948), about an atomic physicist, Barry Harding, who had been involved in top-secret research on the bomb, was one of the first US works of fiction explicitly to urge upon scientists their particular responsibility for persuading their government of the destructive power of atomic weapons. The play opens just before the dropping of the bomb on Hiroshima and includes documentary clips for immediacy and realism. Thus, immediately after the news flash about the Hiroshima bomb, Sinclair introduces the optimistic voice of General Macarthur on the radio, justifying its use as "a holy mission," and thanking "a merciful God that He had given us the faith and the courage and the power from which to mould a victory"[11] while affirming that "the utter destructiveness of war now blots out this alternative. We have had our last chance" (21). This is corroborated by Harding: "There can never be another war like the old ones. The next war will wipe out every city in the first half hour—every city on both sides" (21). However, the play then proceeds to show the hardening of attitudes through the Cold War. Harding, who has joined an association of nuclear physicists trying to educate the public and guide the government away from war, is interviewed by the House Un-American Activities Committee.[12] Eventually an atomic war breaks out, and the characters degenerate to subsistence level in a cave. Harding still hopes to exert influence for good through the physicists' organization, but the only real hope expressed at the end of the play is embodied in his teenage son, who proclaims the need for universal brotherhood and a world state. Thus, although Harding, the representative scientist, is morally more aware than most of his society and agonizes over a possible solution, he is helpless to provide one. The only suggestion Sinclair offers is a hypothetical one, voiced by a child, that seems to have no hope of adoption.

The German playwright Hans Henny Jahnn expresses a similar pessimism in *Die Trümmer des Gewissens oder Der staubige Regenbogen* (The ruins of conscience or The dusty rainbow) (1959). Here the individual scientist, isolated for so long from social issues, recognizes too late the evil potential of his research and the military intentions of his government and is then powerless to avert the threat of mass destruction. Jahnn's protagonist, Jacob Chervat, an atomic physicist living in an Atomstadt, a government-funded science research ghetto, has refused to listen to any discussion of the destructive implications of his research. Even though his son Elia is sexually underdeveloped because of radiation exposure in the Atomstadt and his wife suffocates their next baby rather than expose it to a similar fate, he continues to believe that his discoveries will eventually be for the benefit of humanity and will never be used for destructive purposes. However, his confrontation with Dr. Lambacher, a biologist who is determined to extend his genetic experiments on huge dragonflies to produce a race of supermen, alerts Chervat to the way science is manipulated for destructive purposes by both individuals and the establishment. He realizes that Lambacher's research project is the basis of the government's plan to subdue the nonwhite races by a preemptive nuclear strike, using his research. But although Chervat is transformed from a reclusive scientist into a morally committed activist, his struggle with the authorities is doomed to failure. Acceptance of public funding has made him government property, and his ethical protests are regarded as treason. In despair, Chervat attempts to kill an official and then, appalled at his own violence, commits suicide. Nor does Jahnn leave us with any hope that his sacrifice will have an effect. The last ironic lines of the play emphasize this: "You do wrong! You hope!"[13] There is no good outcome in a world where science is controlled by a totalitarian military power; the lone protesting scientist comes too late.

Ignorance of consequences is no excuse in Dino Buzzati's symbolic short story "Appointment with Einstein" (1958). Buzzati's Einstein has just made his breakthrough in conceiving the idea of curved space. Although usually a humble man, he indulges in a moment of pride, considering himself "above the common herd." Almost immediately he is confronted by a devil (in the person of a Black garage attendant), who announces that his moment of death has arrived. Einstein pleads for time to complete the proofs of his brilliant concept, and after two months, when the work is finished, he returns sadly but honorably to the appointed place to die.[14] At this the devil, laughing uproariously, reveals that the threat of impending death was only a satanic device to make him work harder and finish his work: "God knows how long you would've drawn it out if I hadn't scared you two months ago. . . . The big devils run the show. They say your first discoveries have already been extremely useful. . . . Hell has profited greatly from your ideas." Einstein is irritated and incredulous: "Nonsense! Can you find anything in the world that's more innocent? They're insignificant formulas, pure abstractions, inoffensive, impartial." But he is assured by the devil, "They're satisfied downstairs. Oh, if only you knew!"[15]

Einstein and Rutherford are the two best-known examples of scientists who worked on their seemingly obscure theories believing them to have no sinister significance or destructive potential.[16] However, in his ironic tragicomedy *Die Physiker* (1961) the Swiss playwright Friedrich Dürrenmatt included the physicist Johann Wilhelm Möbius as an exemplar of the powerlessness of even the most brilliant scientists to prevent the misuse of their work. In 1956 Dürrenmatt had reviewed Robert Jungk's book *Heller als tausend Sonnen: Das Schicksal der Atomforscher* (1956) (translated 1958 as *Brighter Than a Thousand Suns*), which detailed the lives of the physicists working on the first atomic bombs and raised the question of their responsibility for the outcome of their research. After the event, it was said by several of the scientists, including Werner Heisenberg, that if even twelve of the physicists had combined to oppose it, they could have prevented the construction of the bomb. Dürrenmatt accepted this but believed that once the first atomic bomb had been constructed, there were no longer any options. The belated warnings of the scientists to society are now useless, merely "demands on an imaginary world, demands not to sin after the Fall."[17]

In *Die Physiker*, Möbius is caught in a paradoxical situation where action and nonaction amount to the same thing, just as the Möbius loop has only one surface and one edge. Dürrenmatt remarked that "the action takes place among madmen and therefore requires a classical framework to keep it in shape."[18] We realize that the concept of a madhouse is relative: a world where atomic physics is directed toward weapons research is also a madhouse. Hence, Möbius reasons, the only place in which to preserve one's sanity is an asylum. As Walter Muschg observes, "For him, the asylum is no poetic metaphor, but the most normal consequence of an actual development. He examines the broadest form of our as-yet-undiagnosed madness, the hunt for atomic physicists."[19]

Through the three representative "patients" in Les Cerisiers asylum, Dürrenmatt explores three alternative approaches available to scientists confronted by a totalitarian system. Möbius has retreated there to prevent his work from being misused; that is, he has sought moral shelter there from an insane world. Kilton, a British physicist and secret agent who claims to be

Newton, and Eisler, an East German secret agent who alleges he is Einstein, have come pretending to be fellow inmates in order to persuade Möbius to work for their respective governments. Like Newton, Kilton is an optimist about science, believing in freedom of research, universal sharing of discoveries, and freedom from political or social responsibilities. He argues, like the amoral scientists of the previous chapter, " I give my services to any system, providing that system leaves me alone. You talk about our moral responsibility. . . . That is nonsense. We have far-reaching pioneering work to do, and that's all that should concern us" (42). However, Dürrenmatt implies that what was appropriate in Newton's time is no longer valid; it can no longer be assumed that scientific research is beneficial to society. Eisler rejects this view, demanding political commitment as the only responsible action, just as Einstein had initially taken a political position and recommended the construction of the atomic bomb. Eisler argues: "We cannot escape our responsibilities. We are providing humanity with colossal sources of power. That gives us the right to impose conditions. If we are physicists, then we must become power politicians" (42).

Möbius has been in the asylum for fifteen years, during which time he has continued to work on his *Weltformel*, his principle of universal discovery. This work, which combines Newton's gravitational theory and Einstein's uncompleted unified field theory, is the greatest breakthrough in modern physics, but Möbius knows that for the sake of humanity it must be suppressed: "There are certain risks which one may not take: the destruction of humanity is one. . . . Fame beckoned me from the university; industry tempted me with money. Both courses were too dangerous. . . . A sense of responsibility compelled me to choose another course" (44). He has therefore burned his manuscripts, and the other two "physicists," their plans thereby subverted, reluctantly agree that to remain in the asylum is the only sanity:

> MÖBIUS: Only in the madhouse can we be free. Only in the madhouse can we think our own thoughts. . . . We ought not to be let loose on humanity. . . .
> NEWTON: Let us be mad, but wise.
> EINSTEIN: Prisoners but free.
> MÖBIUS: Physicists but innocent. (45, 47)

Had the play ended here, Möbius would have been a heroic figure, and the tone one of qualified optimism. It would have implied that physicists at least have a choice, albeit limited, concerning the outcome of their research. However, Dürrenmatt denies the possibility of even such a minimal degree of freedom.

Ironically, the man whose theory promises such power to its possessor has no choice at all. We learn that the director of the asylum, Fräulein Doktor Mathilde von Zahnd, is the only really mad inhabitant. An insane tyrant, she has secretly copied all Möbius's papers as he wrote them and founded a giant cartel ready for world domination through the application of his formulas. Speaking for Dürrenmatt, she says of Möbius: "He tried to keep secret what could not be kept secret. . . . Because it could be thought. Everything that can be thought is thought of some time or other. Now or in the future" (50). Thus the physicists are too late in their decision to take responsibility for their research. Möbius admits, "What was once thought can never be un-thought." Among his "Twenty-One Points to *The Physicists*" Dürrenmatt included the following: "The contents of physics are the concern of the physicists, the consequences are the concern of all"; "What concerns everyone, only everyone can resolve"; "Every attempt by an individual to resolve for himself what concerns everyone, must founder."[20] Here Dürrenmatt suggests as a reason for Möbius's helplessness a certain arrogance underlying his assumption that he has the power to resolve the horrifying predicament of the modern world. But there is no return to the Eden of ignorance, and even the extreme sacrifice Möbius is prepared to undertake by way of reparation is not permitted to achieve absolution.

There are similarities between *Die Physiker* and Michael Frayn's 1992 play *Copenhagen* (film 2002), which retrospectively analyzes a historical event, Werner Heisenberg's visit in 1941 to his former mentor and friend Niels Bohr, a physicist of Jewish descent living precariously in German-occupied Copenhagen. The spirits of Heisenberg, Bohr, and Bohr's wife, Margreth, debate Heisenberg's motives for the visit, which endangers both the physicists and what was said during their walk outdoors. It seems likely that Heisenberg, who had been working in Germany on the technology for producing a nuclear reaction necessary for a bomb, wanted to seek advice from his former mentor about the morality of continuing such research, which would deliver a nuclear bomb to the Nazis. Bohr had not thought the process was feasible, and his shocked reaction ended the meeting inconclusively. Ironically Heisenberg did not proceed with research on the bomb, while Bohr, escaping to the United States, worked at Los Alamos on perfecting the atomic bomb for the Americans. The play draws on the parallels between the indeterminacy of electrons in the so-called Copenhagen interpretation of quantum mechanics and the difficulty of defining the physicists' motives and choices.

Although technically the war ended in August 1945 with the US demonstration of superior military power, the race to devise yet more effective methods of mass destruction continued unabated as the Cold War, with a large proportion of the superpowers' budgets dedicated to "defense." Scientists in appropriate disciplines, especially physics, were readily recruited to work in weapons establishments because they could not secure funding in other areas. However, the Cold War induced a moral quandary for those who were caught up in the superpowers' escalating weapons research but outraged by the barriers to free exchange of scientific information, even between political allies. Klaus Fuchs and other "atom spies," who for various reasons passed information about American atomic research to the Soviets, were only the most publicized representatives of a much larger group of scientists who experienced similar moral anguish about their research into weapons of mass destruction. This theme was explored in fiction before any actual examples of nuclear espionage were known.

This focus on the development of atomic weapons was at first topical but later became symbolic of the increasing scientification of society in the latter half of the twentieth century and the paradoxical situation whereby scientists, because of their unique form of knowledge, became both the most important asset of governments and industry and at the same time disempowered by state secrets acts.

Dürrenmatt's physicists were pitted against a manifest criminal; hence they attained a comparative moral superiority even in their failure. Heiner Kipphardt, on the other hand, is concerned to examine the question of the powerlessness of scientists in a more problematic case: the seemingly highly motivated scientist in conflict with a government that also lays claim to representing the highest values. His play *In the Matter of J. Robert Oppenheimer* (1964) (referred to in a different context in chapter 14) is based on the enquiry in May 1954 into the question of security clearance for Oppenheimer after it had been suspended by the US Atomic Energy Commission (AEC).[21] The frequency with which Oppenheimer featured in the previous chapter and again here suggests that he was an alluring figure for novelists and dramatists. This was partly due to his eloquence and cultured background, unusual among the mostly young physicists working on the Manhattan Project, but also because he was perceived as a tortured soul, providing the necessary agon for dramatic exploration of character in terms other than science itself. Through the questioning of Oppenheimer by State Prosecutor Robb, Kipphardt defines the ironic moral predicament of the physicists: they were trying to develop the bomb because they believed that the German scientists were about to produce it for the Nazis, and they realized too late that the Germans were in fact not anywhere near as far advanced in their development of the bomb. By the time the US atomic bomb was tested, it was also too late, in the face of the demand to end the war in the Pacific as quickly as possible, to prevent its use against Japan.

After the deployment of the atomic bomb against Japan, however, Oppenheimer opposed the immediate undertaking of research into a hydrogen bomb. This led to his arraignment for "lack of enthusiasm" for the hydrogen bomb project, as well as to accusations of disloyalty because of his "objectionable associations" with known communists and left-wing sympathizers. In the final analysis, the charge against Oppenheimer was not that he had undertaken any treasonable action (there was no evidence of any complicity with the atom spies) but that he had committed "ideological treason." Robb sums up his case:

> We could already have had the hydrogen bomb four or five years earlier if Dr. Oppenheimer had supported its development. What is the explanation for such a failure in a man so wonderfully gifted? . . . This is the explanation: Dr. Oppenheimer has never entirely abandoned the utopian ideals of an international classless society, he has kept faith with them consciously or unconsciously, and his subconscious loyalty could only in this way be reconciled to his loyalty to the United States. . . . This is a form of treason which is not known in our code of law, it is ideological treason which has its origins in the deepest strata of the personality and renders a man's actions dishonest, against his own will.[22]

Kipphardt's play is structured around Oppenheimer's inability to resolve the conflict between his fascination with science as an end in itself and his sense of responsibility, which is itself divided between loyalty to a government and loyalty to mankind. Asked about this, Oppenheimer replies: "I would like to put it this way: if governments show themselves unequal to, or not sufficiently equal to, the new scientific discoveries—then the scientist is faced with these conflicting loyalties. . . . In every case I have given undivided loyalty to my government, without losing my uneasiness or losing my scruples, and without wanting to say that this was right" (71). The feeling of the AEC, however, is that Oppenheimer, because of the influence he exerts through his public position, should not be permitted to voice the ethical struggles he may have as a private individual. The right of the government to dictate even the personal attitudes of scientists is also subtly linked with the fact that it pays for their research. This is expressed baldly by Morgan: "When a state spends enormous sums of money on research, is it to be denied the right to do with that research as it thinks fit?" (73–74).[23]

Of the scientists presented in the play, only Oppenheimer, Evans, Bethe, and Rabi (all physicists except for the chemist Evans) have serious moral scruples about the development of the hydrogen bomb, and all support freedom of conscience.[24] Hans Bethe represents the conscientious scientist plagued by conflicting judgments concerning the correct ethical decision to make in an immoral situation. Like most of the team, he had worked on the project believing it necessary to develop the atomic bomb before the Nazis did so, but after Hiroshima he vigorously campaigned for international control and open information about the weapon, believing this to be the safest course. Totally against the hydrogen bomb project, he was one of the twelve scientists who opposed President Truman's decision to press on with it. However, with the outbreak of the Korean War, Bethe, believing again that the United States might be in danger, recapitulated and reluctantly returned to work on the project. Yet, as his testimony at the enquiry indicates, he was never comfortable with this decision: "I have helped to create the hydrogen bomb, and I don't know whether it wasn't perhaps quite the wrong thing to do. . . . After a war with hydrogen bombs, even if we were to win it, the world would no longer be the world which we wanted to preserve, and . . . we would lose all the things we were fighting for."[25]

Oppenheimer's final speech is a reaffirmation of his determination to devote himself to pure research in a belated, futile attempt to return to a prewar, prelapsarian state:

> When I think what might have become of the ideas of Copernicus, or the discoveries of Newton under present day conditions, I begin to wonder whether we were not perhaps traitors to the spirit of science when we handed over the results of our research to the military, without considering the consequences. Now we find ourselves living in a world in which people regard the discoveries of scientists with dread and horror, and go in mortal fear of new discoveries. . . . We, the physicists, find that we have never before been of such consequence and that we have never before been so completely helpless. . . . We have been doing the work of the Devil, and now we must return to our real tasks. We must devote ourselves entirely to research again. We cannot do better than keep the world open in the few places which can still be kept open.[26]

By judicious placement and selection of extracts from the hearing, Kipphardt has effectively investigated the investigating committee and transformed the original "trial" of Oppenheimer into a trial of the US government. Evans, the most sympathetic character among the committee members, clearly implies his personal condemnation of Oppenheimer's opponents Robb and Teller and his exoneration of Oppenheimer and Bethe because of their commitment to humanity.[27] Kipphardt's treatment might seem, therefore, to reinstate Oppenheimer as an idealist, but in this play there are no unambiguous heroes, for although the condemnation of the government agencies, and in particular the FBI, is unequivocal, the real charge against the physicists (as distinct from that being investigated by the committee), namely, that they "have known sin," is never revoked. Oppenheimer's final withdrawal to pure research both challenges and is challenged by Brecht's Galileo and Dürrenmatt's Möbius: in an insane world, can there be any "right" action?

Another of Dürrenmatt's plays, *Der Mitmacher* (The collaborator) (1973), explores the manipulation by industry of a biologist, the significantly nameless Doc, who began his career as an idealist, afire to discover new knowledge: "I wanted to investigate life, to fathom its structure, to pursue its secrets."[28] After a distinguished beginning to his research into the combination of amino acids, the building blocks of proteins, Doc takes a job with a corpse-disposal company and invents a new process, corpse dialysis, which completely disposes of corpses without pollution. Belatedly he realizes that he has produced not only a fortune for the disposal company but a green light for murderers, who cannot now be traced. Like Frankenstein, he has passed from research into the creation of life to dealing in dead bodies, in the process losing control over the use to which his discovery will be put by others. Despite his economic value to the company, Doc has no ethical influence in or on his society. Symbolically, his "laboratory" (also his "home") is an underground warehouse storage room, where water from the sewer system seeps perpetually —a leaky coffin already buried. At the end of the play Doc, now the head of a chemical empire, epitomizes the scientist as conformist, an unintentional and unwilling collaborator with the criminal elements of society and helpless to "uninvent" his process in the interests of justice. Now that he has given society what it wants—possibilities of limitless crime and wealth—his views are disregarded.

The figure of the brilliant scientist rendered helpless by social forces (as opposed to a runaway invention or government decrees) became more common in the 1960s and 1970s. An interesting example is the character of Adam Sedgwick in C. P. Snow's novel *In Their Wisdom* (1974). Formerly a Nobel Prize winner in molecular biology and president of the Royal Society, now a member of the House of Lords, Sedgwick suffers from Parkinson's disease, so that although he retains "one of the most lucid minds extant," his body is racked by uncontrollable spasms and twitches. Sedgwick is Snow's symbol of the scientist in the seventies, a superbly active mind housed in a body that is powerless and uncontrollable, a towering intellect that is ludicrously ineffectual. Thus Snow suggests that importance in the hierarchy of science is of limited worth in the world outside academe. Sedgwick's attempted speech to the House of Lords, for instance, though crisp and coherent in its content and arising from the keenest intellect in the House, nevertheless comes over only as an incoherent sequence of "fumbles of his tongue,"[29] an implicit comment on the breakdown of communication between the scientist and the rest of society and in particular between science and government.

A less obviously helpless scientist, but one who is equally powerless in his world, is Professor Victor Jakob, the protagonist of Russell McCormmach's *Night Thoughts of a Classical Physicist* (1982). Jakob was devised as a composite of several actual scientists living in the Germany of September 1918, but he stands equally well as a representative for the would-be pure scientist today whose work is used in unforeseen and abhorrent ways. Although overtly treated by the university authorities with the respect due to his age and scientific reputation, Jakob is out of tune with his contemporaries, for his real allegiance is to a past age, the age of classical physics. Like Einstein, he finds quantum physics, with its emphasis on chance and statistical results, a travesty of the clarity and order that had attracted him to the nineteenth-century discipline. He also believes in pure, rather than "useful," research, an unpopular stance in 1918, when German physicists were expected to adopt a political role and encourage nationalism in their students. Jakob's long, nocturnal meditation ends on a note of sustained ambivalence about the role and effectiveness of scientists in a world at war. On the one hand, he affirms his belief that "the international spirit of science is evidence that science is stronger than hate, . . . that science is a bearer of understanding between peoples, . . . a reagent of love"; yet he asserts that "it is no betrayal of our scientific creed for us to yield to the dominant pull of patriotism . . . temporarily [to] sacrifice what is essential to our work as scientists."[30] This anachronistic representative of the austere beauty of classical physics (and the analogy with classical

Greece, which Jakob also reveres, is explicit) is seen as irrelevant and helpless to influence his society.

Nearly all the sociomoral positions considered in this chapter in relation to the Cold War are recapitulated in Howard Brenton's provocative play *The Genius* (1983). Brenton's protagonist, the brash young American Leo Lehrer, Nobel Prize winner in mathematics, has solved the equations necessary to establish Einstein's unified field theory and has thereby incidentally discovered the means of making the ultimate atomic weapon. The play examines his successive attempts to live with the knowledge that "no calculation is pure." Lehrer's first reaction is to feign ignorance in the face of the overtures of the US military, whereupon MIT punishes him by allowing him to be "captured" by an English Midlands university, considered by physicists to be the ultimate exile at the end of the world. At this stage he still believes that noninvolvement is possible: "I said—OK, no calculation is pure. Therefore calculate no more. I gave up, . . . I closed down. I exiled me into my own head. If you are shit scared of the damage you can do, do nothing, eh?"[31]

Lehrer soon learns, however, that knowledge cannot be put back in the box. Gilly Brown, a gauche Midlands freshman but also a self-taught mathematical genius, has already solved most of the relevant equations. Seeing her calculations written in the fresh snow of a field, the incredulous Lehrer confronts her with the applications of her work. When Gilly denies any interest in anything but pure math—"I don't want anything to do with politics; I'm here to study mathematics" (21)—the hitherto irresponsible Lehrer adopts the role of teacher implied by his name and in thirty pages of applied physics equations converts her to activism. Together they explore possible alternatives to helplessness: attempting to shock the "world," represented by the university microcosm, into a realization of the nuclear horror; handing the critical calculations to MI6; trying to equalize the balance of power by giving the same equations to the Soviets. All these courses of action fail to avert a nuclear war. Thus the play examines the basic dilemma of modern science: recognition of the terrible potential consequences of so much research and the simultaneous impossibility of not doing it. As Gilly says, "A thought is a thought. You can't not have it" (21). Brenton presents this view in strongly fatalistic terms, and it is this that gives the play its power and pessimism. Thus Leo asserts: "The ideas do not love us. I have come to the conclusion that all the investigations into the atom, discoveries, calculations, formulations, nearer and nearer to the description of the force of nature—the scientific quest of the century—is fundamentally malign" (22). He later adds, "If I were the Pope, I think I would announce that we are forbidden to know the true nature of gravity, electro-magnetism, and the nuclear forces. But—too late, holiness. We're never going to dig that knowledge out of our lives, out of our thoughts, out of our machines" (37).

The helpless scientist, then, may be either reprehensible, insofar as he has failed to foresee the dangerous consequences of his work, or a limited hero. In this latter case he tries to rescue society from the impending evil but fails because the forces against him are too strong. Given the politically unstable decades of the twentieth century, it is hardly surprising that the enemies most commonly depicted as arraigned against the scientist are social forces inimical to his own moral values. In novels and plays of the Cold War period, the most frequent scenario presents the ethically aware scientist (usually a physicist involved in nuclear weapons research) in moral opposition to the militaristic elements of society, which demand the most powerful weapons available. Rarely, if ever, does this figure "win," but like the protagonist of classical tragedy, he assumes a grandeur in defeat.

All the works discussed so far in this chapter have been to some degree tragic in conception: the helpless scientists for one reason or another lose control over the outcome of their discoveries. All the writers considered so far assume that this must inevitably be disastrous, since no one else, no other power, can be trusted to right the situation. However, there are also optimistic variations on the theme. Some writers suggest that although the scientists may be helpless, others may point the way to survival.

Alternative Heroes

During the darkest years of World War II the American humorist Thornton Wilder wrote his fantasy *The Skin of Our Teeth* (1942), exploring the question of survival in a series of time-telescoping situations in which society, both local and global, seems to have broken down. Science is represented in the person of the whimsical and complacent Mr. Antrobus, an inventor in the Edison tradition, who exemplifies what Wilder saw as the optimistic American trust in the future. Antrobus is busy inventing the wheel, the alphabet, and the answer to ten times ten while the rest of the world struggles to survive an ice age, a war, murders, and general social breakdown. Clearly Antrobus, closeted with his precursor of the wheel and alphabet to be, has nothing to offer in such dire straits; science cannot help us in times of global catastrophe.

Wilder's alternative approximates to the instinct for survival by "the skin of our teeth," the innate resourcefulness and endurance of the ordinary individual multiplied over the whole population. This, Wilder believes, is how the human race has hitherto escaped extinction and will perhaps continue to do so. Such optimism, though unrelated to any faith in science, can perhaps still be seen as a characteristically American belief in the power of the ordinary man to create his own future, a frontier faith significantly absent from European writing of the period.

In recent decades the science-based problems under consideration have multiplied to include genetic engineering, robotics, artificial intelligence, nuclear power, and diverse forms of environmental pollution. In the literary treatments of these issues, especially those written from a feminist perspective, the scientist characters are rarely assigned a leading role. The heroic figure who does battle with the business corporations, the military-industrial complex, or the government administration is more likely to be an otherwise ordinary individual with a passion for social justice—a teacher, a farmer, or a mother such as Erin Brockovich in Steven Soderbergh's film of 2000, or Dellarobia Turnbow in Barbara Kingsolver's *Flight Behavior* (2012). This change in focus reflects the general awakening from public apathy that followed the war years. During the 1980s, Western nations experienced the effectiveness of "people power," not only in large-scale operations such as the women's movement, the peace movement, and the environmental movement, but in all areas of social activity. This effect has gained force through social media and volunteer organizations for consciousness raising. The heroic figures of this movement are ordinary citizens rather than scientists or other experts. Interestingly, these new champions of social and environmental justice are more likely to be depicted as successful in their crusade than their predecessors, the noble scientists, were in theirs.

The Responsibility of Society

One of the most interesting emphases in recent representations of the scientist at risk has been the implication that the society that promotes science for the sake of the technological advantages it brings must be held jointly responsible with the scientists for the social and environmental consequences. A striking example of this is Christa Wolf's novel *Störfall* (1987), discussed in chapter 13 in relation to runaway technology. Wolf's nonindividualized presentation of the scientific "experts" who appear on television to extol the merits and potential of nuclear power suggests that they too are merely faceless units of the establishment, as much victims of the disaster as the rest of the society, who want to receive only reassurance. A clear indication of Wolf's view that the scientists are not really in control is the fact that they are not specifically denounced for their part in the process. On the contrary, she asserts that the society that promotes nuclear power in order to create material products and a utopian lifestyle without working for them must share the moral responsibility for the consequences. "Were we monsters when, for the sake of a utopia—justice, equality and humanity for all—that we didn't want to postpone, we fought those whose interests did not (do not) lie in this utopia and, with our own doubts, fought those who dared to doubt that the end justifies the means? That science, the new God, would deliver to us all the answers we sought from it."[32] It seemed to Wolf at that time (prior to the large-scale availability of sustainable energy sources) that a society that demands ever-increasing material wealth is presented with only two alternatives: nuclear power, which produces radioactive waste, or power derived from coal or oil by a process that produces atmospheric pollution and acid rain that destroys the forests (79). Ultimately the responsibility for these consequences lies not only, or even mainly, with the immediate perpetrators of these disasters, the nuclear scientists and the chemical engineers, but with the society that demands the products without acknowledging the cost.

Exploring the responsibility of scientists for the large-scale pollution of the environment led several German writers to distribute the blame more widely. In *Flugasche* (1981) Monika Maron wrote about the deluge of 180 tons of ash falling daily on the city of Bitterfeld as a result of its intense industrialization, while Günter Grass's novel *Die Rättin* (1986) presents a macabre picture of the destruction of the earth by the inundation of garbage resulting from society's rampant consumerism. In these treatments of environmental disasters, the blame has shifted from scientists qua individuals to scientists collectively and thence to the society that commissions their research. Society, by its demands for particular products and by the power it exerts through the selective distribution of funding, directs its scientists and technologists into the initiatives it favors.

Just as the image of the powerful scientist holding a society to ransom has been superseded by that of the helpless scientist, so there has been a change in the causes of helplessness and, correspondingly, the degree of responsibility ascribed to the scientist characters. In the first decades of the twentieth century scientists were most frequently portrayed as victims of their own discoveries or inventions in a playing out of retributive justice. From 1950 to 1980 writers more commonly depicted scientists as being idealistically engaged in a struggle with state authorities or commercial organizations in an effort to ensure that their research was not used for destructive purposes. The 1980s and early 1990s, however, were characterized by more generalized representations of metonymic scientists, who were essentially just obedient civil servants, and the focus of blame shifted to implicate the whole society rather than individuals.

CHAPTER 17

The Scientist as Woman

> I'm not a nineteen-year-old working up a résumé, and I don't expect to be treated as if I were.
> —*Dr. Rae Crane, "The Medicine Man"*

Gender Statistics

Compared with the many and diverse depictions of male scientists in fiction and film, the number of fictional female scientists has been extremely small. In their survey of more than four hundred films about science or scientists, Peter Weingart and colleagues found that only 18 percent of the characters were female and from a study of those female scientists concluded that "science is traditionally a very male world in which women have either no place or 'their' place, i.e. a woman's place."[1] This was confirmed by Eva Flicker's study of sixty feature films from 1929 to 2003 in which female scientists appear. She, too, found that women scientists were rare (18 percent), and when they did appear, their roles differed greatly from those of their male colleagues.[2]

The disparity in numbers noted by Weingart and Flicker reflects the real-life situation up to the 1950s. There were some notable female scientists in the eighteenth century, notably Caroline Herschel, sister of Sir William Herschel and herself a keen astronomer, who discovered several comets and was awarded the Gold Medal of the Royal Astronomical Society, and Ada Lovelace, Byron's daughter, a self-taught mathematician who collaborated with Charles Babbage on his analytical engine, an early mechanical computer.[3] Like Caroline Herschel, whose name was later merged with that of her brother as "The Herschels," Ada Lovelace has been a victim of the "Matilda effect," the female counterpart of the "Matthew effect," whereby the work of junior or lesser-known scientists is attributed to their superiors, partly to gain acceptance and publication, but also because the more famous names are more readily remembered.[4] In hindsight, however, the actual authors of these contributions are usually forgotten. In the nineteenth century there were pioneer women scientists such as Mary Anning (paleontologist), Mary Somerville and Maria Mitchell (astronomers), and Margaret Huggins (spectroscopist). Denied membership in the Royal Society, they nevertheless not only managed to pursue their research but were in the forefront of promoting science education for the public.[5] Even in the twentieth century, when there were many more notable female scientists who made major breakthroughs in their fields, they struggled for acceptance and acknowledgment—Emmy Noether, Dorothy Hodgkin, Nettie Stevens, Barbara McClintock, Lise Meitner, Rosalind Franklin, Rachel Carson, Evelyn Fox Keller, Jocelyn Bell Burnell, Jane Goodall, and Dian Fossey, to name only those whose work has belatedly achieved recognition. The one notable exception in the science hall of fame is Marie Curie. Although before her husband's death she was regarded as his assistant, afterward, chiefly through what has been alleged to be an ingenious campaign waged by Curie herself and her American publicist, the journalist Missey Meloney,[6] she gained international fame and was awarded Nobel Prizes in both physics and chemistry. Yet, ironically, in *Madame Curie* (1943), the best-known film biography, she is presented as subservient to her husband, Pierre, repeatedly asking his advice and opinion, and conforming to the contemporary expectation of the woman scientist as a research assistant.[7]

When, with the rapid advances in science after World War II, there were more opportunities for women to embark on a career in science, films of the 1950s, particularly science fiction films, showed increasing numbers of women engaged in science. Sidney Perkowitz, American physicist and film buff, identified 84 women scientists out of 382 films on the IMDb website featuring scientists (22 percent of the total), which he claimed was close to the percentage of actual women scientists and engineers in the US workforce (25 percent) averaged over that period,[8] and Jocelyn Steinke's survey of 74 science-related films in the period 1991–2001 noted 23 female scientists (31 percent of the total number of scientists).[9] This seemed a strong indication that the sciences were no longer exclusively male territory, although we should note that gender equality in science may have been more apparent in film, in order to further a romantic interest, than was true at the lab bench. In long-running TV series such as *Star Trek* and crime drama series such *NCIS* (2003–), *CSI: Crime Scene Investigation* (2000–15), *CSI: Miami* (2002–), and

Bones (2005–17) women scientists are roughly equal in number to their male colleagues, making for lively gender interactions that sustain the interest of the series.

Characterization

In terms of characterization, female scientists have been portrayed, in both fiction and film, very differently from their male counterparts. The stereotypes of the male scientist described in the preceding chapters rarely apply to fictional female scientists. The most striking difference is that women have never been cast directly in the role of the alchemist, the most prevalent stereotype from medieval times to the twentieth century. This is not surprising, as there are almost no substantiated accounts of females practicing the art during the medieval period.[10] Had there been any, they would almost certainly have been accused of witchcraft and summarily executed. Even in the nineteenth and early twentieth centuries, when the alchemist image resurfaced famously as Frankenstein, Dr. Jekyll, and Dr. Moreau, and in numerous films, the possession of specialized but illicit knowledge, and the power that knowledge entailed, was not attributed to women, who were securely locked in the role of assistants.

In place of the mad, bad male scientist there are a few instances of the mentally unstable female scientist. Esme Charbonneau in Carol Muske-Dukes's novel *Saving St. Germ* (1993) seems to hover on the margins of madness in her volatility and outbursts of anger toward her students and husband when stressed. As an organic chemist, originally working on purifying DNA, she experiences an almost religious sensation on entering the organized arrangement of the laboratory: "I would never compare entering the lab to entering a church, yet the lab glows with an unmistakable aura of prefigurement and awe—the ducts and tall flasks stand like effigies, the distillation apparatus and rubber tubing swing from hooks like ornate censers; it is a sealed-off world touched by grace. A grace of reverse miracles—miracles happening in a controlled state."[11] Her former boyfriend, a doctor, denigrates her research into DNA as "sitting around, splicing and snipping and staining slides, looking at squiggles in microscopes and deducing profundities. Looking for St. Germ. Saving St. Germ. Some tiny organism, some protein sequence that's going to change the quality of human life, but only intellectually, only as a model for thought."[12] Esme suddenly changes career tack, switching to an obsessive search for a Theory of Everything. At the end of the novel it is revealed that Esme was an autistic child, whose father removed her from her mother's influence in order to make her "normal," possibly contributing to her mental instability as an adult.

By the end of the twentieth century, there were sporadic examples in popular fiction and films of the obsessive female scientist, who might be classed as mad. The botanist Dr. Pamela Isley, known as "Poison Ivy" in *Batman and Robin* (1997), becomes a kind of ecoterrorist, vowing to establish the supremacy of her mutant plants at the cost of human lives. "If I can only find the correct dose of Venom, these plants will be able to fight back like animals. I will have given flora a chance against the thoughtless ravages of man."[13] The other examples are less well known[14] or, like Beth Halpern in *Sphere* (1998), have supposedly become mad through some external element of the plot rather than because they are inherently deranged.[15]

In terms of comic women scientists, we have seen that women appeared briefly as foolish virtuosae on the Restoration stage, where they were treated with the same satire as was directed at their male counterparts, such as Sir Nicholas Gimcrack, but with additional ridicule poured on their unfeminine pursuits and conversation. Their real-life model was Margaret Cavendish, Duchess of Newcastle (see chapter 3), whose visit to the Royal Society was lampooned by Samuel Pepys but whose grasp of contemporary science was evident from her *Observations upon Experimental Philosophy* (1668). The virtuosae are redeemed only by abandoning their whimsical hobbies and settling down to marriage and appropriate feminine pursuits. There are few comic women scientists in film, compared with the male counterpart of the absentminded professor, but Steinke has identified two clumsy, absentminded female scientists in films: Dr. Diana Reddin in *Junior* (1994) and Dr. Allison Reed in the comedy *Evolution* (2001).

Eva Flicker has proposed six alternative gender-specific stereotypes of female scientists in feature films:[16]

- The old maid who is only interested in her work. She is professionally competent but lacking in femininity.
- The male woman, assertive, asexual, and lacking in feminine charm but bringing intuition to resolving a problem.
- The naive expert, emotionally and ethically driven, believing in goodness but repeatedly getting into difficulties from which she is rescued by men.
- The evil plotter, sexually attractive but corrupt, using her attractiveness to steal information or trick her opponents.

- The daughter or assistant, attractive and sociable but having no formal scientific qualifications. She has a subsidiary role in science but makes an important contribution in terms of social competence, which the male scientists lack.
- The lonely heroine, who has outstanding qualifications and competence and is focused on her research but who lacks recognition and power, as she is not part of the "boys club." She remains dependent on a male champion for professional recognition.

Of these, the first two have similarities to male stereotypes, though often with variations that mask the connection. Susan Calvin, robopsychologist of Isaac Asimov's robot stories, can be identified with the first of Flicker's stereotypes, but she can equally be seen as a female version of the unfeeling, tunnel-visioned scientist, dedicated to her work and impatient with people who often behave irrationally. Robin Roberts argues that Calvin is intentionally shown as "a failure as a woman" because she is "deformed by her intellect."[17] She embodies the prevailing view of female scientists in the 1960s, in that, being a dedicated scientist, she necessarily loses her femininity. Her only emotions are reserved for the robots, which constitute her substitute children and which she increasingly resembles. Compared with her colleagues, for whom she barely hides her scorn, she unfailingly expresses admiration and affection for the robots: "Susan Calvin talked about [her colleagues] Powell and Donovan with unsmiling amusement, but warmth came into her voice when she mentioned robots."[18] Her only departure from strict rationality occurs in the story "Liar!" (1941), in which she falls prey to vanity and even acquiesces in the effective death of a favorite robot, Herbie, in order to save face about her mistake. Because the robots show human characteristics and some individual personalities, we may not at first realize that Calvin conforms so closely to the stereotype of the unfeeling scientist that emerged during the Romantic reaction to the Enlightenment. Asimov is far less critical than earlier creators of such characters, because Calvin expresses his own belief in the superiority of the robots. Indeed, we might almost infer that Calvin is the voice, even a female persona, for Asimov himself, as she declares her uncritical advocacy of robots and her complete confidence in their inability to act irrationally or to harm humans.

Less authorial favor attaches to Fred Hoyle and John Elliot's female biologist Madeline Dawnay in the TV series *A for Andromeda* (1961, film 2006). The team, led by astronomer and computer scientist John Fleming, is delighted when an encrypted message coming from the Andromeda galaxy proves to contain instructions for the construction of a supercomputer. This in turn directs the scientists to construct ever more complex life forms and eventually a beautiful woman who is then directed by the computer. When Fleming realizes the danger of a computer on track to take over the world, he moves to terminate the program, but Dawnay, who has invested her career in this project, is obsessed with proceeding. Fleming remarks, "Dawnay thinks the machine's given her the power to create life; but she's wrong. It's given itself the power."[19] Ironically Fleming himself is also being directed by the computer, not, as he at first thinks, in the positive sense of being instructed, but literally controlled by it.

Calvin and Dawnay are both examples of Flicker's "male woman" scientist, who is assertive and asexual. In such a category we can also include Sylvia Orloff in Susan Gaines's *Carbon Dreams* (2001). One of the few women scientists to have made it to the top ranking in marine science, she is in charge of her own research vessel and can make or break the careers of early- or mid-career scientists in her department. Aware of her power, she is capricious and quite capable of manipulating their results and assimilating them into her own publications, knowing they have no redress.

Flicker's stereotype of the naive expert can also be identified in fiction. Despite her technical expertise Claire Cyrus in Jennifer Rohn's novel *The Honest Look* (2010) is naive to the point of being immature. Gina Kraymer in Rohn's *Experimental Heart* (2009), although an outstanding researcher, is similarly ingenuous, prone to trusting the wrong people and repeatedly finding herself in physical danger before being rescued by a male colleague.

Physical danger resulting from the superior strength of male colleagues can also be a significant issue, especially for a woman engaged in fieldwork in a remote location. Field stations can be dangerous places, especially for women, as we know from the work of Jane Goodall in Tanzania and Dian Fossey, murdered, presumably by poachers, in her Rwandan camp. In William Boyd's *Brazzaville Beach* (1990) Hope Clearwater has no forewarning of the dangers involved in the research position she undertakes to study chimpanzees at an African research station, Grosso Arvoro, directed by Eugene Mallabar. Hope soon finds that her observations of the violent relations between the chimpanzees, including infanticide, cannibalism, and gratuitous torture, directly oppose the image of benign behavior on which Mallabar has built his international reputation through publications such as *The Peaceful*

Primate. When she tells Mallabar of her observations, he systematically sets out to destroy first her field notes, then her confidence, before orchestrating an attempt on her life. When forced to observe the chimps fighting and willfully killing each other, Mallabar violently attacks Hope's person in actions that mimic those of the chimps:

> He hurled a punch, full force, at my open face. If he had connected he would have broken my nose.... But somehow I managed to jerk my head away and down and the punch hit me on the shoulder.... I fell heavily to the ground. ... Mallabar was some distance off, scrabbling in the undergrowth looking for something.... He stood up, he had a stick in his left hand. He ran over towards me [and] hit me across the back. The stick broke under the blow. He grabbed me and I pushed wildly at his face, scratching.[20]

Women scientists also encounter a range of gender-specific risks and problems in fiction, in film, and arguably in life, as they endeavor to accommodate themselves to life choices entailing two diverse sets of roles and expectations.

Professionalism and Femininity

Weingart has noted that women scientists in films are younger for their professional rank than would be expected in real life and are usually sexually attractive, and Steinke found that in twenty-three films featuring female scientists eighteen were depicted throughout as attractive (physically fit, stylishly dressed, and with fashionable hairstyles) or as having become so after some transformational event,[21] as Ellie Arroway does when she appears in an evening dress for a reception in the film *Contact* (1997). The characters themselves are frequently torn between professional aspirations and the gender roles expected by society that locate them firmly in the domestic sphere.[22] Flicker argues that "the portrayal of women scientists that is oriented on their deficiency—either not a 'real' woman or not a 'proper' scientist—contributes to the formation of myths about women scientists' lack of competence and therefore also to women's experience of social discrimination."[23] Lori Kendall has suggested that "nerdism" in both genders is seen as diminishing sexual attractiveness but that, whereas in male scientists "this is compensated for by the relatively masculine values attached to intelligence and computer skills, in women a lack of sexual attraction is a far greater sin"[24] because it strikes at the perceived feminine role.

In terms of professional capability, older films and novels almost without exception depicted women scientists as subservient to men, either as assistants, as in Flicker's categorization, or as needing to be advised and rescued by male colleagues. Even in the 1996 film about cold fusion *Chain Reaction*, the female physicist Lily Sinclair, supposedly the senior scientist on the team, is unable to make the process work until her lab assistant Eddie Kasalivich discovers the right frequencies. When they are subsequently framed for an explosion in the lab and the murder of one of their colleagues, it is Eddie who contrives their escape and final rescue. More surprisingly, in the lab lit novels of Jennifer Rohn, herself a cell biologist, her female scientists stand in need of repeated rescue by kindly male colleagues from dangers precipitated by their own foolish judgements. In Rohn's first novel, *Experimental Heart* (2008), Gina Kraymer, a beautiful and brilliant young researcher, whose conference presentations on gene therapy are highly acclaimed, appears to have little intelligence outside the lab. Flattered by the attentions of a senior researcher, Richard Rouyle, she becomes his lover without enquiring into his reputation. She is several times rescued by Andy O'Hara, an honorable researcher in a neighboring lab, first from animal rights demonstrators who threaten her for using mice in her experiments, and then from Rouyle, who has kidnapped her to Germany.

In Rohn's second novel, *The Honest Look* (2010), Claire Cyrus, perhaps more technician than scientist, is an expert in her (narrow) field of understanding and operating the Interactrex, a sophisticated, state-of-the-art sampling and analysis machine that, it is hoped, will monitor and vindicate a new technique for curing Alzheimer's disease. Unfortunately, in her overriding desire to be appreciated in the initially hostile atmosphere of the laboratory in which she is employed, and more particularly because of her misjudged romantic entanglement with a wealthy, womanizing director of the company, she makes foolish decisions about nondisclosure of results and has finally to be "rescued" from the serious repercussions of her deceit by a series of male colleagues, including the avuncular head of the research institute.

However, such helpless and dependent females are rare in more recent films and novels. An interesting study of an early-career scientist is Jeannette, an astronomer in Pippa Goldschmidt's *The Falling Sky* (2013). After completing an uncontroversial PhD thesis, Jeannette secures observing time at a high-altitude observatory in Chile, where she and fellow astronomer Maggie observe an extraordinary phenomenon—two galaxies apparently connected in a way that appears to overturn the accepted Big Bang theory. A rerun of the observation confirms the finding that theoretically remains irreconcilable with the Big Bang. As Thomas Kuhn has shown, new scientific ideas are not readily accepted if they threaten the prevailing paradigm,[25] and Maggie prefers to ignore the observation as too controversial. But Jeannette, having considered and excluded alternate explanations (Is the link an aberration? Is it a superimposition of features of the two galaxies?), decides to honor the experimental results by publishing them but without specifically pointing out the inconsistency with the prevailing paradigm. Inevitably, however, other astronomers realize this and the media are intent on setting up a confrontation between alternate theories. Jeannette is horrified to find herself on a TV program with an amateur astronomer, a Big Bang denier, who is jubilant at having now secured, as he thinks, professional support for his position. She herself desperately wants to vindicate the Big Bang, an elegant and beautifully simple explanation for the origin of the universe, while simultaneously realizing the need to capitalize on her temporary celebrity to secure a tenured position. Caught between these forces, she struggles to maintain her integrity. Jeannette has numerous problems in her personal life as well as in astronomy but apparently keeps them separate. Her lesbian relationships do not impinge on her professional life; rather, the latter offers an escape from inconclusive personal issues with her parents and female partners. Yet we never feel that she is in control of her career. *The Falling Sky* offers an unusual insight into the roller-coaster of a research career that nevertheless, overall, offers a bulwark against the even more unpredictable private sphere.

In twenty-first-century fiction female scientists are increasingly shown as not only fully competent researchers but as leaders of projects like Sylvia Orloff in *Carbon Dreams*; Magritte Valorious in *Experimental Heart*; Dr. Grace Augustine, the head of the Avatar Program in the film *Avatar* (2009); and directors of labs, such as Marion Mendelssohn in Allegra Goodman's novel *Intuition* (2006). In cinema, however, romance is an obligatory visual component, and women scientists do not escape the expectation that sexual relations will be a major focus of their lives. They therefore have to be feminized, either by emphasizing their idealism and their moral courage, or by sexualizing or infantilizing them. Kerstin Bergman, who analyzed a sample of three well-known fictional women scientists in fiction, television, and film found that the women were "skilled experts in their fields, appreciated and respected by their peers, and making essential contributions to the solving of crimes. Nevertheless, they are simultaneously treated like children, as well as objects of sexual desire, by their co-workers." This "infantilization and sexualization signifies that the world of science is still far from gender equal."[26] Her sample scientists are nuclear physicist Christmas Jones in the James Bond film *The World Is Not Enough* (1999); forensic anthropologist Temperance Brennan in Kathy Reich's novel *Devil Bones* (2008) and TV series *Bones* (2005); and Abby Sciuto, forensic scientist in the CBS TV drama series *NCIS* (2003–). Although allegedly a physicist, Jones conforms closely to the stereotypical female role in James Bond films, being sexually provocative while posing as naive, innocent, scared, and cautious. "She seems like a lost child, quietly watching the grown-up world with awe-stricken eyes, and she is easy prey for the more experienced Bond," who rescues her twice, disabling an atomic bomb in the process, drags her along by the hand, and predictably takes her off to bed.[27] Brennan indicates her conventional femininity by displays of erratic behavior, including tantrums, and is protected and indulged by her male colleagues; her personality seems to have little connection with her scientific role, which is confined to delivering lengthy explanations of forensic anthropology.

More interesting is Abby Sciuto, who is highly regarded by her colleagues, both male and female, including her boss, Jethro Gibbs, for her forensic skills and her astute intelligence in solving problems. However, to conform to the feminine role, Abby is also endearing and elicits protection from her colleagues in the manner of a younger sister. She is often naive and overtrusting, precipitating danger from which she has to be rescued. Abby is also sexually provocative in an apparently unintentional way, a combination of Goth and schoolgirl, unconsciously arousing sexual fantasies. More interesting, and an integral part of her widespread appeal to viewers, is her essential pleasure in her job, in solving forensic problems. She is never the downcast female, struggling to perform in a man's world; she assumes her equal right to respect, and she has a life outside the lab where her humanitarian values are exercised. As such she stands at the opposite end of the spectrum from her literary ancestor the obsessive, secretive, morally dubious alchemist.

Bergman concludes that all three crime fiction scientists are portrayed as sexualized girls rather than women, standing in need of protection from their male colleagues. While respected for their intellect, they remain objects of patronizing strategies, "both in the embodiment of male fantasies and through their role as ersatz children in need of protection,"[28] and, except in the case of Sciuto, their role as stereotypical women overshadows their role as scientists.

However, there are also women scientists in fiction and film who are passionate about their research. A prime example is Dr. Eleanor (Ellie) Arroway of *Contact* (1997). The film itself is atypical, not merely as a fiction film that includes major elements of futuristic science fiction, but more importantly because Carl Sagan himself supervised the presentation of the science to include his own philosophy about extraterrestrial intelligence. There is no "dumbing down" of the science, though explanations may be delivered obliquely rather than as a lecture. In most cases it is Ellie who explains, sometimes passionately, sometimes scathingly, as when a visiting US senator asks why the extraterrestrial intelligence (ETI) message used prime numbers rather than English to make contact. For Ellie, a committed astrophysicist, the professional objective is paramount, and in the early part of the film she argues and storms her way through objections and rejections, passionately pursing her SETI (Search for Extra-Terrestrial Intelligence) goals when all her colleagues believe she is wasting her time and talent as an astronomer. Denied a continuation of government funding, she spends a year acquiring independent financial support for her project. Yet at other times Ellie responds like the "typical emotional female." She avoids answering when theologian Palmer Joss asks her point-blank at the review meeting if she believes in God and is tearful with frustration when, under attack at the Senate hearings, she is unable to prove satisfactorily that the ETI signal was not a hoax perpetrated by the wealthy Hadden Corporation or that the wormhole encounter was not a delusion, or to account for the "lost" hours when she claims to have traveled to Vega.[29] Most disappointingly, when Palmer Joss tells her that he had deliberately thrown the "God question" at her in order to destroy her chances of selection for the first mission because he did not want to lose her, she does not react angrily to this devious sabotage of her life's goal of searching for ETI; instead she looks tenderly and moist-eyed at him, grateful for his concern. These scenes subvert the intended impression of a strong and dedicated scientist, passionate about making the first extraterrestrial contact. There is also the niggling question of whether Ellie's real desire is to make contact, not with some disembodied ETI, but with her dead father. As Larry Klaes observes, "This is one of the issues I had with *Contact*: Is the film really about searching for ETI, or a woman's quest to regain her father and lost childhood? Granted this event makes Ellie a more three-dimensional character than most of the other roles in the film, but I can equally see the general audience thinking that SETI is not so much about finding what is out there than what is in the human psyche."[30]

Romance and Motherhood

While many of the films featuring female scientists from 1991–2001 emphasized appearance and sexuality and focused on romance, a significant number presented women in professional positions of high status, rather than as assistants or early-career scientists.[31] However, this promotion involves serious difficulties for women. Sandra Hanson writes, "A critical element in the culture of science occupations involves ideas about having to be wedded to one's work—making it difficult for women with families (spouses and/or children), but not men with families, to succeed."[32]

As in real life, fictional women scientists often have to choose between career and romance or motherhood. Both marriage and motherhood are seen as a major hindrance to a career in science, and in the majority of twenty-first-century novels featuring female scientists, most by female authors, these options are rejected or, if they continue to be a possibility, then the career in science is aborted. Claire Cyrus in *The Honest Look* leaves the organization where she has suffered so much stress to write poetry, supported by her wealthy scientist lover.

Steinke concluded that female scientists and engineers in positions of high professional status rarely compromised their career for romance, and this is consistent with the situation in science novels, especially those written by women. At the end of *Carbon Dreams* by Susan Gaines, Tina Arenas accepts the offer of a position at a prestigious research establishment on the other side of the continent, knowing it will mean permanently parting from her gardener-lover. In Goodman's *Intuition* Robin, having realized there is no future in her previous relationship with Cliff, pursues her science career in a new institution.

In *Contact* Ellie Arroway rejects a relationship with theologian Palmer Joss partly because he tries to dissuade her from her search for extraterrestrial life and to make her believe in God, but primarily because she perceives him as a diversion from her main aim. Not only does he not believe in her passionate search for ETI, but he is philosophically incapable of ever doing so: his mind is closed to such a possibility because of its theological implications. If at the end of the film it seems possible that they may renew their relationship, it is on her terms—Joss concedes that she may indeed have had a spiritual experience during the eighteen hours for which the video unit in the spaceship recorded only static. Regardless of this, Ellie continues with her research and public relations, encouraging children to embark on a career in science by honing their curiosity and demanding answers to their questions.

If romance is a distraction, motherhood has a far more severe impact on women's careers in science and is hardly ever an option for female scientists in fiction or film. A notable exception is Esme Charbonneau-Tallich, in Carol Muske-Dukes's novel *Saving St. Germ*. Esme is an organic chemist who, somewhat against her instincts, marries and gives birth to a daughter, Ollie. Shortly afterward Esme and her husband Jay disagree about how to handle the baby, and in time it becomes apparent that Ollie is unusual in her behavior and speech. However, Esme forms a unique bond with her autistic five-year-old daughter, who observes the world from within a cardboard box decorated to resemble a TV and makes seemingly non-sequitur comments that Esme interprets as observations of preternatural wisdom. Her passionate assertion of Ollie's intelligence despite the diagnosis of psychologists and educators alienates Jay and eventually leads to a separation and Esme's loss of custody of Ollie. Closely connected to this personal tumult is her neglect of her teaching duties at her university and her impulsive decision to drop the "flashy" research project on DNA for which the university had hired her to focus on an elusive and wholly abstract Theory of Everything. She loses her university post, along with her husband, daughter, research assistant, and students. Although all are restored to her in the conclusion when her belief in Ollie's gifted intelligence is vindicated, her theory of everything is copublished, and her husband realizes he cannot manage Ollie without her, the main body of the novel depicts the immense difficulty for a woman, albeit an erratic one, of combining science research with family commitments.

Seduction

Women scientists, however dedicated to their research, are not immune to seduction by senior colleagues. Even when they consent to a sexual involvement they are usually unaware of what will be entailed, how far their independence will be eroded, and their work impeded.

In *The Honest Look* Claire Cyrus, rejected by her colleagues in her new position at the biotech startup company NeuroSys, is lonely, vulnerable, and easy prey for the flattering advances of the wealthy Alan Fallengale, one of the directors, for whom she represents only another casual affair. Later she realizes that he assumes ownership of her person, her time, her interest, her thoughts. Yet she accepts the luxury he supplies, even though she feels uncomfortable at the deception she is simultaneously perpetrating against him and the company.

In A. S. Byatt's *A Whistling Woman* (2004) biologist Jacqueline Winwar is fascinated by her research into snail populations as indicators of the genetics versus nurture debate, but first she has to gain a postdoctoral post in a laboratory headed by Lyon Bowman, a known womanizer, who is dismissive of women scientists. "Obsessive women make bad members of teams," he tells her in response to her research proposal. Once in his employ Jacqueline is propositioned by him during their travel to a conference and acquiesces, thinking she is better off taking "the line of least resistance," but also experiencing a certain ambivalence in her own mind: she finds Bowman "at once repellent and irresistible."[33] She is a hostage of her own biological urges, desiring this sexual encounter out of lust, not emotion. In this sense she is using Bowman to promote her career and to assuage her sexual urge as much as he is making use of her.

Gina Kraymer, the seemingly brilliant young scientist in Jennifer Rohn's *Experimental Heart*, has a string of impressive publications and a strong ethical conviction that research should benefit people, particularly the poor and neglected people of Africa. Yet she allows herself to be drawn into a collaboration with Richard Rouyle, a senior scientist in the Geniaxis company for which they both work and to sacrifice her ethical principles by proceeding prematurely to clinical trials in Africa before preliminary tests have been conducted on animals. She also allows herself to become sexually involved with Rouyle, who proves a violent and controlling lover, kidnapping and imprisoning her when she protests. She later explains the hold he had over her sexually and professionally: "He was very charismatic and had power over me, and eventually I couldn't seem to help myself. . . . He had this *force*—it was like I couldn't think straight . . . he just wanted to *subjugate* me. . . . He needed absolute psychological control."[34] In these examples the women are dependent, professionally, on the men who exploit them. That they initially find them sexually attractive is partly a response to power, self-assurance, and the flattery of being chosen, but the situation is more complex: Jacqueline Winwar, Claire Cyrus, and Gina Kraymer are complicit in their seduction and not without a degree of awareness of using their sexuality to achieve some ulterior professional goal.

Discrimination and Plagiarism

Although women scientists are increasingly depicted in senior research positions, they are not, therefore, immune from discrimination, competition, and plagiarism of their work by male colleagues. Their credentials are challenged and criticized and their research ridiculed and dismissed for being unconventional; financial support or research facilities may be withdrawn; yet when they do achieve significant results, their male colleagues quickly step in to claim credit for those same discoveries.

In Marge Piercy's *Small Changes* (1972) Miriam Berg, a PhD student in computer science, has an inspired thought about how to solve a long-standing problem in mathematics. However, when she attempts to make a presentation at a seminar, "she could not even complete setting out her idea before the attacks began. She seemed to have hit a bare nerve in almost everyone. . . . Fred was hardly able to be polite. The idea seemed to affect him personally. . . . Ted . . . said how he thought it was wonderfully fascinating to presume to solve in half an hour what was in essence an insoluble problem. . . . He had been sure the idea was exciting, but the excitement had been fury."[35]

When Claire Cyrus arrives in Amsterdam to take up a senior research position in a start-up firm, she immediately feels the hostility of most of her colleagues, who query the large sum spent by the company on acquiring the Interactrex analyzer at the cost of other projects, and in particular her appointment at such a senior level when, in their eyes, she is merely a technician. Claire takes to hiding away from her colleagues, working mostly at night or arriving early and immersing herself in her work.

In the 1992 film *Medicine Man* Dr. Rae Crane, an established research botanist with a strong reputation, arrives in the Amazonian rain forest bringing supplies and equipment to the brilliant but chauvinist Dr. Robert Campbell, who believes he has found a cure for cancer in an endemic bromeliad but has been unable to synthesize the compound chemically. Annoyed that the expected male assistant has not come, Campbell tells her to go home, questions the extent of her field experience, and orders her around when she is working in the improvised lab. Crane vigorously defends herself, reciting her credentials: "I'm published, and more extensively than Dr. Sealove. I hold degrees from CCNY, Berkeley, and Cambridge. I'm the recipient of the Thurman Award in '82 and '86, the first and only time it's ever been given to the same person twice." But the sexist Campbell is unimpressed, and Crane frequently feels the need to remind him: "I'm not a nineteen-year-old working up a résumé. And I don't expect to be treated as if I were." Only when she learns from the local medicine man that the cure is not derived from the plant but from the ants that live in it does Campbell invite her to continue working with him as codiscoverer of the real source of the cure.

One of the clearest enactments of discrimination and professional jealousy appears in *Contact*. Ellie Arroway is not only a creative and scrupulous researcher but also very competent at gaining funding for, and supervising, a large project. Other scientists acknowledge her expertise, her intelligence, and her competence, but they dismiss as foolish her passion for intercepting signals from extraterrestrial life. In particular Drumlin, the administrator of the National Science Foundation grants, not only cuts the funding for the SETI project, which he considers a waste of time and resources, but publicly ridicules Ellie before other scientists. Yet when she does finally intercept an intelligent radio transmission from space, he quickly steps forward to take credit for her discovery, overrides her in meetings, and answers for her in press interviews. Effectively he plagiarizes her work, relegating her to the role of his assistant, and ensures that he, rather than she, is selected to pilot the spaceship she had designed. This first mission is destroyed by a terrorist bomb. When Ellie eventually receives funding from Hadden Enterprises to build her own spaceship and returns from the mission affirming that she has traveled to Vega and encountered an extraterrestrial intelligence, she is declared delusional until it is revealed that the equipment recorded eighteen hours of static during a time frame that was only a few seconds in terrestrial time. Carol Colatrella has observed that while Carl Sagan's novel (1985) pinpoints "Ellie's gender as a central problem in finding respect for her research, and it acknowledges the difficulty of being a woman in the male-dominated and hostile world of physics," the film version "represents Ellie's marginality as based on her pursuit of 'little green men' rather than on her gender."[36]

Plagiarism is a particular hazard for female scientists who are too far down the hierarchy to demand their rights. In *Experimental Heart* Rouyle appropriates Gina's research for his own purposes. In *Batman and Robin* the male scientist Dr. Woodrue steals Pamela Isley's venom samples to use in his own research into a scheme for world domination. In *Carbon Dreams* Tina Arenas, a geochemist, has devised a method for determining paleotemperatures from marine sediments and thus the ancient climate and the ecology of the oceans—valuable evidence for use in climate change research. But Arenas has her research "stolen" by her own former technician, who leaves her group to work on a project that was started in her lab,[37] and by Sylvia Orloff, who has apparently subverted Tina's request for permanent lab space while incorporating Tina's research into her own paper. In *Brazzaville Beach* the project director, Eugene Mallabar, having tried repeatedly to destroy Hope Clearwater's seminal research, finally incorporates her data into *Primate: The Society of the Great Ape*, his heavily revised volume on behavior of chimpanzees, acknowledging it only in a footnote.

In the past twenty years there has been a marked shift from the traditional depiction of female scientists as subservient assistants to a prominent male figure to their depiction as protagonists with significant research capabilities and, in some cases, a passionate dedication to a project. However, while such portrayals are sympathetic and supportive of these female scientists' goals, they nearly all indicate, even focus on, significant gender differences in the form of risks (seduction, denigration, criticism of their credentials, intimidation, physical violence, manipulation of results, sexist comments, and cultural inequality) and constraints that are not present for male colleagues. Few of these fictional female scientists have continuing relationships, and even fewer combine motherhood with their research: the problems of balancing work and family needs, difficult in life, seem altogether too complex for novelists and filmmakers to contemplate. Unfortunately such imaginary is self-perpetuating, since these images of female scientists in fiction and popular films reinforce cultural and social assumptions. Although these representative examples affirm the capabilities of women to conduct valuable, even cutting-edge research in science and to occupy professional positions of high status, by problematizing the extraprofessional issues besetting women, they indicate, and arguably serve to prolong, a lack of female agency within the culture of science.

CHAPTER 18

Idealism and Conscience

> He was a physicist as Pascal had been; but like Pascal he was also a mystic and a moralist. . . . I saw in him the priest of a new theology.
> —*Heinrich Schirmbeck*

The preceding four chapters have sufficiently indicated the low esteem in which scientists were held by most twentieth-century writers with a background in the humanities. With the exception of the simplified heroes of science fiction, the dominant images of scientists recapitulated the unflattering stereotypes of earlier centuries—the evil scientist, the stupid scientist, the inhuman scientist—or, as a peculiarly twentieth-century contribution, the scientist who has lost control over his discovery. This vote of no confidence by writers was not, however, unanimous. There were some important studies by eminent writers of fictional scientists engaged in a struggle to defend their ideals against the debased values of society or some powerful faction within it. While such figures were rarely, if ever, wholly successful in their quest (such optimism is hardly an option for contemporary writers), in many cases they were used to explore possibilities of improved communication, to offer an example to actual scientists, and in some cases to propose to the wider community an enlarged vision of possibilities beyond the narrow economic and political exploitation of science.

Dedication to Science

Beginning in the 1890s, the novelists and playwrights whose protagonists were idealistic fictional scientists came from a wide variety of backgrounds. Émile Zola had no direct scientific or medical training but was one of the leading figures of French naturalism, a literary movement that attempted to impose scientific procedures and criteria on literature. In his novels *Doctor Pascal* (1893) and *Paris* (1898) the scientifically trained protagonists struggle against the mental oppression, superstition, and violence engendered by the Catholic Church, which Zola portrays as the villainous opponent of science, reason, and humanity. Doctor Pascal has spent a lifetime correlating information about insanity and heredity in generations of his own family and produced scientific data of inestimable value to the world; but his mother, aided by a superstitious, priest-ridden servant, is determined to destroy his work lest it disgrace the family. In Zola's view there is no doubt that the idealistic Pascal, representing medical science, stands for life and health, while the way of religion is the way of death. His novel retains a contemporary relevance, for what Pascal envisaged through his observations of heredity and proposed eugenics is now more immediately accessible through genetic engineering. "If one could only understand it [heredity], master it and make it do one's bidding, one could remake the world at will. . . . No more illness, no more suffering, to be able to limit the causes of death! . . . A new world of perfection and felicity would be hastened by active preventive measures which would make everyone healthy. When all were healthy, strong and intelligent, the human race would be a superior race, infinitely wise and happy."[1] Zola presents this view without irony, yet it resonates strongly with the aims and assumptions of Frankenstein and with the Nazi ideal of a master race.

Zola's other scientist hero, Bertheroy, the celebrated chemist of his novel *Paris*, also affirms his belief that "human happiness can only spring from the furnace of the scientist," that "one single step of science brings humanity nearer to the goal of truth and justice than do a hundred years of politics and social revolt."[2] Bertheroy's research has been in the field of explosives, but his ultimate purpose is not destructive. On the contrary, his dream is to empower society. This dream is realized when his anarchist colleague Guillaume, who had planned to blow up the newly completed Sacré Coeur Basilica at the moment of its consecration, diverts his explosive from purposes of destruction to powering a motor designed by his son. Pointing to the motor, Bertheroy reaffirms his faith in the constructive impact of science as opposed to violence and lack of reason: "That is revolution, the true, the only revolution. . . . It is not by destroying but by creating that you have just done the work of a revolutionary" (485–86).

In the early decades of the twentieth century, the principles espoused by idealized fictional scientists were closely linked to the pursuit of the discipline itself. They embodied a tireless devotion to science, which was usually regarded as being ipso facto for the benefit of society, and as indicating a renunciation of self-interest and personal gain. This idealism was usually expressed in relation to the lifestyle associated with scientists and the way this was evaluated. Long hours given over to research were rated far above social obligations, unless performed by a woman, and were taken as evidence of a noble self-sacrifice on the part of scientists. When the Nobel Prize–winning physicist Robert Millikan asserted that scientists offered an example of superhuman moral probity that religion and society would do well to emulate, he was taken seriously.[3]

During the 1930s this view of scientists was epitomized in the popular perception of Marie and Pierre Curie, who were regarded as exemplary in making the supreme sacrifice for science (see fig. 23). Film versions of their story, notably *Madame Curie* (1943) starring Greer Garson, carefully preempted criticism of Marie Curie as cold and unfeminine by emphasizing her romantic relationship with Pierre and by linking her career in some unspecified way with a passionate Polish nationalism.[4]

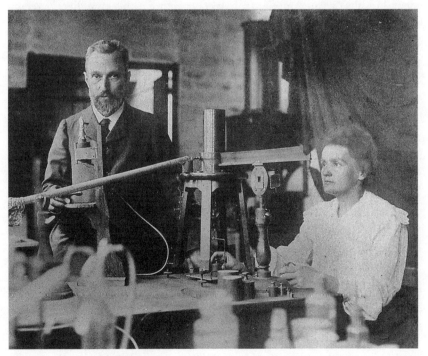

Figure 23. Pierre and Marie Curie in their laboratory, ca. 1904. Wikimedia Commons.

In his novel *Arrowsmith* (1925) the American writer Sinclair Lewis produced a full-length portrait of a doctor involved in bacteriological research. Although Lewis himself had no scientific training, he collaborated in the writing of this novel with a young bacteriologist, Paul de Kruif, and it is to de Kruif's experience that the novel owes its authentic depiction of the laboratory scenes in research institutes and the professional problems and lifestyle of the medical researchers. The major characters are recognizably based on scientists whom de Kruif had known, but most are simplified in order to emphasize the two principal scientists, both idealists in different ways.[5]

Through the charismatic figure of Max Gottlieb, the German pathologist who inspires Arrowsmith to pursue medical research, Lewis depicts vividly the stress and loneliness of the researcher, his isolation from family and social contact as he works through the night until his health breaks down and he suffers temporary neurasthenia from the strain of overwork. Lewis also shows how scientific success brings liabilities as acute as failure. The institute, which needs to publicize its new medical discoveries in order to attract funding for research, pressures Arrowsmith to publish his startling experimental results on pneumonia staphylococci before rechecking them, a recurrent situation when prestige and subsequent large-scale funding are at stake. Remaining true to Gottlieb's principles, Arrowsmith rejects the demand to publish before he is certain, but while he is painstakingly repeating his experiments, a French scientist publishes a description of just such a bacteriophage, thereby rendering Arrowsmith's work effectively redundant.

Although Arrowsmith teeters between devotion to medical research as an end in itself and humanitarian considerations,[6] no such compromises are made by his purist mentor Gottlieb, who enunciates his creed as follows: "To be a scientist—it is not just a different job.... It is ... like mysticism, or wanting to write poetry; it makes its victims all very different from the good normal man.... The scientist is intensely religious—he is so religious that he will not accept quarter-truths because they are an insult to his faith.... He must be heartless.... To be a scientist is like being a Goethe: it is born in you."[7]

Gottlieb is a reflection of de Kruif's deep personal admiration for Jacques Loeb, one of the most vigorous twentieth-century proponents of biological mechanism. Loeb, the pure scientist, despised the methodology of medical research in the 1920s, particularly in America, and Gottlieb reflects this. However, he also represents the spirit of German science in general, which at this time held a special allure for Americans. Gottlieb's devotion to an abstract ideal of scientific research rather than its practical potential and his sacrifice of humanitarian interests in the cause of science might well have suggested at least an ambivalent estimate of such idealism as deleterious to human relationships. But one interesting aspect of Lewis's treatment is the confident transformation of such a character from the unfeeling villain of Romantic literature to heroic status. In fact, the ebb and flow of Arrowsmith's devotion to Gottlieb is used by Lewis as a signifier of his moral worth. His successive battles against the cheap popularization of science, the wealth-oriented research of the fashionable clinic, and the temptation to assume administrative power all dignify his cause, but Lewis (and de Kruif) go further and insist that his final action of leaving his wife and child for a life of monastic research is no less an aspect of his heroism.[8] Gottlieb tells him that "in this vale of tears there is nothing certain but the quantitative method" (36). Humanitarian causes and human emotions are evidently a luxury with which the scientist has no business.

Science and Social Conscience

The contrary view, namely, that social and moral responsibility should take precedence over the drive for research, was argued by H. G. Wells in a number of novels with a utopian subtext. *Marriage* (1912) is a fine study of an idealistic scientist who strives to apply the principles of science in the cause of social reform. The protagonist, Trafford, a crystallographer, is impatient, as all Wells's ideal scientists are, with accepted formulations and theories; he insists on openness and publication rather than secrecy in research, rejecting the ethics of expediency and secrecy practiced by industry. But although he bargains vigorously with company directors for the right to publish, Trafford is also the first of Wells's "mystical" scientists. Not only does he pursue his research into crystals with a religious devotion and articulate it in the language of mysticism,[9] but, like most of Wells's idealists, he rejects the social systems of his time for their failure to engage in "ultimate truth." As he struggles to find a solution to the futility and waste that characterize the society of his time, he concedes that his devotion to science for its own sake is a form of selfishness and vanity. Like Wells himself, he finally leaves his career in colloid chemistry and crystallography to devote his scientific expertise to community reforms, which are implicitly presented as the sociological equivalent of his research.

In Holsten, the atomic physicist of Wells's later novel *The World Set Free* (1914), who discovers how to construct an atomic bomb (a term that Wells coined in this novel), we have the prototype of a recurrent figure in twentieth-century fiction, the scientist who has access to the power that could annihilate civilization and agonizes over where his ethical priorities lie. His scientific principles urge him to publish his results openly, while his awareness of the moral inadequacy of society to cope with such knowledge counsels suppression of those results. Unlike most of his successors, Holsten decides to withhold his results, thereby retarding research in the interest of humanity and avoiding the Pandora's-box problem that confronted the scientists of the preceding chapter. *The World Set Free* proved to be prophetic, not just in predicting, some twenty years before it actually happened, the splitting of the atom and the joint possibilities of atomic energy and atomic weapons,[10] but equally in the description of Holsten's inner struggle.

Integrity versus Nationalism

Bacon's belief that the open exchange of knowledge would foster a spirit of internationalism and avert war, given new emphasis in the first three decades of the century by Wells's utopian writings, was subscribed to by many real-life scientists of the time. Indeed, the scientific world of the 1920s was the closest to an international community of scientists that has yet been seen. Figures such as Ernest Rutherford, James Franck, Niels Bohr, Max Born, Werner Heisenberg, and Peter Kapitza,[11] ignoring the contemporary sociopolitical obsession with the nation-state, remained loyal to an international ideal of science. This brief golden age of internationalism almost certainly contributed to the doubts expressed by many scientists, both real and fictional, about the development of nuclear weapons. It was, however, not strong enough to overcome the nationalist fervor generated by the propaganda machines of the Second World War.

During the 1930s the scientist hero, when depicted at all, was most frequently shown defending ideals that were in conflict with those of his society, but although he might retain a moral advantage, he was rarely seen as achieving a significant victory over the authorities. In the Germany of this period it became increasingly difficult to write anything overtly critical of the government; hence writers commonly resorted to historical parallels to avoid censorship of their contemporary message. Rehberg's play *Johannes Kepler* (1933); Brecht's *Das Experiment* (1939), "Der Mantel des Ketzers" (1939), and *Leben des Galilei* (1938–39); Brod's novel *Galilei in Gefangenschaft* (1948); and Zwillinger's play *Galileo Galilei* (1953) are among the best known of these pseudohistorical treatments of a contemporary issue, the role and responsibility of the scientist in an authoritarian state where society is ignorant of the implications of his work. In the majority of these works the scientist protagonist is, as we have seen in chapters 11 and 16, too flawed to be regarded as heroic, but in *Das Experiment* (1939), "Der Mantel des Ketzers" (1939),[12] and the first version of his *Leben des Galilei*, Brecht portrays his scientists as morally superior to the authorities of their time, whether religious or secular.

Of these examples, *Das Experiment*, based on an idealized reading of Sir Francis Bacon, is the most optimistic, looking forward to a new age when science will not have to contend with reactionary authorities, represented here by the church. Brecht makes use of the traditional story about an incident said to have caused Bacon's death from pneumonia, suggesting the added aura of martyrdom in the cause of science. The experiment of the title, which involves freezing a dead hen by filling it with snow to preserve it, symbolizes the practical application of science. Brecht suggests not only a moral but also a temporal victory for these Marxist ideals, for although Bacon dies before the experiment is concluded, his disciple, a simple stable boy, carries on his master's dedication to socially beneficial science.

Brecht's *Leben des Galilei* is the best known of the works based on a historical scientist. Brecht was writing the first version in 1938–39 in Denmark, where he was living in exile from Nazi Germany because of his communist sympathies, when news of the splitting of the uranium atom by Otto Hahn reached him. He was thereupon moved to dramatize, in the life of Galileo, the problems confronting German scientists, especially physicists, who were being required to work on the construction of the atomic bomb. This first version of the play is a strongly anti-fascist statement, essentially positive in its attitude toward scientists. Brecht himself said that Galileo is portrayed as "an intellectually heroic figure who fights for progress, who deliberately introduces a new age of scientific truth."[13] The main thrust of the drama is the heroic struggle of scientists against authority, symbolized in the play by the Catholic Church. Brecht was concerned, however, to emphasize that the church was only one of many authorities, its twentieth-century counterpart being, obviously, totalitarian governments. He wrote: "The Church functions ... simply as Authority. ... It would be highly dangerous, particularly nowadays, to treat a matter like Galileo's fight for freedom of research as a religious one; for thereby attention would be most unhappily deflected from present-day reactionary authorities of a totally unecclesiastical kind" (1:342). Thus Galileo becomes Brecht's symbol for the noble scientists trapped in Nazi Germany and forced to work on weapons of mass destruction. Temporarily he succumbs to the threat of torture and publicly denies what he knows to be scientific truth (in the play, the Copernican theory; in Brecht's contemporary terms, knowledge of the consequences of an atomic bomb), but later he returns to his research in secret and smuggles his work across the frontier. Despite his other failings, the Galileo of this version is ultimately heroic in the cause of science. In order to continue his research he cheats; risks any danger, including the plague;[14] and recants solely in order to be able to finish his *Discorsi*. The play was not performed until 1943, in Zurich, and it was not produced in the United States until 1947. Ironically, very soon after this first US performance, Brecht was summoned before the House Un-American Activities Committee to answer charges that this play, like several of his earlier plays, had a Marxist bias.[15] Even more ironically, by that time Brecht had reworked the play, in response to the dropping of the atomic bomb on Hiroshima, with an entirely different and much less heroic Galileo, as discussed in chapter 14 above.

In Britain it was possible for writers to be more explicit than their German counterparts in their criticism of government policy, and for this purpose several used the theme of an ethical scientist attempting to preserve his discoveries from misuse by an unprincipled government-military complex. One of the earliest in the field was the dramatist Charles Morgan, who produced two plays exploring this issue of scientists fighting a moral battle against political and military interests. In *The Flashing Stream* (1942) Edward Ferrers, a mathematician working on the development of an aerial torpedo to destroy enemy aircraft, at first cares only about the mathematics of the project. An unusual and interesting aspect of this play is the fact that the only other mathematician capable of working on the problem is a woman, Karen Selby, whose view of science as being only one truth among many is contrasted favorably with Ferrers's strictly objective approach. Here the scientist hero is considerably qualified if not overshadowed by the scientist heroine, whose humanitarian perceptions carry authorial support. In a later play, *The Burning Glass* (1953), Morgan depicts another scientist, Christopher Terriford, who, having made a discovery of some strategic importance, is harassed by the government to pass over the sole rights to the military. Like Wells's Holsten, Terriford is torn between his duty as a scientist (to publish his results openly), his alleged patriotic duty (to divulge the secret only to his own country, England, and its ally America), and his duty as a moral individual (to save the world from the misuse of such power). "Science has never yet kept back its knowledge.... We have always said: 'Ah, but this power has beneficent uses as well. Let's go for it!' We have never yet said: 'We are unfit for it.' The time may have come to say that."[16] As the century progressed, this argument was heard with increasing frequency in relation to many areas of scientific research.

Morgan's protagonists finally achieve their ethical goals, remaining firm in the face of military pressure, and thus both these works end with a certain qualified optimism, but this is an unusual outcome. After the deployment of the atomic bomb at Hiroshima and Nagasaki, a weapon that so clearly owed its existence to the most brilliant physicists of the day, it became progressively more difficult to believe in the moral superiority of scientists and even more difficult to believe in their ability to initiate a new, peaceful society.

The security restrictions of the Second World War and the subsequent Cold War, coupled with the growing public perception of scientists as a country's most valuable resource in time of war, challenged as never before the Baconian tradition of open research and international cooperation in science. This situation elicited a spate of novels and plays about the crisis of conscience experienced by the idealistic scientist caught between two opposing ethical codes. In many of these works, the writers clearly use the scientist character to attack government security policy, which, through increasingly stringent demands for censorship, was affecting their own profession as well.

One of the first uncompromising treatments of this theme of the heroic scientist catapulted into conflict with the governmental edifice and branded a traitor in consequence was *The Big Secret* (1949), by American novelist Merle Colby. Her protagonist, Daniel Upstead, is an idealistic, small-town physics teacher who comes to Washington to hear an eminent physicist speak at a meeting of the National Physical Association. He is incensed to learn that the scientist has been prevented from speaking because his research, although theoretical in nature, is considered to be of military importance and thus to come under national security regulations. Like most of the idealistic scientists considered in this chapter, Upstead is an innocent in the evil world of politics, where atomic weapons represent the new power game and each governmental sector hopes to manipulate the situation for its own profit. To the authorities, scientists, with their ethic of universalism, are ideologically suspect, and there is a move afoot in Washington to push through a new executive order that would require government clearance on all scientific discoveries before publication. Upstead duly becomes involved in a one-person campaign against a network of political intrigue, maneuvering, and corruption. On behalf of the League for the Advancement of Basic Science, he presents a resolution to the White House affirming that new discoveries in basic theory should be freely and universally available to scientists everywhere and that the principle of secrecy should be confined to technical methods alone. It is not surprising that at a time when Oppenheimer and Einstein were campaigning against the continued development of atomic weapons Colby should choose a physicist as her champion of truth and justice and show him accused of un-American activities for just such a stance as Oppenheimer was to make. But she effectively strengthens her case by identifying her physicist, not with a brilliant researcher, to whom few readers could readily relate, but with the more popular image of a morally upright American school teacher. There is never any doubt for the reader that Colby's hero is innocent of the charge of treason, yet his ethic of universalism in science is arguably the same as that which motivated the atom spies during the Cold War.

Other writers of the period raised this controversial issue more explicitly and even dared to present as heroic characters who refused to put patriotism before humanity. This is the case in James Aldridge's novel *The Diplomat* (1949), in which Ivre MacGregor, a British micropaleontologist working in Iran, unwillingly becomes involved in the political machinations of governments during the Cold War. Caught between two extremes, represented on the one hand by the political expediency of his British superior, who tells him that science must serve political ends (nationalism, political face-saving, economic growth), and on the other by a fellow scientist who advocates complete isolation from public affairs, MacGregor's inclination is to opt out. However, he is recalled to a middle way of political responsibility compatible with his scientific principles by his former Iranian professor, Dr. Aqa (a rare example of a Middle Eastern scientist in literature), and exposes the true facts about the Iranian situation despite the threat of being prosecuted for treason.

Aldridge's deliberate rejection of the traditional image of the national hero who puts his country before all else represents a politically subversive and peculiarly twentieth-century treatment of the conflict between individual morality and allegiance to the state. It clearly parallels the almost exactly contemporaneous passing of secret information by the atom spies Allan Nunn May and Klaus Fuchs, although Aldridge could not have known about this when he was writing the novel.[17] Science was probably the only possible justification that could be advanced during the Cold War for rejecting the model of patriotism, and the fact that it could be advanced at all is a measure of the degree to which scientific idealism had been accepted in the first decades of the century.

The theme of betrayal of scientific secrets, usually for idealistic reasons, continued to be treated extensively and with many variations in the first two decades after the Second World War in both literature and film. Given the climate of opinion during the McCarthy period, it is the more significant that although in nearly every case the scientist protagonist is presented as idealistic, the authors' interpretation of what constitutes idealism differs considerably. Three examples from this period suggest the variations in political allegiance that the atom spy plot might be made to serve. In Herman Wouk's play *The Traitor* (1949) an American scientist decides to pass atomic secrets to the Russians—again, in order to secure world peace. He later changes his mind and helps to trap the very espionage ring he had been assisting, so that the title comes to have a double meaning. Although this might seem to damn the protagonist twice over, the authorial assumption is that the second betrayal negates the first, leaving the scientist a born-again patriot whose idealism had been temporarily misapplied but who has seen the error of his ways and been redeemed.

A more provocative situation is explored in Ruth Chatterton's complex novel *The Betrayers* (1953), in which a nuclear physicist is charged with having communist sympathies. Although he is successfully defended by a woman lawyer who is convinced of his idealism and innocence, the physicist has in fact been passing atomic secrets to the Soviets for some time, in the belief that peace can come only from an equal distribution of military power. Chatterton is more daring than most of her contemporaries in her choice of a hero, who is unambiguously a traitor by the political values of the day; but ultimately she can propose no satisfactory moral solution to the antagonism of nationalism and humanitarianism other than the physicist's suicide. Written a decade later, Mitchell Wilson's novel *Meeting at a Far Meridian* (1961) completely reverses the traditional pieties. Although Wilson's nominal protagonist is Nick Rennert, an American cosmic-ray physicist, his moral hero is the Russian atomic physicist Dimitri Gontscharow. Like MacGregor in *The Diplomat*, Rennert is presented with two role models: the cynical, nationalistic Henshel, a character modeled on Edward Teller, and the charismatic Gontscharow, confident, open, and life-affirming. Thus Mitchell's novel effectively vindicates the victims of the McCarthy purges.

Films of this period are more politically conservative in approach. The film industry was quick to capitalize on the paranoia generated by Klaus Fuchs's trial in 1950. *The Atomic City* (1952), *Walk a Crooked Mile* (1952), *Mr. Potts Goes to Moscow* (1953, also released as *Top Secret*), *Escape Route* (1952), *Kiss Me Deadly* (1955), *The Atomic Man* (1956), *The Amazing Transparent Man* (1960), and *Spies, Lies and the Superbomb* (TV miniseries, 2007) all featured nuclear espionage and pandered to the exaggerated level of fear promoted by the CIA and the FBI.

A more humorous but nonetheless serious treatment of the scientist's stand against government security policy is C. M. Kornbluth's short story "Gomez" (1955), in which a brilliant seventeen-year-old, self-educated mathematician, Julio Gomez, is compared to Wiener, Urey, Szilard, and Morrison, all scientists who suppressed knowledge that they knew could be used for atomic weapons. Gomez's brilliance in unified field theory is ripe for exploitation in the production of a new weapon that would confer absolute military power on the United States, and he is effectively imprisoned by the military to solve equations. For Gomez, this is at first merely an intellectual game for which he accepts no ethical responsibility, but his innate morality soon surfaces, and wiping his equations off the blackboard, he demands leave to get married. On his return, he claims to have completely forgotten his equations, and no efforts by the government and military agencies can force him to recall them. Kornbluth clearly implies that this is simply his way of defeating the powers that be, but the overtly humorous conclusion, in which Gomez returns to working in his back-street restaurant, suggests a more pessimistic implication, namely, that there is little hope for even a theoretical mathematician to preserve his moral integrity while pursuing a career in his discipline.

Although these characters carry authorial approval for their antigovernmental stance and in most cases retain their untarnished moral reputation throughout, they are nevertheless essentially powerless to change the political situation. In contrast to the increasing power ascribed to evil scientists before and during the war, the postwar period in general was marked by the rapid decline of the scientist as an effective, that is, an achieving, hero. Zuckmayer's play *Das Kalte Licht* (1955), which closely follows the case history of Klaus Fuchs, Schirmbeck's *Ärgert dich dein rechtes Auge* (1957), Dürrenmatt's *Die Physiker* (1962), and Kipphardt's semidocumentary play *In der Sache J. Robert Oppenheimer* (1964), discussed in chapters 13–15, chart the increasing helplessness of the scientist who tries to retain humanitarian ideals.

Of all these characters, Prince De Bary in Heinrich Schirmbeck's novel is the most outstanding scientist in moral terms. Yet even he has arrayed against him not only political and economic forces but the combined power of technology and scientific research institutions and, more insidiously, his own unwillingness to leave the artistic perfection of pure mathematics for the turmoil of social involvement. Schirmbeck described De Bary, one of literature's last great scientists, as "in some degree derived from the Duc de Broglie

... [although he] bears no resemblance of detail or of circumstance to that great man." Internationally famous for his treatise *Light and Matter*, he is also described in charismatic terms by the younger physicist Thomas Grey: "He was a physicist as Pascal had been; but like Pascal he was also a mystic and a moralist. I loved him deeply as I sat there listening entranced. . . . The big, clear eyes, the elegant restraint of his gestures and the controlled fervour of his voice all combined to invest him with an authority which imposed itself upon his audience with an increasingly mystical effect. I saw in him the priest of a new theology."[18]

But De Bary's very perceptiveness makes him deeply critical of what he sees as the spiritual decay underlying the practice of science—"the lust for knowledge, the furious desire to know—to *know* in the Biblical sense—to master and possess. . . . [The scientist] is a hollow man because in place of knowledge he needs wisdom, contemplation as the Eastern philosophers understand it, but as it is not understood in the West" (409–10). Schirmbeck argues through De Bary that there can be no such thing as "pure" science: "Every accession of knowledge so profoundly affects the observed object that in some cases the measurement is rendered invalid. . . . The recognition of this must in its turn influence the ethical personality of the scientist. It must serve to break down the idealised picture of science, the Pontius Pilate attitude, and set in its place a new conception of science having a definite place in the moral world" (412–13). Yet De Bary himself is guilty of withdrawal from the world through his obsession with the abstract perfection of pure mathematics. Despite his nobility, he remains essentially a figure of a past age, a critic, not a reformer, fleeing from the arena of military-oriented science to his country estate to meditate upon the "original sin of physics," the separation of object and subject. It is the narrator Grey who carries this awareness into the future. In the postscript Grey concludes, "I have learnt one thing at least, that for me there is no escape. I may not pluck out the eye that offends me; for to be worthy of the world is to look it in the face" (445).

This muted heroism and ambivalence about the role of the scientist can also be glimpsed in Douglas Stewart's poem "Rutherford" (1962). Stewart presents his distinguished fellow New Zealander in his study in Cambridge at the height of his scientific career. Well aware of the seductive pleasures of the scientific fraternity,

he had grown to like
This life of power where scientists met together
And felt they were priests and rulers. He liked to talk
With his great peers that language wrapped in mystery.[19]

Rutherford meditates on the nature of power and responsibility and the possible destructive applications of his research into nuclear fission:

You could pay dearly
For probing too deeply into that dark resistance
Where light lay coiled in stone. He had seen clearly
In flashes of the mind each atom exploding the next
To the end of the world, and the light came out of the distance
Like a wave upon him, towering. (176)

Although he tries optimistically to concentrate on the benefits of his work to humanity in medicine and engineering, Stewart's Rutherford cannot quite shut out the awareness of the danger and is on the point of dismantling his apparatus and retiring to a farm in New Zealand. However, in an abrupt volte-face he decides that there is an inner power that, "like force in the atom . . . filled him with its radiance," directs him. Whether or not this is merely self-justification is never finally clear, but Rutherford's decision to continue the work is presented as an act of quiet nobility, the scientist's equivalent of the white man's burden, which the man of integrity must undertake for the greater glory of mankind.

Emotion, Imagination, Relationship

More recent depictions of heroic scientists, and they are few, do not look back to the modes of idealization found in pre-1945 literature but explore new paths whereby scientists might make a contribution to the well-being of society at a level beyond the merely technical. A new wave of optimism concerning the ability of science to save mankind from the ills of the past is to be found in much science fiction, notably that of Arthur Clarke, Isaac Asimov, and James Hogan, all of whom are fascinated by the theoretical possibilities of science and predict a fruitful symbiosis between people and technology. But in contrast to the unalloyed enthusiasm of these science fiction writers, twentieth-century mainstream novelists were more cautious in their reinstatement of the scientist. Rarely was any attempt made to idealize scientists or even to portray them as successful in their aims; at best they were permitted to point a way forward and make a plea for communication and tolerance. Unlike the macho scientists of science fiction tradition, they were unashamed of expressing their emotions and hence could be seen as contributing to the emergence of a new kind of hero. It is perhaps significant that a number of such studies have been produced by women writers.

In her complex novel *The Transit of Venus* (1980) Australian novelist Shirley Hazzard created a young astronomer Ted Tice as her exemplar of humane emotions, independent thought, and social responsibility. Tice, as idealist, is contrasted with an older, established astronomer, Sefton Thrale, who has falsified his "scientific" findings concerning the optimal siting for a telescope in order to accommodate what is politically expedient and who judges other scientists only by their professional success. Thrale's arrogant pragmatism and Tice's concern with intention rather than overt results are contrasted through their respective responses to the story of the eighteenth-century French astronomer Guillaume Legentil. Traveling to India to observe the transit of Venus in 1761, Legentil had arrived in good time at Pondicherry only to find the town fallen to the British. He therefore set out to Mauritius as an alternative observation site but arrived too late to take astronomical measurements of the transit. He waited eight years in the East for the next transit in 1769 and on this occasion reached Pondicherry and built a small observatory. However, when the day of the transit dawned, visibility was so poor that he could make no observations. For Thrale this story is one of humiliating defeat, best forgotten. For Tice, honoring the faith, not the failure, "his story has such nobility that you can scarcely call it unsuccessful."[20]

Whereas Thrale represents the scientific obsession with definition, clarity, and results, an urge Hazzard identifies with the stereotypical male need to conquer and possess, Tice values complexity, seeing art, emotions, and imagination as integral parts of the truth. In this sense he is the ethical hero of the novel, but the price he pays for rejecting conventional standards is relative failure in both science and emotional relationships: his heroism consists essentially in a kind of defeat.

Perhaps the most interesting and innovative depictions of the scientist in twentieth-century literature are those that reject the Cartesian dualism that had characterized Western science since the seventeenth century and urge instead a return to a holistic understanding of the world. One of the first literary presentations of this philosophy was Aldous Huxley's last novel, *Island* (1962), the counterpoint to his earlier, intensely pessimistic *Brave New World*

(1932). In *Island* Huxley explored the possibility of uniting the insights of modern science and ancient yoga, the cognitive and spiritual perceptions, to foster a more creative life. His chief exponent is Dr. Robert MacPhail, surgeon and naturalist, who promotes the holistic view of science as a relationship. On Pala, the island of the title, children learn science from an ecological rather than an analytical base. The school creed is that nothing exists in isolation: "All living is relationship."[21] Bacon's premise of a socially useful science is interpreted as a natural extension of Palan ecological principles rather than as an exploitation of nature. Unfortunately, despite its philosophical interest, *Island* remains a reworking of Huxley's thesis in *Literature and Science* (1963), and the characters fail to come to life, possibly because Huxley has difficulties in imagining how his brave new world could be transformed into Pala. Significantly, the ending of *Island* is far from optimistic about the survival of the utopian community, as its powerful neighbors stand poised to take over and exploit Pala's oil deposits. In the global village of the twentieth century, islands, however ideologically sound, cannot exist in isolation.

Potentially more optimistic and innovative as a portrait of the idealistic scientist searching for wholeness is Shevek, protagonist of Ursula Le Guin's futuristic novel *The Dispossessed* (1974). Le Guin draws parallels between the Taoist search for the unity of opposites as the path to understanding and the fundamental paradoxes inherent in quantum physics. Like Hazzard's character Ted Tice, Shevek, a brilliant physicist of a future world, Anarres, rejects easy simplification and false clarity. Instead, he applies to his study of time the unifying concepts that Einstein tried to establish in his search for a theory of general relativity. Shevek's General Theory of Temporality involves a synthesis of his earlier Principles of Simultaneity with the Principles of Sequency, concepts as difficult to hold together as the wave and particle theories of light. This search for a theoretical unity in temporal physics parallels Shevek's struggle to bring unity and tolerance to the polarized political systems of Urras and its moon, Anarres. In socialist Anarres, his theoretical physics is regarded as a self-indulgent substitute for real work; in capitalist Urras, scientists are highly respected, but social conscience is taboo. Shevek finds his idealist principles, his defense of openness in scientific research, and the free sharing of ideas with the inhabitants of other planets are regarded by both societies as treachery.

Shevek returns to Anarres as an ambassador for the idea of "unbuilding walls" between individuals and societies and, although a much modified hero, he represents in Le Guin's terms the only hope for the future. Through his rejection of violence to achieve his purposes, he stands in contrast to the scientist heroes popular at the beginning of the century, who never doubted the ethics of nationalism and aggression, whether against the alien creatures of another planet or the equally abhorrent fellow humans of a different political stripe. Shevek's goal (symbolized by his unifying theory of temporality and by the ansible, the communication device he has constructed as a practical spin-off from his research) is to facilitate interplanetary communication and overcome the many different kinds of walls that divide people. The symbol of the wall is a clear reference to the Berlin Wall, but it is no less applicable to the many political barricades that have succeeded it: "For seven generations there had been nothing in the world more important than that wall. Like all walls it was ambiguous, two-faced. What was inside it and what was outside it depended upon which side of it you were on."[22]

Le Guin has replaced the traditional image of the physicist associated with disintegration (the atomic bomb or some destructive precursor) with a character involved in a project that is essentially creative in both the scientific and the social sense. Through the figure of Shevek she explores new interpretations of utopia, rejecting both the nineteenth-century ideal of material well-being and the self-negating ideal of uniformity for the possibility of creative freedom, where the individual is voluntarily in harmony with the society. This newly defined heroic role for the scientist represents both a radical rethinking in terms of ideology and, arguably, the most realistic hope for the future.

From the discussion of the characters in this chapter it is apparent that the concept of the idealized scientist has been a rapidly changing one. Early in the century it signified the single-minded pursuit of the discipline as an end in itself. Nothing else was required of the ideal scientist but objectivity, rationality, and dedication to science to the exclusion of all else. Toward the end of the Second World War some writers put forward the radical proposal that integrity might be identified with noncompliance with the orders of the government-military establishment, that is, with treason as conventionally defined. This interpretation shifted the criterion from within the discipline itself to an external principle, in this case rejection of militarism and totalitarianism in some form or other. Later it acquired an increasingly sociological component. By the 1970s the idealism of the scientist (in literature as opposed to life) was assessed very little in relation to devotion to research and very considerably in relation to social conscience. More recently, the noble scientist was characterized by an ability to conceive of a viable future society in which the claims of the individual and the state were integrated, allowing scope for emotional and spiritual wholeness without sacrificing rationality and efficiency. By the late 1970s the ideal scientist was required to be a philosopher and an effective communicator with nonscientists. This latter role in particular was emphasized in literature long before the need was acknowledged by real-life scientists, most of whom considered the popularization of science an activity to be denigrated, if not actively opposed. During the 1980s and early 1990s, as the perception of the world's problems changed from a political to an environmental focus, the ideal scientist was required to be committed to restoring the ecological health of the planet, channeling technological expertise into the solving of environmental problems, and combating the economic imperative of multinational corporations. This particular emphasis in literature predated, in many cases, the parallel integrated vision of much contemporary science.

A number of the writers considered in this chapter have contributed to evolving a new role model for the scientist hero, replacing the crude, macho Martian-basher with a multidisciplinary and socially aware communicator who, like Shevek, "unbuilds walls." Whether that ideal is compatible with the pressures placed on scientists in the competitive world of peer review and limited research funding is arguable, and most writers have suggested that it may be necessary for society to change those requirements before these fictional examples can eventuate in real life. The message from literature would seem to be that a society produces the scientists it deserves.

CONCLUSION

New Images of Scientists

Throughout this study it has been apparent that the portrayals of unattractive scientists, whether as suspicious, foolish, arrogant, inhuman, amoral, mad, evil, dangerous, or helpless, predominate in both fiction and film. We have traced the diverse cultural, political, and religious reasons for this unflattering perception, many of which are grounded in deep-rooted fears of the new, of a loss of emotional roots, and even of extinction of the entire human race; fears concerning loss of individuality and the stability engendered by accepted values; fears of the cargo cult of technology, bringing with it immense power and unanswered questions about its control. Cultural historian Theodore Roszak spoke for the majority of the writers whose scientist characters are discussed in this book: "Dr. Faustus, Dr. Frankenstein, Dr. Jekyll, Dr. Cyclops, Dr. Caligari, Dr. Strangelove. The scientist who does not face up to the warning in this persistent folklore is himself the worst enemy of science. In these images of our popular culture resides a legitimate public fear of the scientist's stripped-down depersonalized conception of knowledge—a fear that our scientists, well-intentioned and decent men and women all, will go on being titans who create monsters. What is a monster? the child of knowledge without gnosis, of power without spiritual intelligence."[1]

However, the last two decades have marked a cultural watershed, a distinct attitudinal change toward scientists, in the media, in fiction, and (though somewhat less marked) in film. There are several reasons driving this more benign appraisal. One of these is the preoccupation with environmental issues. Until recently scientists were held accountable for destroying the environment by producing pesticides, radioactive contamination from nuclear weapons and power-station accidents, chemical waste, monocultures, hormonal feed supplements, etc. But these evils are now more commonly laid at the door of big business—pharmaceutical companies, medical research laboratories, gene banks, agribusiness. Scientists, by contrast, are in the vanguard of attempts to rescue nature from these enemies and to devise new forms of ecopower generation. These efforts are well reported in the media, and ecofiction, or cli-fi, is one of the fastest-growing subgenres of the twenty-first-century novel. In film the number of altruistic, well-intentioned scientists is growing: Dr. Robert Campbell of *The Medicine Man* (1992) and the virologists of *Outbreak* (1995) were only the forerunners. Female scientists, of whom there are increasing examples, are rarely either mad or evil but often resolve the problems: Dr. Grace Augustine in *Avatar*, Diane Fossey in *Gorillas in the Mist*, Dr. Ellie Sattler in *Jurassic Park*.

There has also been a move toward exploring the complexities of the private mental lives of scientists rather than the implications of their actions. In particular, following the publicity given to the solution of Fermat's Last Theorem by British mathematician Andrew Wiles in 1993, there has been a revived interest in mathematicians, once considered too abstruse for presentation in film. *Breaking the Code* (1995) and *The Imitation Game* (2014) are two dramatizations of the life of Alan Turing, a British mathematical genius and designer of the computer that broke the German Enigma code during World War II. These films explored Turing's personal problems in admitting his homosexuality at a time when it was illegal and raised questions about national security risks. *Enigma* (2001) follows the Bletchley code breakers, and *A Beautiful Mind* (2001) follows the mental journey of mathematician John Forbes Nash, who arrives at Princeton with great mathematical ability but few people skills and develops paranoid schizophrenia. *Ramanujan* (2014) and *The Man Who Knew Infinity* (2015) are biographical films about the Indian mathematical genius Srinivasa Ramanujan, who spent painful but intellectually challenging years in Cambridge with British mathematician G. M. Hardy; *A Brief History of Time* (1991) and *The Theory of Everything* (2014) document the life and hardships of theoretical physicist Stephen Hawking.

Uncertainty about brilliance and madness is posed in Rebecca Goldstein's novel *Properties of Light* (2000), which explores the mystical mathematical perceptions of a father and daughter, and in David Auburn's play *Proof* (2000, film 2005), in which, when a once brilliant mathematician Robert dies, the proof to a paradigm-changing theorem on prime numbers is discovered in his notebooks. Robert's daughter Catherine asserts that she, not her father, who was already suffering from dementia, had devised the proof, but almost no one believes her and she herself begins to doubt her own sanity. Proof is never definite. All these films bypass the mathematical problem to focus on the motivation and sometimes anguish of the mathematicians themselves, engendering empathy for these mentally isolated figures.

The most significant change in recent times has been the demise of the mad scientist, since there is now less reliance on scientists to provide the situations and objects of dread that inspired horror films. Since 2001 we have been conditioned to fear most the terrorism and fanaticism arising from political and religious systems and, underpinning them, the madness of despotic or fanatical leaders. As before, the psychology of the unbalanced, evil mind is the real and abiding source of fear, but this is no longer associated with scientists. The mad scientist is in recession because we no longer need him.

The new face of terror is—the terrorist.

In the modern "world risk society" scientists are perceived as having three roles, one malevolent and two benign. They have long been characterized as risk producers, and we have seen seven centuries of such attribution. Now they are also seen as risk monitors, our best hope for realizing in time the signs and implications of impending dangers, whether from earthquakes, volcanoes, avalanches, and other natural disasters or from more insidious risks of genetic engineering, environmental causes of cancer and other health issues, endangered species, and climate change. Potentially they may also become the risk averters of the future, devising means to save endangered habitats and species; discovering medical cures for pandemics; developing printer-cut body parts that remove the "necessity" for organ transplants, stem cells, or clones; and opening communications channels for those isolated through age or a disability.

In a forthcoming volume I look at the new scientists of twenty-first-century literature and film, asking why and how they have overturned the comfortable, well-worn stereotypes and cliché responses of centuries to engage our compassion as fellow human beings.

APPENDIX

Films and TV Series with Scientist Characters

2001: A Space Odyssey, 1968
A for Andromeda (TV series), 1961
A for Andromeda, 2006
Atomised, 2006
Avatar, 2009
Back to the Future, 1985
Batman and Robin, 1997
Battlestar Galactica (TV series), 2004–9
A Beautiful Mind, 2001
Blade Runner, 1982
Blueprint, 2003
Bones (TV series), 2005–present
The Boys from Brazil, 1978
Breaking the Code, 1995
A Brief History of Time, 1991
Chain Reaction, 1996
The Cloning of Joanna May, 1992
Contact, 1997
CSI: Crime Scene Investigation (TV series), 2000–2015
CSI: Miami (TV series), 2002–present
The Day after Trinity, 1980
Die Frau im Mond, 1928
Dr. Cyclops, 1940
Dr. Jekyll and Mr. Hyde, 1931
Dr. Renault's Secret, 1946
Dr. Strangelove, or: How I Learned to Stop Worrying and Love the Bomb, 1964
Dr. X, 1931
Enigma, 2001
Enthiran, 2010
Erin Brockovich, 2000
Evolution, 2001
Ex Machina, 2015
Fail-Safe, 1964
Fat Man and Little Boy, 1989 (also known as *The Shadow Makers*)
The Fly, 1958
The Fly, 1986
Frankenstein, 1910
Frankenstein, 1931
Frankenstein, 2007
Frankenstein: The Real Story, 1992
Gattaca, 1997
The Genius of Marie Curie: The Woman Who Lit Up the World (TV documentary), 2013
The Godsend, 2004
Godzilla, 1998

Godzilla, 2014
Godzilla, King of Monsters!, 1954
Gorillas in the Mist, 1995
Honey, I Shrunk the Kids, 1989
Humans (TV series), 2015–
I, Robot, 2004
The Imitation Game, 2014
The Invisible Ray, 1936
The Island, 2005
The Island of Doctor Moreau, 1977
The Island of Doctor Moreau, 1996
The Island of Lost Souls, 1932
Junior, 1994
Jurassic Park, 1993
Jurassic Park III, 2001
Jurassic World, 2015
Le cabinet de Méphistophélès, 1897
The Lost World: Jurassic Park, 1997
L'uomo meccanico, 1921
Madame Curie, 1943
The Man Who Knew Infinity, 2015
Maria Sklodowska-Curie, 2016
Marie Curie (TV miniseries), 1977
Marie Curie, 2016
The Martian, 2015
Mary Shelley's Frankenstein, 1994
The Matrix, 1999
The Medicine Man, 1992
Metropolis, 1927
The Murders in the Rue Morgue, 1932
My Sister's Keeper, 2009
NCIS (TV series), 2003–present
Never Let Me Go, 2010
Orphan Black (TV series), 2013–16
Outbreak, 1995
Outer Limits (TV series), 1963–65
Proof, 2005
Ramanujan, 2014

RoboCop, 1987
Rodan, 1956
Runaway, 1984
Seven Days to Noon, 1970
The Sixth Day, 2000
Sphere, 1998
Splice, 2009
Son of Frankenstein, 1939
Star Trek movies, 1979–2016
Star Trek (TV series), 1966–present
Star Wars, 1977
Stargate, 1994
Stargate (TV series), 1994–2011
The Stepford Wives, 1975
The Stepford Wives, 2004
Surrogates, 2009
Them!, 1954
The Terminator, 1984
The Theory of Everything, 2014
The Thing, 1951
The Thing, 1982
The Thing, 2011
A Trip to Mars, 1910
The Vampire Bat, 1933
The Walking Dead, 1936
The World Is Not Enough, 1999

NOTES

Introduction • Myths of Science

1. Ulrich Beck (2007), *World at Risk* (Cambridge: Polity Press).
2. Jon Turney (1998), *In Frankenstein's Footsteps: Science, Genetics and Popular Culture* (New Haven, CT: Yale University Press), 8.
3. Andrew Tudor (1989), *Monsters and Mad Scientists: A Cultural History of the Horror Movie* (Oxford: Oxford University Press), 17–47.
4. Peter Weingart, Claudia Muhl, and Petra Pansegrau (2003), "Of Power Maniacs and Unethical Geniuses: Science and Scientists in Fiction Film," *Public Understanding of Science* 12, no. 3: 279–87.
5. Douglas Chapman (1988), "To a New World of Gods and Monsters: Mad Scientists and the Movies," *Strange Magazine*, no. 2, retrieved Oct. 2016, www.strangemag.com/madscientists/madscientists.html.
6. H. G. Wells (1934), *Experiment in Autobiography* (London: Jonathan Cape), 497–98.
7. For a full discussion of the words *science* and *scientist* see Sydney Ross (1962), "Scientist: The Story of a Word," *Annals of Science* 18: 65–85.
8. Francis Bacon (1857–74), "The New Organon; or, True Directions Concerning the interpretation of Nature" [1620], in *The Works of Francis Bacon*, ed. James Spedding, Robert Leslie Ellis, and Douglas Denton Heath, 14 vols. (London: Longman), 4:47–248.
9. Ross, "Scientist," 82.
10. [W. Whewell] (1834), review of *On the Connexion of the Physical Sciences*, by Mrs. Somerville, *Quarterly Review* 51: 58–61. In this article Whewell made the analogy with *sciolist, economist*, and *atheist*, thereby sabotaging the suggestion from the start, but seven years later he made the suggestion again, comparing the word with *artist* and *physicist*. *The Philosophy of the Inductive Sciences* (1847), vol. 1 (London: Parker), cxiii).
11. Ross, "Scientist," 66.
12. See chapter 8 below.
13. Karl R. Popper (1963), "Science as Falsification," in *Conjectures and Refutations: The Growth of Scientific Knowledge* (London: Routledge & Kegan Paul), 33–39.

Chapter 1 • Evil Alchemists and Doctor Faustus

1. J. C. Cooper (1990), *Chinese Alchemy: The Daoist Quest for Immortality* (New York: Sterling), 55–90.
2. The idea of a drug that could confer immortality was mentioned in the ancient Indian Vedas, dating from before 1000 BCE, and this was possibly the source of the Chinese belief.

3. Fabrizio Pregadio (2014), *The Way of the Golden Elixir: An Introduction to Taoist Alchemy* (Mountain View, CA: Golden Elixir Press). Retrieved from www.goldenelixir.com/files/The _Way_of_the_Golden_Elixir.pdf. Chinese medicine, which emphasizes the harmony of body and soul, derives from this aspect of alchemy.

4. See, e.g., R. Cummings (1966), *The Alchemists: Fathers of Practical Chemistry* (New York: David McKay), 17–27.

5. *Chem* or *kmt* was the ancient Egyptian name for the black land of the Nile delta and, by extension, for Egypt itself. *Chyma* was the Greek word for fusing or casting metals. The Arabic name was thus probably a combination of both derivations.

6. Cummings, *Alchemists*, 42.

7. Lyndy Abraham (1990), *Marvell and Alchemy* (Aldershot: Scolar), 4–6.

8. Ibid., chap. 1.

9. These acts culminated in the edict *Spondent* of Pope John XXII, denouncing all alchemists as tricksters and *falsarii* (counterfeiters). The edict is quoted in translation in E. H. Duncan (1968), "The Literature of Alchemy and Chaucer's *Canon's Yeoman's Tale*," *Speculum* 43: 636–37.

10. Betty Jo Teeter Dobbs (1991), *The Janus Faces of Genius: The Role of Alchemy in Newton's Thought* (Cambridge: Cambridge University Press), 1.

11. See John Cohen (1966), *Human Robots in Myth and Science* (London: George Allen & Unwin), 43–48.

12. The notion of the golem is first found in the story of Rabbah in *Sanhedrin* 65b of the *Talmud*. In Talmudic tradition Adam was designated as a golem, made from dust. See Cohen, *Human Robots*, 41–43, 48. The Golem of Prague was allegedly created in 1580.

13. J. M. Keynes (1947), "Newton the Man," *Proceedings of the Royal Society: Newton Tercentenary Celebrations, 15–19 July 1946* (Cambridge: Cambridge University Press).

14. There is a brief, allegedly autobiographical account of the adventures of Solomon Trismosin and his search for the philosopher's stone in a collection of alchemical tracts entitled *Aureum Vellus oder Guldin Schatz und Kunstkammer*, printed in Rorschach in 1598.

15. Geoffrey Chaucer (1957), *The Canon's Yeoman's Prologue and Tale*, in *The Canterbury Tales*, trans. Neville Coghill (Harmondsworth: Penguin).

16. S. Foster Damon has argued that Chaucer was not only defending true alchemists in this last part of the tale but that he must himself have been an initiate in alchemy. "Chaucer and Alchemy," *PMLA* 39 (1924): 782–88. This view is supported by Duncan, "Literature of Alchemy," 656, and by Chaucer's unfinished *Treatise on the Astrolabe* (1391), which indicates a comprehensive knowledge of contemporary science.

17. See J. W. Smeed (1975), *Faust in Literature* (New York: Oxford University Press), 13.

18. Christopher Marlowe (1963), *The Tragical History of Doctor Faustus*, in *Marlowe's Plays and Poems*, ed. M. R. Ridley (London: Dent), 122.

19. H. G. Wells's treatment of Griffin, the Invisible Man, owes much to Faustus. See chapter 10.

20. John Dee is alluded to in act 2, sc. 6. Edmund Kelley's association with Rudolph II of Bohemia is mentioned in act 4, sc. 1, 89–91. Simon Forman, self-styled astrologer and alchemist, was alive when the play was written and was mentioned explicitly in Jonson's earlier play *Epicoene* (1609).

21. See Roslynn D. Haynes (2006), "The Alchemist in Fiction: The Master Narrative," *HYLE: The International Journal for the Philosophy of Chemistry* 12, no. 1: 5–29.

22. Mary Shelley (1980), *Frankenstein, or the Modern Prometheus* [1818], ed. J. Kinglsey and M. K. Joseph (Oxford: Oxford University Press), 32.

Chapter 2 • Bacon's New Scientists

1. Francis Bacon (1857–64), *The Interpretation of Nature*, in *The Works of Francis Bacon*, ed. James Spedding, Robert Leslie Ellis, and Douglas Denon Heath, 14 vols. (London: Longman), 3:505–20.
2. Sir Thomas More had argued in similar terms in *Utopia*. See, e.g., Robert P. Adams (1949), "The Social Responsibilities of Science in *Utopia*, *New Atlantis*, and After," *Journal of the History of Ideas* 10: 374–98.
3. Bacon, *Interpretation of Nature*, ,3:222. Bacon was familiar with the writings of Paracelsus, who believed that in the biblical millennium mankind would regain the domination over nature that Adam had enjoyed before the Fall. See Peter Heimann (1979), "The Scientific Revolution," in *The New Cambridge Modern History* (Cambridge: Cambridge University Press), 13:252.
4. Francis Bacon, *The Advancement of Learning*, bk. 1, in *Works*, 3:274.
5. Ibid., 3:294.
6. Francis Bacon (1857–74), *New Atlantis*, in *Works*, 3:145.
7. This college for research incorporated ideas that Bacon had evolved some years earlier when he had hoped to be appointed master of some foundation at Oxford or Cambridge, but it is also possible that he drew on reports of the Accademia dei Lincei in Rome, founded in 1603 for research into physics and chemistry.
8. Francis Bacon, *De Augmentis Scientiarum*, bk. 7, chap. 1, in *Works*, 5:8.
9. Francis Bacon, *Magna Instauratio*, preface, in *Works*, 4:20–21 (my italics).
10. Francis Bacon, *The New Organon*, Aphorism 63, in *Works*, 4:64–65.
11. Ibid., Aphorism 13, 4:49.
12. Bacon, *New Atlantis*, in *Works*, 3:156–66.
13. Every twelve years, six of the Brethren of Salomon's House embark on a fact-finding mission "whose errand was only to give us knowledge . . . especially of the sciences, arts manufactures, and inventions of all the World; And withal to bring unto us, books, instruments, and patterns." Ibid., 164.
14. Robert K. Merton (1942), "The Normative Structure of Science," in *The Sociology of Science: Theoretical and Empirical Investigations* (Chicago: University of Chicago Press), 270–78. *The New Atlantis* appears to have been written about 1623 although it was not published until 1627, the year after Bacon's death (Gutenberg eBook, 2008, introductory note).
15. Bacon described his philosophy as mediating "between the presumption of pronouncing on everything and the despair of comprehending anything,." Bacon, *New Organon*, preface, in *Works*, 4:1.
16. "He held up his bare hand, as he went, as blessing the people" and two attendants bear before his litter "the one a crozier, the other a pastoral staff." Bacon, *New Atlantis*, in *Works*, 3:155.
17. "We have certain hymns and services, which we say daily, of laud and thanks to God, for his marvellous works: and forms of prayers, imploring his aid and blessing for the illumination of our labours, and the turning of them into good and holy uses." Bacon, *New Atlantis*, in *Works* 3:166.
18. Ibid., 3:165.
19. Bacon, *New Organon*, Aphorism 129, in *Works*, 4:115.
20. Rudolph Metz writes: "In the great process of setting learning free from shackles of faith and secularizing it, which begins with the Renaissance and continues its advance with humanism, Bacon's teaching plays an important part. . . . His efforts led to an essential strengthening of the secular at the expense of the spiritual. . . . The course which he adopted in religious matters led directly to the deism of the Enlightenment, and further to the skepticism of Hume." Rudolph Metz (1967), "Bacon's Part in the Intellectual Movement of his Time," trans.

Joan Drever, in *Seventeenth-Century Studies Presented to Sir Herbert Grierson*, ed. J. Dover Wilson (New York: Octagon), 32.

21. "It is very probable that the *Motion of Gravity* worketh weakly, both farre from the Earth, and also within the Earth; The former because of the Appetite of Union of Dense Bodies with the Earth, in respect of the distance is more dull." Francis Bacon, *The History of Dense and Rare*, in *Works*, 5:340.

22. "As for that Region of Fire our Philosophers talke of, I heard no news of it, my eyes having sufficiently informed me there can be no such thing." Francis Godwin (1971), *The Man in the Moone: or a Discourse of a Voyage thither by Domingo Gonzales, the Speedy Messenger* (Menston, Yorkshire: Scolar), 66.

23. Ibid., 101.

24. Abraham Cowley (1661), *A Proposition for the Advancement of Learning* (Ann Arbor, MI: Text Creation Partnership). In this work Cowley attempted to translate Bacon's House of Salomon into practical contemporary terms. He envisaged a college of twenty philosophers or teachers and sixteen young scholars accommodated and equipped to carry out research and "give the best education in the world gratis" to some two hundred pupils. See Dorothy Stimson (1968) *Scientists and Amateurs* (New York: Greenwood), 21–22.

25. Thomas Sprat (1959), *History of the Royal Society of London for the Improving of Natural Knowledge*, ed. J. L. Cope and H. W. Jones (London: Routledge), 35.

26. Henry Oldenburg (1665), "The Introduction," *Philosophical Transactions*, March 6.

27. Robert Hooke (1848), manuscript papers, quoted in C. R. Weld, *History of the Royal Society*, 2 vols. (London: John W. Parker), 1:146.

28. William Wotton (1705), *Reflections Upon Ancient and Modern Learning* [1697] (London: Tim Goodwin), 30.

29. See Betty Jo Teeter Dobbs (1975), *The Foundations of Newton's Alchemy; or, "The Hunting of the Greene Lyon"* (Cambridge: Cambridge University Press), 194–95.

30. Denis Diderot (1875–77), *L'Encyclopédie, ou Dictionnaire raisonné des sciences, des arts et des métiers, dirigé par Diderot & d'Alembert*, prospectus, in *Oeuvres complètes de Diderot*, 20 vols. (Paris: Garnier), 13:133 (my translation).

31. See Dobbs, *Foundations of Newton's Alchemy*, 59.

32. According to Plato's account Atlantis was located west of the Pillars of Hercules (the promontories that flank the Strait of Gibraltar). The title page of Bacon's *Instauratio Magna* features the two pillars in stylized form and the motto *Multi pertransibunt et augebitur scientia*, "Many will pass through and knowledge will be the greater."

33. Francis Bacon (1620), *New Organon*, Aphorism 129, in *Works*, 4:114.

34. Bacon, *New Atlantis*, in *Works*, 3:156.

35. See, e.g., Brian Easlea (1981), *Science and Sexual Oppression: Patriarchy's Confrontation with Women and Nature* (London: Weidenfeld & Nicholson); Evelyn Fox Keller (1996), *Feminism and Science* (Oxford: Oxford University Press), 1–40; Genevieve Lloyd (1984), "Francis Bacon: Knowledge as the Subjugation of Nature," in *The Man of Reason: "Male" and "Female" in Western Philosophy* (London: Routledge), 1–15; Carolyn Merchant (1980), *The Death of Nature: Women, Ecology and the Scientific Revolution* (San Francisco: Harper & Row), 164–90.

36. Kate Aughterson (2002), "'Strange Things So Probably Told': Gender, Sexual Difference and Knowledge in Bacon's *New Atlantis*," in *Francis Bacon's New Atlantis: New Interdisciplinary Essays*, ed. Bronwen Price (Manchester: Manchester University Press), 156–79.

37. Bacon, *New Organon*, Aphorism 2, in *Works*, 4:47.

38. Aughterson, "'Strange Things So Probably Told,'" 163.

39. Ibid., 175–76.

Chapter 3 • Foolish Virtuosi

1. The term *virtuoso* was originally applied to those who claimed a special interest in, and aspired to knowledge of, art and science. However, as we see in this chapter, it became debased and used to refer to foolish pretenders to scientific knowledge.

2. See, e.g., R. H. Syfret (1950), "Some Early Critics of the Royal Society," *Notes and Records of the Royal Society of London* 8: 45.

3. These factors included the emphasis on classical writers and Greek philosophy, the influence of Descartes, and, perhaps most immediately, the reaction against Puritan fanaticism.

4. Robert K. Merton (1938), "Science, Technology, and Society in Seventeenth-Century England," *Osiris* 4: 360–632. Cartwright and Baker suggest that the Puritan ideal of the "priesthood of all believers" is commensurate with the scientific study of nature using fresh observations and a mind purged of classical and medieval dogma. See John H. Cartwright and Brian Baker (2005), *Literature and Science: Social Impact and Interaction* (Santa Barbara, CA: ABC-CLIO), 79.

5. R. F. Jones (1949), "Background of the Attack on Science in the Age of Pope," in *Pope and His Contemporaries*, ed. J. L. Clifford and L. A. Landa (Oxford: Oxford University Press), 104.

6. Thomas Sprat (1959), *History of the Royal Society of London for the Improving of Natural Knowledge*, ed. J. I. Cope and H. W. Jones (London: Routledge), 417.

7. Edmund Halley gained a reputation for mechanistic views from which he had to extricate himself. See Michael Hunter (1981), *Science and Society in Restoration England* (Cambridge: Cambridge University Press), 174.

8. Ibid., 175. Joseph Glanvill, a devoted Baconian, encouraged by Robert Boyle, the acclaimed physicist, was almost obsessive in collecting allegations of witchcraft, which he published in *Saducismus Triumphatus: A full and plain Evidence Concerning Witches and Apparitions*, a work that ran into four editions from 1681 to 1726. Boyle himself published *The Excellencey of Theology, Compared with Natural Philosophy* (1674) to counter the fact that "under-valuation . . . of the study of things sacred . . . is grown so rife among many (otherwise ingenious) persons, especially studiers of physicks." Robert Boyle (1772), "Author's Preface," in *The Works of the Honourable Robert Boyle*, 4:2.

9. Richard S. Westfall (1958), *Science and Religion in Seventeenth-Century England* (New Haven, CT: Yale University Press), 145.

10. See Hunter, *Science and Society in Restoration England*, 66–67.

11. Ken Arnold (2006). *Cabinets for the Curious: Looking Back at Early British Museums* (Aldershot: Ashgate), 17–18.

12. The entrance fees and weekly dues were such that at least 50 percent of the members were wealthy gentlemen virtuosi. Londa Schiebinger (2006), "Women of Natural Knowledge," in *The Cambridge History of Early Modern Science*, ed. Katherine Park and Lorraine Daston (Cambridge: Cambridge University Press), 3:192–205.

13. See, e.g., Hunter, *Science and Society in Restoration England*, 68–69; and J. A. Bennett (1987), *The Divided Circle: A History of Instruments for Astronomy, Navigation, and Surveying* (Oxford: Phaidon), 88–92.

14. See Bennett, *The Divided Circle*, 65–68.

15. *The Diary of Samuel Pepys* (1669), transcribed by Mynors Bright, ed. Henry B. Wheatley (New York: Random House), 2:536–37.

16. See Sylvia Bowerbank and Sara Mendelson (2000), *Paper Bodies: A Margaret Cavendish Reader* (Peterborough, ON: Broadview), 26.

17. *The Blazing World* was originally appended as a fictional illustration to Cavendish's *Observations upon Experimental Philosophy*.

18. Margaret Cavendish, Duchess of Newcastle (1666), *The Description of a New World, Called The Blazing World* (London: A. Maxwell), 27–28.

19. Ibid., 28–29.

20. "Ballad of Gresham Colledge," quoted in Dorothy Stimson (1932), "Ballad of Gresham Colledge," *Isis* 18: 108–17.

21. Factors affecting the production of woolen cloth, a cheaper method of tanning leather, a diving bell, aids to navigation, etc.

22. S. V. Bruun (1969), "Who's Who in Samuel Butler's 'The Elephant in the Moon,'" *English Studies* 50: 381–89.

23. Samuel Butler (1854), "The Elephant in the Moon,", in *The Poetical Works of Samuel Butler*, ed. Rev. George Gilfillan, 2 vols. (Edinburgh: James Nicol), 2:135, ll. 527–30, 537–38.

24. Samuel Butler (1854), "A Satire upon the Royal Society," in *The Poetical Works*, 2:138–39, ll. 85–87, 91–92, 97–98.

25. Henry Fielding (1957), *The History of Tom Jones*, 2 vols. (London: Dent), vol. 2, bk. 12, chap. 5, 171.

26. See, e.g., Joseph M. Gilde (1970), "Shadwell and the Royal Society: Satire in *The Virtuoso*," *Studies in English Literature, 1500–1900* 10: 469–90; Marjorie H. Nicolson (1966), introduction to *The Virtuoso*, by Thomas Shadwell, ed. M. H. Nicolson and D. S. Rodes (London: Edward Arnold).

27. Thomas Shadwell (2003), *The Virtuoso, A Comedy* [1676], http://name.umdl.umich.edu /A59463.0001.001 (Ann Arbor, MI: Text Creation Partnership). "The College" is Gresham College, the precursor of the Royal Society, a name that continued to be applied to the latter institution. 2.1, p. 43.

28. Lisa Jardine (2003), *The Curious Life of Robert Hooke: The Man Who Measured London* (London: HarperCollins), 322.

29. The blood transfusion mentioned, between a spaniel and a bulldog, was described by Thomas Cox at a meeting of the Royal Society in May 1667 and published in the *Philosophical Transactions*, 2: 451. See Jessie Dobson (1969), "Doctors in Literature," *Library Association Record* 71: 270.

30. Although in his *History of the Royal Society* Thomas Sprat had affirmed its "constant resolution to reject all amplifications, digressions, and swellings of style: to return to the primitive purity and shortness" (113), by the eighteenth century the style of the *Transactions* had departed considerably from this ideal. C. J. Horne (1973), "Literature and Science," in *The Pelican Guide to English Literature*, ed. Boris Ford (Harmondsworth: Penguin), 4:196.

31. Susannah Centlivre (1982), *The Basset Table* [1705], in *Plays*, ed. Richard Frushell (New York: Garland), 97–98.

32. James Miller (1730), *The Humours of Oxford: A Comedy in Five Acts and in Prose* [1726], 2d ed. (London: J. Watts), 4.

33. Other famous scientists have been incorporated into fiction, notably Galileo, the Curies, Thomas Edison, and Albert Einstein, but they were treated as fictional characters rather than depicted in their own person.

Chapter 4 • Newton

1. John Maynard Keynes (1946), "Newton the Man," in *The Royal Society Newton Tercentenary Celebrations: 15–19 July* (Cambridge: Cambridge University Press), 27.

2. Betty Jo Teeter Dobbs (1991), *The Janus Faces of Genius: The Role of Alchemy in Newton's Thought* (Cambridge: Cambridge University Press), 230, 5.

3. See Michael White (1999), *Isaac Newton: The Last Sorcerer* (Boston: Da Capo).

4. Karin Figala (2004), "Newton's Alchemy," in *The Cambridge Companion to Newton*, ed. I. Bernard Cohen and George Edwin Smith (Cambridge: Cambridge University Press), 375.

5. Isaac Newton (1946), *Sir Isaac Newton's Mathematical Principles of Natural Philosophy*

and His System of the World, 2d ed. [1729], trans. Andrew Motte, ed. Florian Cajori (Berkeley: University of California Press), xviii.

6. Isaac Barrow became the first Lucasian Professor of Mathematics while Newton was at Cambridge.

7. See, e.g., Henry Guerlac (1965), "Where the Statue Stood: Divergent Loyalties to Newton in the Eighteenth Century," in *Aspects of the Eighteenth Century*, ed. E. R. Wasserman (Baltimore: Johns Hopkins University Press), 325–28.

8. Jean le Rond D'Alembert (1821), *Eléments de philosophie* [1759], in *Oeuvres philosophiques, historiques, et littéraires de d'Alembert*, ed. J. F. Bastien, 2d ed., 8 vols. (Paris: A. Belin, Readex Microprints), 1:121; Pierre Simon, Marquis de Laplace (1830), *The System of the World* [1796], trans. J. Pond (London: R. Phillips, Readex Microprints), vol. 2, bk. 5, 324.

9. Isaac Newton (1704), *Opticks: or, A Treatise of the Reflexions, Refractions, Inflexions and Colours of Light* (London: Sam. Smith and Benjamin Walford, Printers to the Royal Society), bk. 1, part 1.

10. Richard G. Olson ed. (1971), *Science as Metaphor: The Historical Role of Scientific Theories in Forming Western Culture* (Belmont, CA: Wadsworth), 71.

11. Voltaire to Cideville, in *Voltaire's Correspondence* (1977–88), ed. Theodore Besterman, 10 vols. (Paris: Gallimard), vol. 4, no. 838, 48–49.

12. *A Philosophic Ode on the Sun and the Universe* (1963), quoted in W. P. Jones, "Newton Further Demands the Muse," *Studies in English Literature* 3: 297.

13. Newton, *Mathematical Principles of Natural Philosophy*, 398.

14. James Thomson (1961), "A Poem Sacred to the Memory of Sir Isaac Newton," ll. 82–84, in James Thomson, *The Castle of Indolence and Other Poems*, ed. A. D. McKillop (Lawrence: University of Kansas Press).

15. William Tasker (1963), "An Ode to Curiosity," in *Poems* (London, 1779), quoted in Jones, "Newton Further Demands the Muse," 304.

16. Newton himself had not aspired to the certainty accorded to him. He admitted only that the propositions arrived at by induction were "very nearly true" and that "although arguing from Experiments and Observations to Induction be no Demonstration of general conclusions; yet it is the best way of arguing which the Nature of things admits of." Newton, *Opticks*, Query 31, p. 404.

17. Olson, *Science as Metaphor*, 72.

18. Newton's original explanation of gravity as the result of the pressure of a descending shower of ether had to be modified when, as the result of an experiment with pendulums, he was forced to conclude that the ether did not exist. He then had recourse to the concepts of attraction and repulsion, which he called "Powers" and "Virtues," whereby particles of matter act upon each other at a distance. This was strikingly similar to the theories of sympathies and antipathies to be found in the mystical literature of the Renaissance. See Richard S. Westfall (1971), *Force in Newton's Physics: The Science of Dynamics in the Seventeenth Century* (London: Macdonald), 375–77. The same ideas expressed as elective affinities remained current in chemistry into the nineteenth century as an explanation for chemical reactions. Goethe's popular novel *Die Wahlverwandtschaften*, or *Elective Affinities* (1809), is based on this metaphor.

19. Samuel Johnson, 30 July 1763, quoted in James Boswell (1961), *The Life of Samuel Johnson, L.L.D.* (New York: Oxford University Press), 276.

20. In 1714 the British Parliament passed the Longitude Act, offering a reward for solutions to the problem of calculating longitude at sea. It was Harrison's chronometer, tested aboard Captain James Cook's ship *Resolution* on a voyage to the South Pacific (1772–75), that first provided sufficient accuracy to determine longitude. See Dana Sobel (1995), *The Illustrated Longitude* (New York: Walker).

21. E. N. da C. Andrade (1943), "Newton and the Science of His Age," *Proceedings of the Royal Society, Series A* 181, no. 986 (May): 241–42.

22. Wright emphasizes the logic of the mechanism (an eidouranion, or transparent orrery, attributed to Adam Walker in the 1780s, a precursor of the planetarium) with its elliptical metal bands encircling space and catching the light from the hidden candle representing the sun. The figure of the natural philosopher standing at the center and dominating the group has an air of remoteness, amplified in the figure of the alchemist-like demonstrator in *An Experiment on a Bird in the Air Pump* (see fig. 11 below).

23. Olson, *Science as Metaphor*, 73.

24. Sir Thomas Browne (1962), *Religio Medici* (London: Dent), 17.

25. Roger Cotes to Newton, in *Correspondence of Sir Isaac Newton and Professor Cotes*, ed. J. Edleston (London: John Parker, 1850), 153.

26. See, e.g., E. W. Strong (1952), "Newton and God," *Journal of the History of Ideas* 13: 146–47. Cotes's own preface to the second edition of the *Principia* also stresses the implications for religious faith: "Therefore we may now more nearly behold the beauties of Nature and ... adore the great MAKER and LORD of all." Roger Cotes (1729), preface to Newton, *Mathematical Principles of Natural Philosophy*, xxiii.

27. Newton, *Opticks*, Query 28, 345.

28. The lectures, endowed by Robert Boyle of the Royal Society, were designed to consider the relationship between Christianity and the new natural philosophy (now called science).

29. Allan Ramsay (1974), "Ode to the Memory of Sir Isaac Newton: Inscribed to the Royal Society," in *Poems by Allan Ramsay and Robert Fergusson*, ed. Alexander Manson Kinghorn and Alexander Law (Edinburgh: Scottish Academic Press).

30. James Thomson (1961), "A Poem Sacred to the Memory of Sir Isaac Newton" [1727], in *The Castle of Indolence and Other Poems*, edited by A. D. McKillop (Lawrence: University of Kansas Press), 181.

31. Henry Grove (1714), *Spectator*, no. 635 (20 Dec.).

32. David Mallett (1810), *The Excursion: A Poem in two Books*, bk. 2, in *The Works of the English Poets*, ed. Alexander Chalmers, 21 vols. (London: J. Johnson), 14:16–24.

33. Abraham Cowley (1959), "To the Royal Society," in Thomas Sprat, *History of the Royal Society of London for the Improving of Natural Knowledge* [1667], ed. J. I. Cope and H. W. Jones (London: Routledge).

34. James Thomson (1860), *The Seasons*, "Summer," ll. 1531–63, in *The Poetical Works of James Thomson* (London: Bell and Daldy), 1:92. Ironically, William Blake was later to link Bacon, Locke, and Newton as the infernal trinity who had reduced a spiritual universe to a mechanical one (see chapter 6 below).

35. Halley's successful prediction that the Great Comet of 1682 would return in 1758 was important in vindicating Newton's theories of motion.

36. Richard Savage (1905), *The Wanderer*, in *Works of the English Poets*, ed. Samuel Johnson, 68 vols. (London, 1779–81), ed. G. B. Hill (Oxford: Oxford University Press), 45:12.

Chapter 5 • Arrogant and Godless

1. Andrew Pyle, ed. (2000), "Meric Casaubon," in *Dictionary of Seventeenth-Century British Philosophers* (London: Thoemmes Continuum), 162–63.

2. The much repeated, but disputed, anecdote concerns a meeting between Napoleon and Laplace. Napoleon allegedly queried why Laplace had written a "large book on the system of the universe and never even mentioned its Creator," to which Laplace is said to have replied, "I have no need of that hypothesis." The astronomer Hervé Faye wrote that this was an incorrect account and that it was not God, but rather divine intervention at any particular point, that

Laplace had declared to be a hypothesis. Hervé Faye (1884), *Sur l'origine du monde: théories cosmogoniques des anciens et des modernes* (Paris: Gauthier-Villars), 109–11.

3. Alexander Pope (1962), *The Dunciad*, bk. 4, ll. 469–76, in *Selected Poetry and Prose*, ed. W. K. Wimsatt Jr. (New York: Holt, Rinehart & Winston), 443.

4. Alexander Pope, *Epistle I*, ibid., 33–34.

5. Pope was himself a student of Newtonian astronomy and Cartesian cosmogony. See Marjorie Hope Nicolson and George S. Rousseau (1968), *This Long Disease, My Life: Alexander Pope and the Sciences* (Princeton, NJ: Princeton University Press).

6. Jonathan Swift (1957), *The Mechanical Operation of the Spirit*, in Swift, *Collected Works*, 16 vols. (Oxford: Basil Blackwell), 1:174. "Squaring the circle," or constructing a square with the same area as a circle, was a mathematical preoccupation of the time, but Swift's mention of it involves a specific reference to Newton, whose invention of the calculus provided the closest approximation to this long-standing riddle.

7. At this time Newton, now knighted, was not only a national hero but, as Master of the Mint and a visitor at court, he also had considerable political power. The manifest success of England's productivity as a result of her scientific inventions made it difficult to dismiss scientists as foolish, bumbling virtuosi.

8. Marjorie Nicolson has explored in detail the parallels between Swift's Academy of Projectors in Lagado and the Royal Society. See Marjorie Nicolson (1956), "The Scientific Background of Swift's *Voyage to Laputa*," in *Science and Imagination* (Ithaca, NY: Cornell University Press), 134–35.

9. "Terrella" (little Earth) was the name given by William Gilbert to a magnetized sphere, with which he demonstrated to Queen Elizabeth his theory of the earth's magnetism. He showed that a small compass moving around the terrella always pointed north–south. Gilbert argued that the same effect was seen in relation to the earth, indicating that it, too, was a magnetized sphere. William Gilbert (1991), *De Magnete* (New York: Dover), facsimile of 1893 English translation by Paul Fleury Mottelay.

10. For the former see Marjorie Nicolson (1960), *Voyages to the Moon* (New York: Macmillan), 193; and R. Philmus (1970), *Into the Unknown* (Berkeley: University of California Press), 10–12. For the latter see Marjorie Nicolson and Nora Mohler (1937), "Swift's 'Flying Island' in the *Voyage to Laputa*," *Annals of Science* 2: 299–335.

11. Jonathan Swift (1956), *Gulliver's Travels* [1726], ed. Harold Williams (London: Dent), bk. 3, chap. 2, 169.

12. Isaac Newton, "Axioms, or Laws of Motion," in *Mathematical Principles of Natural Philosophy*, Law III, Corollary II.

13. Samuel Johnson (1958), *The History of Rasselas, Prince of Abissinia*, in Johnson, *Rasselas; Poems and Selected Prose*, ed. B. H. Bronson (New York: Holt, Rinehart & Winston), chap. 40, 590.

Chapter 6 • Inhuman Scientists

1. The first calculating machine was created by the mathematician-philosopher Pascal. Leibniz also dreamed of a reasoning machine able to derive a complete mathematical system of the universe, possibly as an answer to Newton.

2. Julien Offray de La Mettrie (1960), *L'Homme machine* [1747] in *La Mettrie's L'Homme Machine: A Study in the Origins of an Idea*, ed. Aram Vartanian (Princeton, NJ: Princeton University Press), 190.

3. See Joseph Ben-David (1971), *The Scientist's Role in Society: A Comparative Study* (Englewood Cliffs, NJ: Prentice-Hall); and Bernard Smith (1988), *European Vision and the South Pacific* (Sydney: Harper & Row).

4. See Roslynn D. Haynes (2006), "The Alchemist in Fiction: The Master Narrative," *HYLE: International Journal for Philosophy of Chemistry* 12: 5–29.

5. Friedrich Schiller, Johann Gottlieb Fichte, Georg Wilhelm Hegel, August Schlegel and Karl Wilhelm Friedrich Schlegel, Clemens Brentano, Ludwig Tieck, Wilhelm von Humboldt, and Friedrich Schelling were all part of the Jena group.

6. Goethe's observation of colors in shadows, which was later to fascinate the impressionist painters, formed the basis of his most important scientific work, *Die Farbenlehre* (1810). See Karl J. Fink (1992), *Goethe's History of Science* (Cambridge: Cambridge University Press).

7. Gotthold Ephraim Lessing's *Faust* (of which only a fragment survives) portrays a pure and noble character, devoted to wisdom, and although he fell prey to Satan, he was eventually saved. This is the first statement in a Faust play that the quest for knowledge cannot be intrinsically evil but is the noblest human quality.

8. See J. W. Smeed (1975), *Faust in Literature* (New York: Oxford University Press), 8.

9. Johann Wolfgang von Goethe (1987), *Faust*, part 1 [1808], trans. David Luke (Oxford: Oxford University Press), p. 54, ll. 382–84.

10. Smeed, *Faust in Literature*, 19. They thus embody the more general Romantic longing for transcendence. In the subsequently added "Prolog im Himmel," Goethe has God affirm from the outset that Faust, despite his errors and sins, will never finally fall victim to Mephistopheles but will eventually return to the right road.

11. The major exception was Percy Bysshe Shelley, whose training in science and particular fascination with chemistry have been well documented in Carl Grabo (1968), *A Newton among the Poets: Shelley's Use of Science in Prometheus Unbound* (New York: Cooper Square).

12. Byron's character Manfred has been cited as closely parallel to Goethe's Faust. See Mary Anna Hahn, "A Comparison of Goethe's Faust and Byron's Manfred," BA thesis, University of Illinois, 1912, retrieved from https://archive.org/stream/comparisonofgoetoohaan/comparisonofgoetoohaan_djvu.txt. (But see Smeed, *Faust in Literature*, 226–27.) Manfred does not desire more knowledge or even more experience, only oblivion and escape from his guilt.

13. For a detailed discussion of the English Romantic poets and science see A. N. Whitehead (1953), *Science and the Modern World* (Cambridge: Cambridge University Press), chap. 5; M. H. Abrams (1958), *The Mirror and the Lamp* (New York: W. W. Norton), 184; and Richard Holmes (2008), *The Age of Wonder: How the Romantic Generation Discovered the Beauty and the Terror of Science* (New York: HarperCollins).

14. In *Phaedrus* (245a) Plato distinguished four types of divine madness—prophetic, initiative, poetic, and erotic. In *Ion* (533d–536d) he elaborated on poetic inspiration as arising from ecstasy or divine frenzy, insisting that it was prior to consciousness and unrelated to the art or skill of the poet. Rather, the poet was divinely possessed in order to utter his rhapsodic words.

15. See, e.g., Donald Ault (1974), *Visionary Physics: Blake's Response to Newton* (Chicago: University of Chicago Press).

16. Paley's famous analogy of the watch and the watchmaker, published in 1802, exemplified this teleological argument, recently revived as intelligent design.

17. William Blake (1966) to Thomas Butts, 22 Nov. 1802, 2d letter, ll. 83–84, 87–88, in *Blake: Complete Writings*, ed. Geoffrey Keynes (London: Oxford University Press), 818–19.

18. William Blake, *Jerusalem*, plate 54 in *Blake: Complete Writings*, ll. 15–18, ibid., 685.

19. William Blake, "Mock on, Mock on, Voltaire, Rousseau," ibid., 418.

20. Ault, *Visionary Physics*, 2–3.

21. S. Foster Damon (1979), *A Blake Dictionary* (London: Thames & Hudson), 299.

22. Blake, *Jerusalem*, plate 98, line 9, 745.

23. Blake, like Swift, also feared that scientists would embark on irresponsible and dangerous experiments. Underlying the comedy of the unfinished satire, *An Island in the Moon* (1784–85) is the serious assertion that science is destructive and scientists are irresponsible.

24. William Wordsworth (1978), *Prelude*, bk. 3, ll. 61–63, in *Wordsworth Poetical Works*, ed. Thomas Hutchinson (Oxford: Oxford University Press), 509.

25. William Wordsworth, preface to the second edition of *Lyrical Ballads*, ibid., 738.

26. William Wordsworth, "The Tables Turned," ibid., 377.

27. The air pump, invented in 1654 by Otto von Guericke of Magdeburg, and further developed by Boyle and Hooke in England, became an item of popular entertainment after the mid-eighteenth century to demonstrate the effect of a near vacuum on living organisms.

28. Percy Bysshe Shelley (1970), *A Defence of Poetry*, ed. Roland A. Duerksen (New York: Appleton-Century-Crofts), 190.

29. Walt Whitman (1958), "When I Heard the Learn'd Astronomer" [1865], in *The Portable Walt Whitman*, ed. Mark Van Doren (New York: Viking), 266.

30. W. B. Yeats (1994), "The Song of the Happy Shepherd" [1889], in *The Collected Poems of W. B. Yeats* (Ware, HERTS: Wordsworth Editions), 3–84.

31. Thomas Carlyle (1915), "Signs of the Times" [1829], in *English and Other Critical Essays* (London: Dent), 228.

32. Boz [Charles Dickens] (1837), "Full Report of the First Meeting of the Mudfog Association for the Advancement of Everything," *Bentley's Miscellany* 2: 397–413.

33. Reprinted in *New Scientist*, 9 Dec. 2011, retrieved 7 Oct. 2015 from www.newscientist.com/article/dn21266-the-mathematician-in-love/.

34. Honoré de Balzac (n.d.), *The Quest of the Absolute* [1834], trans. Ellen Marriage (London: Newnes), 84.

35. Ignaz Denner is a physics professor whose *Wunderdoktor* father has entered into a pact with the devil to obtain knowledge about nature, in order to acquire wealth. His demonic deeds bring only suffering and destruction. "Der Sandmann" appears in English as "The Sandman" (1983), trans. J. T. Bealby, in *Isaac Asimov Presents: The Best Science Fiction of the Nineteenth Century*, ed. I. Asimov, M. Greenberg, and C. G. Waugh (New York: Beaufort Books).

36. Hoffmann may have been influenced by an exhibition of automata in Dresden in 1813. See Patricia S. Warrick (1980), *The Cybernetic Imagination in Science Fiction* (Cambridge, MA: MIT Press), 34.

37. Quoted in Terence Martin (1983), *Nathaniel Hawthorne* (Boston: Twayne), 99.

38. Nathaniel Hawthorne (2001), "The Birthmark," in *Mosses from an Old Manse* (Mechanicsville, VA: Electric Book), 12.

39. Nathaniel Hawthorne (1986), "Rappaccini's Daughter," in *The Portable Hawthorne*, edited by Malcolm Cowley (Harmondsworth: Penguin Books), 204.

40. Chillingworth's laboratory is equipped with a "distilling apparatus and the means of compounding drugs and chemicals which the practiced alchemist knew well how to turn to purpose." Nathaniel Hawthorne (1986), *The Scarlet Letter* [1850], in *The Portable Hawthorne*, 413.

41. This was a preoccupation of Edward Bulwer-Lytton, in *Zanoni* (1842), *A Strange Story* (1861), and *The Coming Race* (1871).

Chapter 7 • *Frankenstein and the Creature*

1. Jon Turney (1998), *Frankenstein's Footsteps: Science, Genetics and Popular Culture* (New Haven, CT: Yale University Press).

2. Ken Russell's film *Gothic* (1986); the opera *Mer de Glace* (1991), libretto by David Malouf; and the novel by Molly Dwyer (2008), *Requiem for the Author of Frankenstein* (Fort Bragg, CA: Lost Coast).

3. Mary Shelley (1980), introduction to the 1831 edition of *Frankenstein, or The Modern Prometheus*, ed. J. Kinglsey and M. K. Joseph (Oxford: Oxford University Press), 9–10.

4. Shelley would certainly have known part 1 of Goethe's *Faust*, published in 1808. Godwin's novel includes a sequence in which a stranger, Francesco Zampieri, arrives at the home of Count Reginald St. Leon, to whom he reveals the secrets of the alchemists—immortality (an herbal elixir that brings youth and vigor and heals the sick) and the art of multiplying gold, both ascribed to the philosopher's stone.

5. Benjamin Franklin's most famous discovery—that lightning is caused by the discharge of static electricity in the atmosphere—was made as a result of a highly dangerous experiment whereby he tied a metal key to a silk conducting thread attached to a kite, which he flew in a lightning storm. By touching the key Franklin showed that electrical sparks would leap from the metal. Several scientists who attempted to repeat his experiment were electrocuted; however, the experiment led to the development of the lightning rod.

6. Percy Shelley, preface to 1818 edition of *Frankenstein*, 5.

7. The monster responds to the beauties of nature, to the joys of domesticity, to the ideas in great books—all things that Frankenstein has put aside for his research. But the monster also kills Frankenstein's younger brother William, his fiancée Elizabeth, and his friend Clerval, the people whom Frankenstein is duty-bound to love but whom he subconsciously wishes to be rid of because they attempt to distract him from his obsession.

8. Many critics have discussed the doppelgänger relationship between Frankenstein and his creation. See, e.g., Harold Bloom (1965), "*Frankenstein*, or the New Prometheus," *Partisan Review* 32, no. 4: 611–18; George Levine and U. C. Knoepflmacher, eds. (1979), *The Endurance of Frankenstein* (Berkeley: University of California Press); and Masao Miyoshi (1969), *The Divided Self: A Perspective on the Literature of the Victorians* (New York: New York University Press), 79–89.

9. Note the word play on "confinement," since Frankenstein is abrogating the female as well as the male role in presuming to create a man by himself.

10. As Harold Bloom pointed out, the novel's subtitle, "The Modern Prometheus," would have had connections with Napoleon for Shelley's contemporaries. Byron associated Napoleon with Prometheus and the sequence in the novel of Victor's pursuit of the monster across Russia would have recalled Napoleon's recent desperate retreat from Moscow. See Harold Bloom (2007), *Mary Shelley's Frankenstein* (New York: Chelsea House), 188. For a detailed study of this political reference see Lee Sterrenburg (1979), "Mary Shelley's Monster: Politics and Psyche in *Frankenstein*," in Levine and Knoepflmacher, *Endurance of Frankenstein*, 143–71.

11. William Godwin (1976), *Enquiry Concerning Political Justice and Its Influence on Morals and Happiness* (London: Penguin Books).

12. "Remember that I am thy creature; I ought to be thy Adam; but I am rather the fallen angel whom thou drivest from joy for no misdeed. Everywhere I see bliss, from which I alone am irrevocably excluded. I was benevolent and good; misery made me a fiend." *Frankenstein*, 100; cf. Milton's Adam: "Did I request thee, Maker, from my clay / To mould me Man? did I solicit thee / From darkness to promote me?" John Milton (1950), *Paradise Lost*, in *Complete Poetry and Selected Prose* (New York: Modern Library), 339.

13. Frankenstein ignores the warning of the monster that he will be with him on his wedding night. The neglect of this warning is so incredible, given the previous murder, that we are surely intended to see it as a suppressed wish by Frankenstein.

14. See, e.g., Anthony Frewin (1988), *One Hundred Years of Science Fiction Illustration, 1840–1940* (London: Bloomsbury Books). The menacing monsters range from robots to Martians but the point is the same. The prevalence of the image suggests that the threatening monsters represent what the scientist-heroes, rendered temporarily helpless and thus reduced to voyeurs, would like to do—attack the helpless female—but are prevented by the restrictions of their decent, rational civilization.

15. George Levine, "The Ambiguous Heritage of *Frankenstein*," in Levine and Knoepflmacher, *The Endurance of Frankenstein*, 15.

16. Carlos Clerens (1967), *An Illustrated History of the Horror Film* (New York: Capricorn), 64.

17. For the titles see http://en.wikipedia.org/wiki/Category:Frankenstein_films.

Chapter 8 • Victorian Scientists

1. For a comprehensive discussion of these issues see, e.g., J. Hillis Miller (1963) *The Disappearance of God* (Cambridge, MA: Harvard University Press); Walter E. Houghton (1957), *The Victorian Frame of Mind, 1830–1870* (New Haven, CT: Yale University Press); David Oldroyd (1980), *Darwinian Impacts: An Introduction to the Darwinian Revolution* (Kensington, NSW: UNSW Press); and Iain McCalman (2009), *Darwin's Armada* (Camberwell, Vic.: Penguin).

2. Joachim Schummer (2007), "Historical Roots of the 'Mad Scientist': Chemists in Nineteenth-Century Literature," in *The Public Image of Chemistry*, ed. J. Schummer, B. Bensaudse-Vincent, and B. Van Tiggelen (Singapore: World Scientific), 37–79.

3. University College London, founded 1826, and King's College, founded in 1828, provided a number of paid positions for scientists but without research facilities.

4. The mathematician Charles Babbage, designer of the first digital computer and a founding member of the BAAS, was prominent in the reform of science. His book *Reflections on the Decline of Science in England* (1830) attacked the traditionalism of the Royal Society.

5. The realist novels of Flaubert and Balzac in France; Thackeray, George Eliot, Hardy, and George Moore in England; and Joseph Furphy in Australia are literary expressions of biological and social determinism.

6. "So ist der Mensch die Summe von Eltern und Amme, von Ort und Zeit, von Luft und Wetter, von Schall und Licht, von Kost und Kleidung" (Humans are the sum total of parents and nurses, place and time, air and weather, sound and light, food and clothing); Jacob Moleschott, *Kreislauf*, letter 19, quoted in J. W. Smeed (1975), *Faust in Literature* (New York: Oxford, University Press), 201.

7. Many novels dealt with the dilemma facing clergymen who felt they could no longer honestly subscribe to a fundamentalist understanding of Creation as described in Genesis or of the miracles ascribed to Jesus. See, e.g., Margaret M. Maison (1961), *Search Your Soul, Eustace* (London: Sheed & Ward).

8. George Henry Kingsley studied medicine at Edinburgh and Paris and was wounded on the Paris barricades in the 1848 rebellion before setting out to explore the South Seas and the Rocky Mountains. See Susan Chitty (1974), *The Beast and the Monk: A Life of Charles Kingsley* (London: Hodder & Stoughton).

9. Charles Kingsley (1889), *Two Years Ago* [1857], (London: Macmillan), 74.

10. Charles Kingsley wrote and lectured extensively on what he called the science of health. See Charles Kingsley (1880), *Sanitary and Social Essays*, in *Collected Works*, vol. 18 (London: Macmillan).

11. Tertius Lydgate, usually cited as the first such character, is in fact a later creation (*Two Years Ago*: 1857; *Middlemarch*: 1871) although the later novel is set twenty years earlier. There are, as Harold Gillespie points out, many parallels between the two characters and their situations. See H. R. Gillespie (1964), "George Eliot's Tertius Lydgate and Charles Kingsley's Tom Thurnall," *Notes and Queries* 2, no. 1: 226 27.

12. An interesting later vignette of a female doctor is Dr. Mary Prance in Henry James's *The Bostonians* (1866). Although Dr. Prance is only a minor character, she is morally above reproach. She appears to have been based on Dr. Mary Walker, an eminent but eccentric doctor. See R. E. Long (1964), "A Source for Dr. Mary Prance," *Nineteenth-Century Fiction* 19: 87–88.

13. Kingsley to F. D. Maurice (1877), 29 Mar. 1863, in *Charles Kingsley: His Letters and Memories of His Life*, ed. Frances Eliza Grenfell Kingsley (London: Henry S. King), 2:181.

14. In *The Water Babies* Kingsley refers respectfully to Owen, Huxley, Muchison, Darwin, Sedgwick, and Faraday, although he pokes gentle fun at a Professor Ptthmllspths, who is so busy classifying species that he cannot see them for what they are.

15. Charles Kingsley (1949), *Glaucus; or, The Wonders of the Shore* (London: Dent).

16. Elizabeth Gaskell (1969), *Wives and Daughters* (Harmondsworth: Penguin), 75.

17. We know from one of Elizabeth Gaskell's letters that the parallel with Charles Darwin, though anachronistic in terms of the time frame of the novel, was intended. See Elizabeth Gaskell to George Smith, 3 May 1864, in *Letters of Mrs. Gaskell*, ed. J. A. V. Chapple and Arthur Pollard (Manchester: Manchester University Press, 1966), 732.

18. Thomas Arnold and Thomas Henry Huxley both delivered speeches to the Royal Academy of Arts (Arnold in 1881 and Huxley in 1883) in response to the toast "Science and Literature." See David A. Roos (1977), "Matthew Arnold and Thomas Henry Huxley: Two Speeches at the Royal Academy, 1881and 1883," *Modern Philology* 74, no. 3: 316–24.

19. Gustav Flaubert's *Madame Bovary* (1856) also depicts a spectrum of medical practitioners in early-nineteenth-century France, from Homais, the local pharmacist, to Larivière, the Paris surgeon with an international reputation. Charles Bovary, complacent, unambitious, failing to read the medical journals or even his own former notebooks, lacks all the qualities necessary for a country doctor.

20. Anna T. Kitchel, ed. (1949–1950), "Quarry for *Middlemarch*," *Nineteenth-Century Fiction* 4 (suppl.).

21. "This morning I finished the first chapter of *Middlemarch*. . . . I am reading Renouard's *History of Medicine*." Journal entry, Aug. 1869, in *George Eliot's Life as Related in Her Letters and Journals*, ed. J. W. Cross (1885), 3 vols. (Edinburgh and London: Blackwood), 3:97. A month later she commented, "I have achieved little during the last week except reading on medical subjects—the Encyclopedia about the Medical Colleges, Cullen's *Life*, Russell's *Heroes of Medicine*, etc." (3:99).

22. This is made explicit when Lydgate's reactionary colleagues discuss the program of reform publicized by Wakley, proprietor of the *Lancet*.

23. See, e.g., Charles Newman (1957) *The Evolution of Medical Education in the Nineteenth Century* (London: Oxford University Press).

24. George Eliot (1959), *Middlemarch: A Study of Provincial Life* (London: Dent), 1:128.

25. In Paris Lydgate had met Broussais and Pierre-Charles Louis, whose "new book on Fever" he studies so avidly. See C. L. Cline (1974). "Qualifications of the Medical Practitioners in *Middlemarch*," in *Nineteenth-Century Perspectives: Essays in Honor of Lionel Stevenson*, ed. C. de L. Ryals (Durham, NC: Duke University Press), 271–82.

26. Lydgate diagnoses Casaubon's heart condition with the help of a stethoscope. The details of this instrument had been published by René Laënnec only ten years earlier in 1819.

27. See Kitchel, *Quarry for Middlemarch*, 35–36.

28. See Cline, "Qualifications of the Medical Practitioners of *Middlemarch*," 275, 280.

29. George Lewes, Eliot's de facto husband, had also worked through the concept of a subjective element in scientific procedure and concluded that "perception is the assimilation of the Object by the Subject. . . . Hence the search after *the thing in itself* is chimerical: the thing being a group of relations, it is what these are." George Henry Lewes (1874–75), *Problems of Life and Mind*, quoted in George Levine (2008), *Realism, Ethics and Secularism: Essays on Victorian Literature and Science* (Cambridge: Cambridge University Press) 33.

30. Bichat can be seen as the inheritor of the "vitalistic materialism" of Harvey's disciple Francis Glasson, modified by von Haller, La Mettrie, and Diderot.

31. Chapter 45 gives a vivid description of the marshaling of medical and lay forces against Lydgate, as inaccurate rumors and gossip are spread at all levels from the mayoral dinner table to the local public house.

32. Lydgate is soon popularly regarded as Bulstrode's man, so that the latter's disgrace destroys Lydgate's reputation also.

33. G. H. Lewes had written an article advocating the institution of facilities and grants for scientist-doctors to pursue research full time, unhampered by time-consuming practice. After his death Eliot endowed just such a scheme in his memory—a scholarship in physiology for a young man eager to carry out research but without the financial means to do so. See Eliot to Mme. Bodichon, 8 Apr. 1879, quoted in Cross, *George Eliot's Life*, 3:355.

34. François-Vincent Raspail (1833), *Nouveau système de chemie organique, fondé sur des méthodes nouvelles d'observation* (Paris: Baillière), 205.

35. W. J. Harvey (1967), "The Intellectual Background of the Novel: Casaubon and Lydgate," in *Middlemarch: Critical Approaches to the Novel*, ed. B. Hardy (London: Athlone), 25–37.

36. Harvey (ibid.) points to a further irony. Eliot herself was limited in her biological knowledge by her time. Biochemistry has reverted to something not unlike Lydgate's notion of a primitive tissue.

37. Alfred Tennyson (1971), *In Memoriam A.H.H.*, in *Tennyson's Poetry*, edited by Robert W. Hill Jr. (New York: W. W. Norton), pp. 121–22, stanza 3, ll. 5–12; p. 147, stanza 55, ll. 5–8; p. 148, stanza 56, ll. 1–4.

38. See J. R. Lucas (1979), "Wilberforce and Huxley: A Legendary Encounter," *Historical Journal* 22, no. 2: 313–30, doi:10.1017/S0018246X00016848; and Keith Stewart Thomson (2000), "Huxley, Wilberforce and the Oxford Museum," *American Scientist* 88, no. 3: 210, doi:10.1511/2000.3.210.

39. Benjamin Disraeli, a British member of Parliament, later to be prime minister, asked during a speech at the Oxford Diocesan Conference (25 Nov. 1864), "What is the question now placed before society with the glib assurance which to me is most astonishing? That question is this: Is man an ape or an angel? I, my lord, I am on the side of the angels. I repudiate with indignation and abhorrence those new fangled theories." *Times*, 26 Nov. 1864, quoted in Raymond Corbey (2005), *The Metaphysics of Apes: Negotiating the Animal-Human Boundary* (Cambridge: Cambridge University Press), 34.

40. Chambers had tried to convey the ideas of embryologist Karl Ernst von Baer, who showed the parallels between stages of embryonic development, the Linnaean classification system, and the sequence of fossils. This was incorrectly interpreted (and ridiculed) by many of his contemporaries to mean a rapid recapitulation of evolution during gestation.

41. Edmund Gosse (1986), *Father and Son* [1907], ed. Peter Abbs (Harmondsworth: Penguin), 103.

42. *Omphalos* is Greek for "navel." This esoteric title relates to the necessary conclusion of Gosse's theory, namely, that the creation was designed to look as though it had evolved, in order to test our faith. Hence Adam, although created de novo, would have been given a navel to look as though he had undergone a fetal existence.

43. Gosse, *Father and Son*, 104.

44. Charles Kingsley wrote *The Water Babies* specifically to introduce children to evolutionary theory, which he believed entirely compatible with Christianity, without recourse to Gosse's sophistry.

45. See, e.g., Constance Areson Clark (2000), "Evolution for John Doe: Pictures, the Public, and the Scopes Trial Debate," *Journal of American History* 87, no. 4: 1275–1303. Jerome Lawrence and Edwin Robert Lee's 1955 play *Inherit the Wind* and the 1960 film of the same name were loosely based on the trial.

46. Thomas Hardy (1986) *A Pair of Blue Eyes* [1873], ed. Roger Ebbatson (Harmondsworth: Penguin), 271.

47. Hardy applied to the Astronomer Royal on 26 Nov. 1881 for permission to visit the Greenwich Observatory. F. E. Hardy (1928), *The Early Life of Thomas Hardy* (London: Macmillan), 195.

48. In his preface to the 1895 edition, Hardy explained that his intention was "to set the emotional history of two infinitesimal lives against the stupendous background of the stellar universe, and to impart to readers the sentiment that of these contrasting magnitudes the smaller might be the greater to them as men." Thomas Hardy (1922), *Two on a Tower* (London: Macmillan), i.

49. In 1834 the British astronomer Sir John Herschel, having reexamined the double stars cataloged by his father, William Herschel, traveled to South Africa to catalog the stars and nebulae of the southern sky. Hardy was certainly influenced not only by Herschel's discoveries but also by the descriptive style in which the astronomer wrote. In writing the "astronomical passages" of *Two on a Tower*, he had similar descriptions available to him from both Herschel and Proctor. See Pamela Gossin (2007), *Thomas Hardy's Novel Universe: Astronomy, Cosmology, and Gender in the Post-Darwinian World* (Aldershot: Ashgate), 164. He would also have known Richard Proctor's *Essays in Astronomy* (1872).

50. George Gissing (1970), *Born in Exile* [1892] (London: Gollancz), 67.

Chapter 9 • The Scientist as Adventurer

1. Alfred North Whitehead (1953), *Science and the Modern World* (Cambridge: Cambridge University Press), 120.

2. Many wonderful-journey stories in the nineteenth century employed scientific or pseudoscientific details to transport the adventurers to the unknown place without creating scientist characters. Edgar Allan Poe's "The Unparalleled Adventure of One Hans Pfaall" (1938), and *The Narrative of Arthur Gordon Pym of Nantucket* (1938), both in *Complete Tales and Poems of Edgar Allen Poe* (New York: Random House), are of this kind. They were followed in the twentieth century by space travel stories in which heroes exemplified by the Buck Rogers and Flash Gordon stereotypes.

3. Lewis Mumford (1934), *Technics and Civilization* (London: Routledge), 184.

4. Pierre-Jules Hetzel's morally didactic intention was for his journal to "augment both the knowledge and the wholesome ideas, the logic and the taste, the compassion and the wit, which constitute the moral and intellectual fabric of French youth." Quoted in Arthur B. Evans (1988), *Jules Verne Rediscovered: Didacticism and the Scientific Novel* (New York: Greenwood), 34.

5. Roland Barthes (1972), "The *Nautilus* and the Drunken Boat," in *Mythologies*, trans. Annette Lavers (London: Jonathan Cape), 65–67.

6. Ibid., 65.

7. The best-known example is Captain Nemo's *Nautilus*. At the 1867 Exposition Universelle Verne had studied a model of the newly developed French navy submarine, *Plongeur*, the inspiration for the *Nautilus*. Verne the engineer delights in detailing its innovatory technicalities. It is powered by electricity provided by sodium/mercury batteries and has floodable tanks to achieve buoyancy and control depth. The crew gathers and farms food from the sea and distils sea water for drinking. For the comfort and delight of its passengers the *Nautilus* is also furnished with an extensive library, a pipe organ, and the best culinary offerings.

8. Verne was so fascinated by this idea that he failed to see the extreme danger that a gas leak would incur. This was demonstrated in 1937 when the German passenger airship *Hindenburg* caught fire and was totally destroyed.

9. Jules Verne (1970) *Journey to the Centre of the Earth*, trans. Robert Baldick (Harmondsworth: Penguin), 39–40.

10. *Lycoperdon* is a family of puff-ball mushrooms first described in 1790 by the French botanist Jean Baptiste François Pierre Bulliard, whom Verne references.

11. Barthes, "The *Nautilus* and the Drunken Boat," 65.

12. Jules Verne (1976), *The Mysterious Island* [1875], trans. Lowell Blair (London: Corgi), 208.

13. This dream recalls Alton Locke's feverish dream in Charles Kingsley's *Alton Locke: Tailor and Poet*, chap. 36 (1850) and looks forward to the journey of Wells's Time Traveller.

14. See, e.g., Peter Costello (1978), *Jules Verne, Inventor of Science Fiction* (London: Hodder & Stoughton), 108.

15. In *The Mysterious Island* (1875), we learn that Nemo was formerly an Indian prince, Dakkar, which sufficiently accounts for his hatred of British imperialism.

16. Jules Verne (1968), *20,000 Leagues under the Sea* [1870], trans. H. Frith (London: Dent), 135.

17. Percy Greg's *Across the Zodiac* (1880) employs the interplanetary voyage in the manner of Verne, hypothesizing about gravitational effects and the topography of Mars. Verne's *The Mysterious Island* may also have been the inspiration for Edward Bellamy's *Looking Backward* (1888), often cited as the first example of the modern technological utopia.

18. H. G. Wells (1961), "The Chronic Argonauts," reprinted in Bernard Bergonzi, *The Early H. G. Wells: A Study of the Scientific Romances* (Manchester: Manchester University Press), 190.

19. Arthur Conan Doyle (1979), *The Lost World* [1912] (London: John Murray & Jonathan Cape), 89.

20. H. G. Wells's novella *In the Days of the Comet* (1906) recounts a similar story of the earth passing through a stream of gas from the tail of a passing comet, though in this case the gas induces a state of euphoria and benevolence.

21. Arthur Conan Doyle (1913), *The Poison Belt* (London: Hodder & Stoughton), 47.

22. With his conversion to spiritualism Challenger's personality changes, becoming "gentler, humbler and more spiritual." Ibid., 516.

23. See chapter 11 below.

24. The latter nineteenth century saw a surge in popularity of various forms of spiritualism, often presented as fringe sciences—mesmerism, clairvoyance, levitation, spirit photography, phrenology, parapsychology, and physiognomy. Nathaniel Hawthorne and W. B. Yeats were among the other writers attracted to these pursuits. Even the biologist Alfred Russel Wallace saw no incompatibility between science and spiritualism. See Malcolm Jay Kottler (1974), "Alfred Russel Wallace, the Origin of Man and Spiritualism," *Isis* 65, no. 2: 144–92.

Chapter 10 • *Efficiency and Power*

1. John Tyndall (1892), "The Belfast Address," in *Fragments of Science*, 2 vols. (New York: Appleton), 2:201.

2. However, Wöhler's contemporaries Justus von Liebig and Louis Pasteur continued to believe in vitalism.

3. See, e.g., Charles Darwin (1887) to Asa Gray, 22 May 1860, in *Life and Letters of Charles Darwin*, ed. Francis Darwin, 2 vols. (London: Murray), 2:312.

4. Samuel Butler (1967), *Erewhon* [1872] (New York: Airmont Classics), 123.

5. H. L. Sussman (1968), *Victorians and the Machine* (Cambridge, MA: Harvard University Press), 155.

6. E. M. Forster (1977), "The Machine Stops," in *Collected Short Stories* (Harmondsworth: Penguin), 131.

7. Poe was almost certainly drawing on rumors about von Kempelen's bogus "automatic" chess player in the form of a life-size Turk in which was concealed a real master chess player. See John Cohen (1969), *Human Robots in Myth and Science* (London: George Allen & Unwin), 90–91.

8. Ambrose Bierce (1954), "Moxon's Master" [1899], in *Science Fiction Thinking Machines: Robots, Androids, Computers*, ed. Groff Conklin (New York: Vanguard), 8.

9. Edward Bellamy's short story "Dr. Heidenhoff's Process" (1880), H. G. Wells's "Story of

the Days to Come" and *The Sleeper Awakes*, Aldous Huxley's *Brave New World*, and George Orwell's *1984* employ hypnotism and hypnopedia to instill values compatible with those of a mechanistic society and to remove those contrary to it.

10. G. K. Chesterton (1927), *Robert Louis Stevenson* (London: Houghton & Stoughton), 72–73.

11. It seems likely that the novel is an autobiographical wish list. The aristocratic Count Villiers de l'Isle-Adam was unsuccessful in finding a suitable wife, and the title of his novel, while symbolic, also connects with his name.

12. Auguste Villiers de L'Isle-Adam (1982), *Tomorrow's Eve*, trans. Robert Martin Adams (Urbana: University of Illinois Press), 64.

13. Jon Stratton (2001), *The Desirable Body: Cultural Fetishism and the Erotics of Consumption* (Urbana: University of Illinois Press), 215.

14. For possible causes, both personal and social, see Arthur B. Evans (1988), *Jules Verne Rediscovered: Didacticism and the Scientific Novel* (New York: Greenwood), 79–82.

15. Jules Verne (1886), *Robur the Conqueror* (Paris: Hetzel), 246.

16. In his novel *Sybil, or The Two Nations* (1845), published the same year as Friedrich Engels's *The Condition of the Working Class in England in 1844*, Disraeli explores the plight of the English working classes—the "Condition of England question" as it was known.

17. H. G. Wells (1970), "The Time Machine," in *Selected Short Stories* (Harmondsworth: Penguin), 63.

18. See Bernard Bergonzi (1961), *The Early H. G. Wells: A Study of the Scientific Romances* (Manchester: Manchester University Press), chap. 1.

19. See Roslynn D. Haynes (1980), *H. G. Wells: Discoverer of the Future* (London: Macmillan), 25.

20. H. G. Wells (1970), "The Lord of the Dynamos," in *Selected Short Stories*, 184.

21. See, e.g., R. D. French (1975), *Antivivisection and Medical Science in Victorian Society* (Princeton, NJ: Princeton University Press), 71.

22. H. G. Wells (1967), *The Island of Doctor Moreau* (London: Penguin), 114.

23. Anne Stiles (2009), "Literature in Mind: H. G. Wells and the Evolution of the Mad Scientist," *Journal of the History of Ideas* 70, no. 2: S317–45, S324–25.

24. Alfred Tennyson (1850), *In Memoriam A.H.H.*, in *Tennyson's Poetry*, edited by Robert W. Hill Jr. (New York: W. W. Norton), p. 147, stanza 55, ll. 6–7.

25. Thomas Henry Huxley (1947), "Evolution and Ethics" [1893], in J. S. Huxley and T .H. Huxley, *Evolution and Ethics* (London: Pilot), 80, 82.

26. See Roslynn D. Haynes (1981), "Wells's Debt to Huxley and the Myth of Doctor Moreau," *Cahiers Victoriens et Édouardiens* 13 (April): 31–41.

27. Wells, *The Island of Doctor Moreau*, xxxiii.

28. H. G. Wells (2010), *The Invisible Man* (New York: Penguin Putnam), 110.

29. In Fritz Lang's film *Die Frau im Mond* (1928) Professor George Mannfeldt, an astronomer, largely based on Wells's Cavor, leads an expedition to mine gold on the moon. Mannfeldt conforms closely to the mad scientist stereotype, being old, with long gray hair and wild eyes. On arriving on the moon, he becomes obsessive about the gold before falling to his death in a crevasse. The film was famous for its presentation of a multistage rocket launch with countdown to zero. The first successful A-4 rocket, launched from Peenemünde, had the *Frau im Mond* logo painted on its base. See Christopher Frayling (2005), *Mad, Bad and Dangerous? The Scientist and the Cinema* (London: Reaktion Books), 82–83.

30. H. G. Wells (1924), *The First Men in the Moon*, in *The Works of H. G. Wells*, Atlantic ed. (London: T. Fisher Unwin), 6:110.

31. H. G. Wells (1909), *Tono-Bungay* (London: Unwin), 329.

32. In *First and Last Things* (1908) Wells describes a nightmare vision of misrule spreading

over the earth. "I see humanity scattered over the world, dispersed, conflicting, unawakened. ... I see human life as avoidable waste and curable confusion. ... Their disorder of effort, the spectacle of futility, fills me with a passionate desire to end waste, to create order, to develop understanding." H. G. Wells (1926), *First and Last Things* (London: Unwin), 277.

33. H. G. Wells (1941), "Religion and Science," in *Guide to the New World* (London: Gollancz), 103.

Chapter 11 • *The Scientist as Hero*

1. Carl Becker (1965), *Progress and Power* (New York: Vintage Books), 2.

2. Rush Welter (1955), "The Idea of Progress in America," *Journal of the History of Ideas* 16: 401–5.

3. See Frank M. Turner (1993), "Public Science in Britain, 1880–1919," in *Contesting Cultural Authority: Essays in Victorian Intellectual Life* (Cambridge: Cambridge University Press), 201–30.

4. Edison, dismissed from school as being "retarded" and homeschooled by his mother, came to despise theorists of science as useless. See Wyn Wachhorst (1981), *Thomas Alva Edison: American Myth* (Cambridge, MA: MIT Press).

5. The only other scientist of modern times with a comparable popular status was Einstein, whose reputation was problematic because of his association with the atomic bomb.

6. John Clute (1993), "Edisonade," in *The Encyclopedia of Science Fiction*, ed. John Clute and Peter Nicholls, 2d ed. (New York: St Martin's), 368–70.

7. Chesney's novel, written just after the Prussian victory in the Franco-Prussian War, was intended to galvanize Britons into improving their defenses.

8. Robert Cromie (1976), *A Plunge into Space* [1890] (Westport, CT: Hyperion), 70.

9. Garrett P. Serviss (1947), *Edison's Conquest of Mars* (Los Angeles: Carcosa House), 179–80.

10. Weldon J. Cobb (1901), *A Trip to Mars: Or, The Spur of Adventure* (London: Smith & Street). Cobb had considered "To Mars with Tesla" as an alternative title. See Everett Franklin Bleiler and Richard Bleiler (1990), *Science Fiction: The Early Years* (Kent, OH: Kent State University), 967. Tesla claimed to have detected coherent signals originating on Mars and developed his Teslascope for the purpose of communicating with Mars, so his inclusion was not improbable. N. Tesla (1901), "Talking with the Planets," *Collier's Weekly*, March, 359–61.

11. The covers of these pulp magazines characteristically display a mix of horror, violence, and helpless Anglo-Saxon women being attacked by aliens. See, e.g., Anthony Frewin (1988), *One Hundred Years of Science Fiction Illustration, 1840–1940* (London: Bloomsbury Books), chapters 4, 5.

12. Ursula K. Le Guin (1975), "American SF and the Other," *Science Fiction Studies* 2, no. 3: 210, 209.

13. Bram Stoker (1965), *Dracula* (New York: Airmont), 99.

14. Stanley Waterloo (1898), *Armageddon* (Chicago: Rand McNally), 258–59.

15. See, e.g., the discussions reported in William Broad (1985), *Star Warriors* (New York: Simon & Schuster).

16. Simon Newcomb (1900), *His Wisdom, the Defender* (New York: Harper), 110–11.

17. Garrett P. Serviss (1948), "The Second Deluge," *Fantastic Novels Magazine* 2 (July): 8, 115.

18. Bernhard Kellermann (1913), *Der Tunnel* (Berlin: S. Fischer), 398.

19. Wells's novel proved to be prophetic in several ways. Wells was originally inspired by Frederick Soddy's interpretation of radium, but in 1913, when *The World Set Free* was written, physicists had no proposals for splitting the atom, yet 1933, the year in which the scientists of the novel first begin constructing fission bombs to be used in the 1956 holocaust, was actually the year in which Frédéric and Irène Joliot-Curie first produced radioactive phosphorus, the first

step in tapping atomic energy. In the novel the scientists discover a substance called caroliniun, which has the same properties and uses as plutonium, an element that was isolated only much later and that, with uranium, is the most important source of nuclear fuel and component of nuclear weapons. The physicist Leo Szilard acknowledged a debt to Wells's novel, both for the idea of how to liberate atomic energy on a large scale and for the realization of the probable consequences. He therefore did what Holsten in the novel had done: he attempted to stop the spread of the knowledge by applying for patents to cover his invention: "Knowing what this would mean—and I knew it because I had read H. G. Wells—I did not want these patents to become public." Leo Szilard (1968), "Reminiscences," *Perspectives in American History* 2: 102.

20. Pax's first message makes this clear: "To all mankind. I am the dictator of human destiny. I command the cessation of hostilities and the abolition of war . . . I appoint the United States as my agent for this purpose." Arthur Train and Robert Williams Wood (1915), *The Man Who Rocked the Earth* (New York: Doubleday, Page), 11.

21. For example, G. Powell, *All Things New* (1926) (London: Hodder & Stoughton); B. Newman, *Armoured Doves* (1931) (London: Jarrolds); and H. Edmonds, *The Professor's Last Experiment* (1935) (London: Rich & Cowan).

22. H. G. Wells, *The Open Conspiracy* (1930) (London: Hogarth); *The Shape of Things to Come* (1933) (London: Hutchison); and *World Brain* (1938) (London: Methuen).

23. Warren Wagar (1982), *Terminal Visions: The Literature of Last Things* (Bloomington: Indiana University Press), 27, 110.

24. Broad, *Star Warriors*.

25. Rudolf Heinrich Daumann (1940), *Protuberanzen. Ein utopischer Roman* (Berlin: Schützen), 122.

26. Robert Millikan's *Science and the New Civilization* (1930) was a national best seller.

27. Arthur Conan Doyle (1955), "The Final Problem" [1893], in *A Treasury of Sherlock Holmes Stories*, ed. Adrian Conan Doyle (Garden City, NY: Hanover House), 259.

28. Stuart Edward White (1912), *The Sign at Six* (Indianapolis: Bobbs-Merrill), 250.

29. William MacHarg and Edwin Balmer (1910), foreword to *The Achievements of Luther Trant* (Boston: Small, Maynard), 1.

30. Lasswitz was greatly influenced by his teacher Gustav Fechner, who incorporated the idealistic philosophy of Goethe and Schiller into his scientific studies to arrive at a particular neo-Romantic vitalist stance. See Edwin J. Kretzman (1938), "German Technological Utopias of the Pre-War Period," *Annals of Science* 3: 418.

31. "This ignorance of our great men tends to create ignorance in our future leaders; is hurtful to the strength of the nation now, and retards our development in all ways." John Perry, quoted in Turner, "Public Science in Britain," 215.

32. Turner, "Public Science in Britain,", 221.

33. H. G. Wells (1926), *A Modern Utopia* [1905] (London: Unwin), 92.

34. H. G. Wells (1926), *The World Set Free* (London: Unwin), 141–42. Wells repeated this concept in *The World Brain* (London: Methuen, 1938) and *The Fate of Homo Sapiens* (London: Secker & Warburg, 1939).

35. See, for example, William J. Fanning Jr. (2015), *Death Rays and the Popular Media, 1876–1939: A Study of Directed Energy Weapons in Fact, Fiction and Film* (Jefferson, NC: McFarland), 143–44.

36. B. F. Skinner (1948), *Walden Two* (New York: Macmillan), 211–12.

37. Eugene Rabinowitch (1956), "History's Challenge to Scientists," *Bulletin of the Atomic Scientists* 12: 238. This attitude is discussed in more detail in chapter 14 below.

38. The first US artificial satellite, Explorer I, was launched in January of the following year (1958). The vilification of scientists in America began only after the first Russian atomic test, on

29 Aug. 1949, when it was realized that they were leading the way in space research and atomic weaponry.

39. Ian Fleming's creation James Bond, Agent 007, became the most famous prototype of this new hero in films of the sixties and seventies. See Mick Broderick (1988), *Nuclear Movies: A Filmography* (Northcote, Vic.: Post-Modem), 22–23.

Chapter 12 • Mad, Bad, and Dangerous to Know

1. See Andrew Tudor (1989), *Monsters and Mad Scientists: A Cultural History of the Horror Movie* (Oxford: Oxford University Press), 17–47.
2. Peter Weingart, Claudia Muhl, and Petra Pansegrau (2003), "Of Power Maniacs and Unethical Geniuses: Science and Scientists in Fiction Films," *Public Understanding of Science* 12: 279–87.
3. See www.youtube.com/watch?v=7JDaOOwoMEE.
4. See http://thebehindthescenes.blogspot.com.au/2012/01/trip-to-mars-1910.html.
5. For a video clip of the Edison film see https://www.youtube.com/watch?v=w-fM9meqfQ4, accessed 10 Nov. 2016.
6. See, e.g., Ulrich Beck (2009), *World at Risk*, trans. Ciarin Cromer (Malden, MA: Polity), chaps. 7 and 11.
7. Sir William Crookes, quoted in Frederick Soddy to Sir Ernest Rutherford, 19 Feb. 1903. See Richard Moore (2001), *The Royal Navy and Nuclear Weapons* (London: Frank Cass), 6.
8. Quoted in Spencer R. Weart (1979), *Scientists in Power* (Cambridge, MA: Harvard University Press), 294n11.
9. Frederick Soddy (1932), *The Interpretation of the Atom* (London: John Murray), 4.
10. Spencer R. Weart (1988), "The Physicist as Mad Scientist," *Physics Today*, June, 35.
11. Freeman Dyson (1981), *The Day after Trinity: J. Robert Oppenheimer and the Atomic Bomb* (Kent, OH: Transcript Library), 14, 30.
12. W. H. Rhodes (1974), "The Case of Summerfield," in *Caxton's Book*, ed. Daniel O'Connell (Westport, CT: Hyperion), 19. Kurt Vonnegut's *Cat's Cradle* [1963] uses a similar scenario but substitutes ice for fire.
13. Robert Cromie (1895), *The Crack of Doom* (London: Digby, Long), 85–86.
14. Arthur Conan Doyle (1976), "The Disintegration Machine," in *The Complete Professor Challenger Stories* (London: John Murray & Jonathan Cape), 531.
15. Georg Kaiser (1953), *Gas I*, trans. Hermann Scheffauer, in *Twenty-Five Modern Plays*, ed. S. M. Tucker and A. S. Downer (New York: Harper), 605.
16. Georg Kaiser (1953), *Gas II*, trans. Winifred Katzin, in *Twenty-Five Modern Plays*, ed. S. M. Tucker and A. S. Downer (New York: Harper), 639.
17. Priestley may have had in mind Einstein's resistance to the statistical implications of quantum theory and his much-quoted dictum "God does not play dice."
18. J. B. Priestley (1938), *The Doomsday Men* (London: Heinemann), 289.
19. F. H. Rose (1946), *The Maniac's Dream* (London: Duckworth), 158–59.
20. L. Sprague de Camp (1955), "Judgement Day," *Astounding Science Fiction* 55, no. 6: 72, 73.
21. See Paul Brians (1987), *Nuclear Holocausts: Atomic War in Fiction, 1895–1984* (Kent, OH: Kent State University Press), 277.
22. Christopher Frayling (2005), *Mad, Bad and Dangerous? The Scientist and the Cinema* (London: Reaktion Books), 103.
23. Ibid., 105.
24. Philip K. Dick (1977), *Dr. Bloodmoney; or, How We Got Along after the Bomb* (Boston: Gregg), 165.

25. Fredric Jameson (1975), "After Armageddon: Character Systems in *Dr. Bloodmoney*," *Science Fiction Studies* 2, no. 1: 40.

26. J. Tyndall (1875), *Address Delivered before the British Association Assembled at Belfast* (New York: Appleton), 1–65.

27. Another Faustian biologist who in human breeding experiments intends to outdo God by producing stronger types is Lambacher in Hans Henny Jahnn's novel *Trümmer des Gewissens (Der staubige Regenbogen)*, (The ruins of conscience [The dusty rainbow]) (1961), ed. Walter Muschg (Frankfurt am Main: Europäische Verlagsanstalt).

28. André Gide (1952), *The Vatican Cellars* [1914], trans. Dorothy Bussy (London: Cassell), 5.

29. A. Hyatt Verrill (1927), "The Plague of the Living Dead," *Amazing Stories*, Apr. 20.

30. A. Hyatt Verrill (1927), "The Ultra-Elixir of Youth," *Amazing Stories*, Aug., 483.

31. Alexander Snyder (1926), "Blasphemers' Plateau," *Amazing Stories*, Oct., 663.

32. See Leland Sapiro (1964), "The Faustus Tradition in the Early Science Fiction," *Riverside Quarterly* 1, no. 2: 53–55.

33. In 2008 an opera *The Fly* was performed in Paris and Los Angeles, directed by David Cronenberg, who cowrote and directed the 1986 film.

34. Aldous Huxley (1948), *Along the Road: Notes and Essays of a Tourist* [1925] (London: Chatto & Windus), 223.

35. Aldous Huxley (1926), *Jesting Pilate* (London: Chatto & Windus), 273.

36. Huxley prefaced *Brave New World* with a quotation from Nicholas Berdiaeff: "Les utopies apparaissent bien plus réalisables qu'on ne le croyait autrefois. Et nous nous trouvons actuellement devant une question bien autrement angoissante: Comment éviter leur réalisation définitive?" (Utopias appear more feasible than previously thought. And we are now confronted with a more agonizing question: How to avoid their final completion?) Aldous Huxley (1971), *Brave New World* (Harmondsworth: Penguin Books), n.p.

37. Other examples include Bernard Wolfe, *Limbo* [1952]; Kurt Vonnegut, *Player Piano* [1952]; Isaac Asimov, *The Caves of Steel* [1954]; Ray Bradbury, *Fahrenheit 451* [1954]; and Frederick Pohl, *Drunkard's Walk* [1960].

38. See, e.g., J. B. S. Haldane (1969), "Auld Hornie, F.R.S.," reprinted in *Shadows of Imagination*, ed. M. R. Hillegas (Carbondale: Southern Illinois University Press), 15–25.

39. C. S. Lewis (1944), *Perelandra* [1943] (New York: Macmillan), 81.

40. C. S. Lewis (1946), *Out of the Silent Planet* [1938] (London: John Lane, the Bodley Head), 21–22.

41. Cf. Gulliver's account of Europe to the King of Brobdingnag and Cavor's to the Grand Lunar in H. G. Wells's *The First Men in the Moon* (see chapter 10 above).

42. *Perelandra*, 81.

43. H. G. Wells (1902), "Discovery of the Future," *Nature* 65 (6 Feb.): 331.

44. C. S. Lewis (1946), *That Hideous Strength* (London: John Lane, the Bodley Head), 45.

45. Christa Wolf (1982), "Ein Brief," in *Mut zur Angst. Schriftsteller für den Frieden*, ed. Ingrid Krüger (Darmstadt: Luchterhand), 54–55, 157.

Chapter 13 • *The Impersonal Scientist*

1. See Uwe Schimank (1992), "Science as a Societal Risk Producer: A General Model of Intersystemic Dynamics, and Some Specific Institutional Determinants of Research Behavior," in *The Culture and Power of Knowledge: Inquiries into Contemporary Societies*, ed. Nico Stehr and Richard V. Ericson (Berlin: De Gruyter), 218.

2. See Gyorgy Markus (2011), *Culture, Science, Society: The Constitution of Cultural Modernity* (Leiden: Brill), 216.

3. Louis MacNeice (1979), "The Kingdom," in *Collected Poems of Louis MacNeice*, ed. E. R. Dodds (London: Faber & Faber), 252–53.

4. D. H. Lawrence (1966), *The Rainbow* (Harmondsworth: Penguin), 350.

5. During eighteen months in England (1916–17) supervising the construction of Russian icebreakers, Zamyatin became familiar with Wells's work and praised his scientific romances as "social-scientific fantasy," the new urban fairy tales of the industrial age. Yevgeny Zamyatin (1970), "H. G. Wells," in *A Soviet Heretic: Essays by Yevgeny Zamyatin*, ed. and trans. Mirra Ginsburg (Chicago: University of Chicago Press), 259–90.

6. Yevgeny Zamyatin (1924), *We*, trans. Gregory Zilborg (New York: E. P. Dutton), 4.

7. Aldous Huxley (1971), *Brave New World* [1932] (Harmondsworth: Penguin), 37.

8. Thomas Pynchon (1973), *Gravity's Rainbow* (New York: Viking), 434.

9. Walter Gratzer suggests that Dr. Obispo bears some resemblance to Serge Voronoff, who became famous for grafting monkey testicle tissue onto human testicles allegedly to reinvigorate older recipients. But Huxley may have had in mind Casimir Funk, who tried to procure from Abyssinian tribesmen the testicles of Italian prisoners in order to isolate testosterone. Walter Gratzer (1989), *The Longman Literary Companion to Science* (Harlow, Essex: Longman), 245.

10. Aldous Huxley (1947), *Point Counter Point* (London: Chatto & Windus), 25.

11. Dr. Poole, the botanist of *Ape and Essence* (1951), is another emotionally undeveloped scientist.

12. Aldous Huxley (1955), *The Genius and the Goddess* (London: Chatto & Windus), 23.

13. C. P. Snow (1959), *Death under Sail* (London: Heinemann), 121–22. Snow published this detective story anonymously lest it adversely affect his career as a scientist.

14. C. P. Snow (1979), *The Search*, rev. ed. (Harmondsworth: Penguin), 69.

15. "If false statements are to be allowed, if they are not to be discouraged by every means we have, science will lose its one virtue, truth.... If we do not penalize false statements in error, we open up the way, don't you see, for false statements by intention. And of course a false statement of fact, made deliberately, is the most serious crime a scientist can commit." Ibid., 257.

16. The fictional incident parallels a similar one in Snow's own career. Having worked with P. V. Bowden on the chemistry of vitamin A and its derivatives, Snow published a paper that was well received. However, it was later found that the experimental work was seriously flawed and the conclusions drawn from it unfounded. Snow published no retraction and gave no satisfactory reason for the errors, but he did not publish another scientific paper.

17. Irving Fineman (1939), *Dr. Addams* (New York: Random House), 206–7.

18. Max Frisch (1969), *Don Juan; or, The Love of Geometry*, in *Max Frisch: Four Plays*, trans. Michael Bulloch (London: Methuen), 118–19.

19. Erich Nossack (1956), "Die Schalttafel," in *Spirale* (Frankfurt: Suhrkamp), 47–113.

20. Heinz von Kramer (1964), "Aufzeichnungen eines ordentlichen Menschen," in *Leben wie im Paradies* (Hamburg: Hoffmann & Campe), 15.

21. Hermann Broch (1986), *Die unbekannte Größe* in *Gesammelte Werke*, ed. Ernest Schönwiese (Zurich: Rhein), 10:76.

22. Stephen Sewell (1983), *Welcome the Bright World* (Sydney: Alternative Publishing Cooperative and Nimrod Theatre Press), 24.

23. Robert Musil *Der Mann ohne Eigenschaften* [1930–43], trans. as *The Man Without Qualities* by Eithne Wilkins and Ernst Kaiser (1953–60), 3 vols. (London: Secker & Warburg), 1:296.

24. Christa Wolf (1978), "Self-Experiment: Appendix to a Report" [1973], trans. Jeanette Clausen, *New German Critique* 13 (Winter): 115.

25. Friederike Eigler (2000), "Rereading Christa Wolf's 'Selbstversuch': Cyborgs and Feminist Critique of Scientific Discourse," *German Quarterly* 73, no. 4: 410.

26. Christa Wolf (1988), *Störfall. Nachrichten eines Tag* (Frankfurt am Main: Luchterhand), 70.

27. For an account of Klaus Fuchs see Alan Moorehead (1952), *The Traitors* (New York: Charles Scribner's Sons).

28. Carl Zuckmayer (n.d.), *The Cold Light*, trans. Elizabeth Montagu (typescript, Columbia University Libraries), act 1, sc. 1, 7.

29. The other scientists in the play span a wide spectrum of types and attitudes. The militaristic Ketterick is troubled by no scruples about the use of the atomic bomb against any enemy, real or potential; the sensitive Löwenschild, a man of conscience, profoundly regrets the destructive potential of the research he does but believes it is the lesser of two evils.

Chapter 14 • "Scientia Gratia Scientiae"

1. Quoted by Barton J. Bernstein (1975) in "Shatterer of Worlds: Hiroshima and Nagasaki," *Bulletin of the Atomic Scientists* 31, no. 10 (December): 21.

2. Shiv Visvanathan (1984), "Atomic Physics: The Career of an Imagination," *Alternatives: A Journal of World Policy* 10, no. 2: 208.

3. Quoted in Spencer R. Weart's (1982) review of the film, in *Bulletin of the Atomic Scientists* 38 (August/September): 41–42.

4. See David Irving (1967), *The German Atomic Bomb: The History of Nuclear Research in Nazi Germany* (New York: Simon & Schuster), 67.

5. See Brian Easlea (1983), *Fathering the Unthinkable: Masculinity, Scientists and the Nuclear Arms Race* (London: Pluto).

6. Robert Merton enunciated a set of scientific norms, which he believed to be consistent with the goals of science and binding on scientists. In his original version these were communalism, universalism, disinterestedness, and organized skepticism (CUDOS). R. K. Merton (1942), "The Normative Structure of Science," in *The Sociology of Science: Theoretical and Empirical Investigations* (Chicago: University of Chicago Press). This list was later expanded to include originality.

7. Russell McCormmach (1982), *Night Thoughts of a Classical Physicist* (Cambridge, MA: Harvard University Press), 148–49.

8. Albert Einstein (1985), quoted in John S. Weltman, "Trinity: The Weapons Scientists and the Nuclear Age," *SAIS Review* 5, no. 2: 39.

9. Freeman Dyson (1984), "Weapons and Hope," pt. 2, *New Yorker*, 13 Feb., 77.

10. Joseph Rotblat (1985), "Leaving the Bomb Project," *Bulletin of the Atomic Scientists* 41: 18. A signatory of the Russell-Einstein Manifesto of 1955, protesting against the further development of nuclear weapons, Rotblat was secretary-general of the Pugwash Conferences on Science and World Affairs and shared with them the 1995 Nobel Peace Prize for efforts toward nuclear disarmament.

11. Carl Sandburg (1950), "Mr. Attila," in *Complete Poems* (New York: Harcourt Brace).

12. Bertolt Brecht (1965), *The Life of Galileo* [1943], trans. Desmond I. Versey, in *Plays*, 2d ed. (London: Methuen), 8.

13. Quoted in Käthe Rütlicke (1963), "Bemerkungen zur Schluß-Szene," in Bertolt Brecht, *Materialen zu Brechts "Leben des Galilei"* (Frankfurt: Suhrkamp), 97, 106.

14. Bertolt Brecht (1965), notes on *The Life of Galileo*, in *Plays*, 2d ed. (London: Methuen), 340.

15. Arthur Koestler (1988), *The Sleepwalkers* (London: Penguin).

16. Brecht held that the audience should never surrender its separation from the events taking place on the stage. He believed that a sense of alienation (*Verfremdungseffekt*) was essential to provoke the audience to think objectively about the issues presented in the drama. See Frederic Jameson (1998), *Brecht and Method* (London: Verso).

17. C. P. Snow (1954), *The New Men* (London: Macmillan), 46.

18. US Atomic Energy Commission (1954), *In the Matter of J. Robert Oppenheimer* (Washington, DC: USGPO), 251.

19. Philip M. Stern in collaboration with Harold P. Green (1969), *The Oppenheimer Case: Security on Trial* (New York: Harper and Row).

20. Under the Baruch Plan the United States proposed turning over all its weapons if all other countries agreed not to produce them and agreed to inspection by the UN. However, the Soviet Union refused to comply on the grounds that the UN was controlled by the United States and its allies and therefore could not be impartial.

21. Heinrich Schirmbeck (1960), *Ärgert dich dein rechtes Auge*, trans. as *The Blinding Light* by Norman Denny (London: Collins), 341–42.

22. Heinz von Cramer (1961), *Die Konzessionen des Himmels* (Hamburg: Hoffmann & Campe). The major atomic tests were held at Emu Field in 1953 and Maralinga in 1956. The bombs ranged in size from one kiloton to twenty-five kilotons, the largest being greater than the bomb dropped on Hiroshima.

23. Judy Nunn's novel *Maralinga* (2009) is set in 1956 at a British air base in the South Australian desert, where atomic weapons are being tested. The tests, assented to by the Australian government on a need-to-know basis, which excluded Australians from any knowledge of what was happening, took no account of the Aboriginal people on whose lands they were occurring and who, with their descendants, continue to suffer from the radiation effects.

24. Daniel Lange (1972), "Ex-Oracles," *Harper's* 245 (Dec.): 34.

25. Pearl S. Buck (1959), *Command the Morning* (New York: John Day).

26. The reference here, as in the title, is to Job 38:12: "Hast thou commanded the morning since thy days?"

27. Dexter Masters (1955), *The Accident* (London: Cassell), 18.

28. Heiner Kipphardt (1967), *In the Matter of J. Robert Oppenheimer*, trans. Ruth Speirs (London: Methuen), pt. 2, sc. 2, 82.

29. Martin Cruz Smith (1992), *Stallion Gate* (Glasgow: HarperCollins), 167.

30. The film title references in inverse order the code names of the two nuclear weapons detonated over Hiroshima and Nagasaki. These code names were based on characters in the works of detective novelist Dashiell Hammett.

31. Kurt Vonnegut (1965), *Cat's Cradle* (London: Gollancz). The title refers to a children's game in which the strings can form elaborate and ever-changing patterns or become hopelessly knotted.

32. However, lest we should begin to gain faith in scientists, Vonnegut tells us that von Koenigswald has been an official at Auschwitz and thus has a "terrible deficit . . . in his kindness account" (159).

33. William Broad (1985), *Star Warriors* (New York: Simon & Schuster), 13, 85.

34. Christa Wolf (1988), *Störfall. Nachrichten eines Tag* (Frankfurt am Main: Luchterhand), 70.

35. Saul Bellow (1978), *Mr. Sammler's Planet* (Harmondsworth: Penguin), 176.

36. Wells's treatise *The Open Conspiracy: Blueprints for a World Revolution* (1928) offered a program for setting up a utopian society based on efficiency and order by means of a science-based education program.

37. For example, *Cybernetics: Or Control and Communication n the Animal and the Machine* [1948], *The Human Use of Human Beings: Cybernetics and Society* [1950], and *I Am a Mathematician* [1956].

38. Rorbert Wiener (1971), *The Human Use of Human Beings: Cybernetics and Society* (New York: Avon Books), 244.

Chapter 15 • Robots, Androids, Cyborgs, and Clones

1. See Jacques Ellul (1964), *The Technological Society* (New York: Knopf); and Stephen Hawking, "Do Black Holes Have No Hair?," BBC Reith Lecture, 26 Jan. 2016, transcript retrieved from www.bbc.com/news/science-environment-35354313.
2. Charles Dickens (1998), *Hard Times* [1854] (Ware, HERTS: Wordsworth Classics), chap. 11.
3. Čapek's robots were not the first mechanical servants in literature. In *The Iliad* Homer writes that Achilles the warrior had new armor made by Hephaestus, god of fire, assisted by "handmaidens wrought of gold in the semblance of living maids. In them is understanding in their hearts, and in them speech and strength." Homer (1924), *The Iliad*, trans. A. T. Murray (Harvard: Harvard University Press), bk. 18, 410.
4. Klaus Benesch (1999), "Technology, Art, and the Cybernetic Body: The Cyborg as Cultural Other in Fritz Lang's 'Metropolis' and Philip K. Dick's 'Do Androids Dream of Electric Sheep?'" *Amerikastudien/American Studies*, 44, no. 3: 385.
5. Čapek's robots are not automata. They are not mechanical devices but artificial biological organisms that appear humanlike and have independent thought. They are more like androids.
6. Karel Čapek (1923), *R.U.R. (Rossum's Universal Robots: A Fantastic Melodrama)* [1921], trans. Paul Selver (Garden City, NY: Doubleday, Page), 14.
7. See www.youtube.com/watch?v=NDNi37Qlrn4.
8. Christopher Frayling (2005), *Mad, Bad and Dangerous? The Scientist and the Cinema* (London: Reaktion Books), 60.
9. This is the situation first described by H. G. Wells in "The Time Machine" (1895). In some future time, the Eloi, living a carefree existence aboveground, are supported by the subterranean Morlocks, who drive the machines that provide for their existence. However, the Time Traveller later learns that the Morlocks emerge at night and drag the Eloi belowground to be eaten like cattle. In the case of *Metropolis* it also represents an interwar society split by capital/labor conflicts.
10. Thea von Harbou (1927), *Metropolis* (London: Readers Library), 61; Fritz Lang (1973), *Metropolis* (Letchworth: Garden City Press).
11. Andreas Huyssen (1981–82), "The Vamp and the Machine: Technology and Sexuality in Fritz Lang's *Metropolis*," *New German Critique* 24/25: 221–37, 227.
12. Quoted in Gray Kochhar-Lindgren (1998), "Ethics, Automation, and the Ear: Capitalism, Technology, and the Suspension of Animation in Ernst Jünger's *The Glass Bees*," Ctheory.net, https://journals.uvic.ca/index.php/ctheory/article/view/14623/5489.
13. Isaac Asimov (1979), introduction to *I, Robot*, in *Isaac Asimov Presents the Great SF Stories* (New York: Daw Books). Eando (E and O) was a pen name of two brothers, Earl and Otto Binder.
14. Asimov was considerably influenced at the start of his career by John W. Campbell, the editor of *Astounding Stories*, where most of his robot stories were published. See Isaac Asimov (1973), *The Early Asimov, or Eleven Years of Trying* (London: Gollancz).
15. Isaac Asimov, "On Computers," *DP Solutions*, Apr. 1975, 2.
16. The Three Laws of Robotics devised by Asimov were introduced in the short story "Runaround." They state, "One, a robot may not injure a human being or, through inaction, allow a human to come to harm." "Two, . . . a robot must obey the orders given it by human beings except where such orders conflict with the First Law." "Three, a robot must protect its own existence as long as such protection does not conflict with the First or Second Laws." Isaac Asimov (1989), "Runaround," in *I, Robot* (London: Collins), 50–51.
17. Isaac Asimov (1989), "Evidence," in *I, Robot*, 206.
18. Philip K. Dick (1972), "The Android and the Human," accessed 10 Nov. 2016, http://1999pkdweb.philipkdickfans.com/The%20Android%20and%20the%20Human.htm.

19. The Turing test is a game devised by British mathematician Alan Turing to decide, from answers delivered through a computer terminal, whether a respondent is a human or a machine.

20. Benesch, "Technology, Art, and the Cybernetic Body," 381. Benesch's use of the word *cyborg* does not conform to the modern meaning. Maria in *Metropolis* is a female android or gynoid, not a cyborg.

21. See, e.g., David Levy (2009), "The Ethical Treatment of Artificially Conscious Robots," *International Journal of Social Robotics* 1: 210.

22. Stanislaw Lem (1971), "Robots in Science Fiction," in *SF: The Other Side of Realism: Essays on Modern Fantasy and Science Fiction*, ed. Thomas D. Clareson (Bowling Green, KY: Bowling Green University Popular Press), 320.

23. Donna Haraway (1991), "A Cyborg Manifesto," in *Simians, Cyborgs and Women: The Reinvention of Nature* (New York: Routledge), 150.

24. Immanuel Kant's categorical imperative is best phrased as "Act according to that maxim which you would will to become a universal law."

25. See Mark Coeckelbergh (2009), "Personal Robots, Appearance, and Human Good: A Methodological Reflection on Roboethics," *International Journal of Social Robotics* 1: 221.

26. Reuters (2016), "Europe's Robots to Become 'Electronic Persons' under Draft Plan," *Sydney Morning Herald*, 22 June, retrieved from www.smh.com.au/world/-gpotdk.html.

27. H. Lim (1999), "Caesareans and Cyborgs," *Feminist Legal Studies* 7, no. 2: 133–73.

28. Abby Lippman (1991), "Prenatal Genetic Testing and Screening: Constructing Needs and reinforcing Inequities," *American Journal of Law and Medicine* 17: 15–50, 19.

29. See also chapters 6 and 10 above.

30. Dorothy Nelkin and M. Susan Lindee (1995), *The DNA Mystique: The Gene as a Cultural Icon* (New York: W. H. Freeman), 41.

31. Michael Lemonick (1993), "Cloning Classics," *Time*, 8 Nov., 68.

32. The name is derived from Genesis 1:26–27, which describes the creation of man on the sixth day.

33. Christopher Rose (2003), "How to Teach Biology Using the Movie Science of Cloning People, Resurrecting the Dead, and Combining Flies and Humans," *Public Understanding of Science* 12: 289–96, 294.

34. Fay Weldon (1989), *The Cloning of Joanna May* (London: Collins), 122.

35. This possibility is now largely discredited by scientists who found that the potential for DNA survival in resin is minimal. See David Penney et al. (2013), "Absence of Ancient DNA in Sub-Fossil Insect Inclusions Preserved in 'Anthropocene' Colombian Copal," *PLoS ONE* 8, no. 9: e73150, doi:10.1371/journal.pone.0073150, retrieved from https://www.researchgate.net/publication/256614113_Absence_of_Ancient_DNA_in_Sub-Fossil_Insect_Inclusions_Preserved_in_'Anthropocene'_Colombian_Copal. There was a similar proposal by Michael Archer to revive the extinct thylacine, or Tasmanian tiger, by cloning from DNA of a thylacine embryo preserved in the Australian Museum. However, the project was abandoned when the DNA was declared to be too degraded. See Judy Skatsoon (2005), "Thylacine Cloning Project Dumped," *ABC Science*, 15 Feb., retrieved from www.abc.net.au/science/articles/2005/02/15/1302459.htm.

36. Quoted in A. McGregor (1993), "The Crichton Factor," *Time Out*, 7 Apr., 21.

37. Kim Newman (2001), interview in *Celluloid Scientists*, BBC, Radio 4.

38. Spencer Weart (2001), interview, *Celluloid Scientists*, BBC, Radio 4.

39. Jon Turney (1998), *Frankenstein's Footsteps: Science, Genetics and Popular Culture* (New Haven, CT: Yale University Press), 213.

40. The word *gattaca* is derived from the first letters of guanine, adenine, thymine, and cytosine, the four nucleotide bases of DNA.

41. David Kirby (2004), "Extrapolating Race in GATTACA: Genetic Passing, Identity, the New Eugenics and the Science of Race," *Literature and Medicine* 23, no. 1: 184–200, 190.
42. Caryl Churchill (2002), *A Number* (New York: Theatre Communications Group), 60–61.
43. Steven Spielberg's DreamWorks purchased the film rights for *Spares* but never made a film of the novel. When the rights lapsed, DreamWorks produced the film *Island*, which had many similarities to *Spares*.
44. L. Suchman, R. Trigg, and J. Blomberg (2002), "Working Artefacts: Ethnomethods of the Prototype," *British Journal of Sociology* 53, no. 2: 163–79, 164.
45. David Kirby (2011), *Lab Coats in Hollywood: Science, Scientists, and Cinema* (Cambridge, MA: MIT Press), 196.
46. Stephen J. Burn (2008), "An Interview with Richard Powers," *Contemporary Literature* 49 (Summer): 163–79, 178.
47. Jennifer Doudna (2015), "Genome-Editing Revolution: My Whirlwind Year with CRISPR," *Nature* 528 (24 Dec.): 469–71, doi:10.1038/528469a.

Chapter 16 • Pandora's Box

1. Joe Kane (1969), "Nuclear Films," *Take One* 2 (July/August): 10.
2. As early as 1939 the work of Enrico Fermi and Leo Szilard on nuclear chain reactions was being repeated by von Weizsäcker and others at the Kaiser Wilhelm Institut in Berlin.
3. Kepler is persecuted by the Protestants for supporting the Gregorian (and hence Catholic) calendar and by the Catholics for advocating the Copernican view of the universe, which was interpreted as undermining religious authority.
4. Hans Rehberg (1933), *Johannes Kepler: Schauspiel in drei Akten* (Berlin: S. Fischer), 78.
5. Max Brod (1948), *Galilei in Gefangenschaft* (Winterthur: Mondial), 774.
6. In another sense, however, Zwillinger's play was prophetic, since, in 1992 a papal edict absolved Galileo of heresy.
7. Nigel Balchin (1949), *The Small Back Room* (London: Gollancz), 190.
8. William Shakespeare (1916), *King Richard II*, ed. W. J. Craig (London: Oxford University Press), act 4, sc. 1, ll. 244–48.
9. Freedom to publish research has been regarded as a fundamental right of science but became a critical issue in the Second World War. More recently it is threatened from other directions by commercial secrecy requirements.
10. Nigel Balchin (1949), *A Sort of Traitors* (London: Collins), 39.
11. Upton Sinclair (1948), *A Giant's Strength* (London: T. Werner Laurie), 20.
12. Sinclair's scenario predates the actual enquiry into the activities of J. Robert Oppenheimer.
13. Hans Henny Jahnn (1961), *Trümmer des Gewissens (Der staubige Regenbogen)* (Frankfurt am Main: Europäische Verlagsanstalt), 199.
14. Buzzati indicates Einstein's pride in wishing to complete his work before he dies: "Here my work has considerable interest." Dino Buzzati (1984), "Appointment with Einstein," in *Restless Nights*, trans. Lawrence Venuti (Manchester: Carcanet), 12.
15. Dino Buzzati (1984), "Appointment with Einstein" [1958], in *Restless Nights*, trans. Lawrence Venuti (Manchester: Carcanet), 14.
16. Compare Douglas Stewart's presentation of Rutherford, aware of the destructive potential of his discovery of nuclear fission, but persuading himself that it would be used only for good.
17. Friedrich Dürrenmatt (1966), *Theaterschriften und Reden* (Zurich: Im Verlag der Arche), 275.
18. Friedrich Dürrenmatt (1963), *Die Physiker*, trans. James Kirrup as *The Physicists* (London: French's Acting Edition), 1.

19. Walter Muschg (1961), "Dürrenmatt und *Die Physiker*," *Moderna Sprak* 56: 280.
20. In Dürrenmatt, *Die Physiker*, 60.
21. The secret hearing lasted from 12 Apr. to 14 May, and the play is distilled from the three thousand pages of typed records published by the US Atomic Energy Commission in May 1954.
22. Heiner Kipphardt (1967), *In the Matter of J. Robert Oppenheimer*, trans. Ruth Speirs (London: Methuen), 102–3.
23. Kipphardt also explores the relative importance of public and private roles of scientists. Oppenheimer sees no inconsistency in having active communists as personal friends while being an advisor to the government on atomic matters, but it is clear that this is an anomaly for the committee of enquiry.
24. In 1983, at a reunion of Los Alamos scientists, in a paper entitled "We Meant Well," Isador Rabi lamented that the "nations are now lined up like people before the ovens of Auschwitz while we are trying to make the ovens more efficient. . . . We meant well. . . . We gave the power away to people who didn't understand it and now it's gotten out of our hands." I. Rabi, quoted in Greg Herken (1983), *Counsels of War* (New York: Knopf).
25. Kipphardt, *In the Matter of J. Robert Oppenheimer*, 88–89.
26. Ibid., 106–7.
27. See Evans's soliloquy: "Perhaps I should have turned down this appointment. . . . I cannot reconcile these interrogations with my idea of science." Ibid., 24.
28. Friedrich Dürrenmatt (1973), *Der Mitmacher. Komodie* (Basel: Reiss), 14.
29. C. P. Snow (1974), *In Their Wisdom* (London: Macmillan), 145.
30. Russell McCormmach (1982), *Night Thoughts of a Classical Physicist* (Cambridge, MA: Harvard University Press), 148.
31. Howard Brenton (1983), *The Genius* (London: Methuen), 21.
32. Christa Wolf (1988), *Störfall. Nachrichten eines Tag* (Frankfurt am Main: Luchterhand), 37.

Chapter 17 • *The Scientist as Woman*

1. Peter Weingart (2003), with assistance from Claudia Muhl and Petra Pansegrau, "Of Power Maniacs and Unethical Geniuses: Science and Scientists in Fiction Film," *Public Understanding of Science* 12: 279–87, 283.
2. Eva Flicker (2003), "Between Brains and Breasts—Women Scientists in Fiction Film: On the Marginalization and Sexualisation of Scientific Competence," *Pubic Understanding of Science* 12: 307–18.
3. Although there is some dispute about authorship, it has generally been conceded that Lovelace was largely responsible for the "Notes on the Engine," which include the first algorithm designed to be carried out by a machine—that is, the first computer program.
4. Robert K. Merton (1968), "The Matthew Effect in Science," *Science* 159, no. 3810: 56–63; Margaret W. Rossiter (1993), "The Matilda Effect in Science," *Social Studies of Science* 23: 325–41.
5. See, e.g., Richard Holmes (2010), "The Royal Society's Lost Women Scientists," *Guardian*, 21 November, retrieved from https://www.theguardian.com/science/2010/nov/21/royal-society-lost-women-scientists?CMP=twt_gu.
6. See Eva Hammungs Wirtén (2015), *Making Marie Curie: Intellectual Property and Celebrity Culture in an Age of Information* (Chicago: University of Chicago Press).
7. A. Elena (1997), in "Skirts in the Lab: Marie Curie and the Image of the Woman Scientist in the Feature Film," *Public Understanding of Science* 6: 269–76, emphasizes this subservient role, but two more recent films produced by the BBC, the four-part *Marie Curie* (1977) and *The Genius of Marie Curie—The Woman Who Lit Up the World*" (2007) in the Great Lives series,

portray Curie as more independent and assertive, as does the French miniseries, *Marie Curie, une femme honourable* (1991).

8. Sidney Perkowitz (2006), "Female Scientists on the Big Screen," *Scientist*, 21 July, www.the-scientist.com/?articles.view/articleNo/24170/title/Female-scientists-on-the-big-screen/.

9. Jocelyn Steinke (2005), "Cultural Representations of Gender and Science: Portrayals of Female Scientists and Engineers in Popular Films," *Science Communication* 27, no. 1: 27–63.

10. The doubtful figure of Maria the Prophetess or Maria the Jewess was alleged in some hermetic writings to have invented several pieces of alchemical apparatus and even to have been the first true alchemist of the Western world. See Raphael Patai (1995), *The Jewish Alchemists: A History and Source Book* (Princeton, NJ: Princeton University Press).

11. Carol Muske-Dukes (2014), *Saving St. Germ: A Novel* (New York: Open Road Intgerated Media), chap. 5, Kindle locus 871.

12. Ibid., Kindle locus 849.

13. ITT: Screenplay for *Batman and Robin*, 118, retrieved from www.ign.com/boards/threads/itt-the-screenplay-for-batman-and-robin.454619122/.

14. Jess Nevins (2011), "From Alexander Pope to 'Splice': A Short History of the Female mad Scientist," retrieved from http://io9.gizmodo.com/5794436/from-alexander-pope-to-splice-a-short-history-of-the-female-mad-scientist.

15. In *Sphere* (1998), based on Michael Crichton's novel of the same name (1987), marine biologist Beth Halpern is one of a group of scientists sent to investigate a spacecraft discovered on the floor of the Pacific Ocean. The spacecraft contains a mysterious, seemingly impenetrable, floating fluid sphere. The fears of the scientists are actualized by the sphere, and Beth develops suicidal thoughts, which are interpreted by her colleagues as madness.

16. Flicker, "Between Brains and Breasts."

17. Robin Roberts (1993), *A New Species: Gender and Science in Science Fiction* (Urbana: University of Illinois Press).

18. Isaac Asimov (1968), "Catch That Rabbit" [1944], in *I, Robot* (St. Albans, HERTS: Panther), 90.

19. Fred Hoyle and John Elliot (1961), *A for Andromeda*, BBC TV, episode 7.

20. William Boyd (2009), *Brazzaville Beach* (London: Penguin), 265–66.

21. Jocelyn Steinke (2005), "Cultural Representations of Gender and Science," *Science Communication* 27, no. 1: 27–63, 39–41.

22. Bonnie Noonan (2005), *Women Scientists in Fifties Science Fiction Films* (Jefferson, NC: McFarland), 49.

23. Flicker, "Between Brains and Breasts," 316–17.

24. Lori Kendall (2000), "Oh No! I'm a Nerd: Hegemonic Masculinity on an Online Forum," *Gender and Society* 14, no. 2: 256–74, 265.

25. Thomas Kuhn (1962), *The Structure of Scientific Revolutions* (Chicago: University of Chicago Press).

26. Kerstin Bergman (2012), "Girls Just Wanna Be Smart? The Depiction of Women Scientists in Contemporary Crime Fiction," *International Journal of Gender, Science and Technology* 4, no. 3: 313–29, 313, retrieved from http://genderandset.open.ac.uk.

27. Ibid., 318.

28. Ibid., 324.

29. Only later is she told that in fact there were eighteen hours of static unaccounted for on the video recorder aboard the spaceship.

30. Larry Klaes (1997), "'Contact' Film Review," retrieved from www.coseti.org/klaescnt.htm#Eleanor%20Arroway%20. I am indebted to Klaes's article for clarifying several of the scientific issues in *Contact*.

31. Steinke, "Cultural Representations of Gender and Science."
32. Sandra L. Hanson (2000), "Gender, Families and Science: Influences on Early Training and Career Choices," *Journal of Women and Minorities in Science and Engineering* 6: 169–87, 170.
33. A. S. Byatt (2004), *A Whistling Woman* (New York: Vintage Books), 56.
34. Jennifer Rohn (2009), *Experimental Heart* (New York: Cold Spring Harbor Laboratory Press), 356–58.
35. Marge Piercy (1974), *Small Changes* (Robbinsdale, MN: Fawcett Crest), 364.
36. Carol Colatrella (2011), *Toys and Tools in Pink: Cultural Narratives of Gender, Science and Technology* (Columbus: Ohio State University Press), 120.
37. Susan M. Gaines (2001), *Carbon Dreams* (Berkeley, CA: Creative Arts Book Company), 327.

Chapter 18 • Idealism and Conscience

1. Émile Zola (1957), *Doctor Pascal*, trans. Vladimir Kean (London: Elk Books), 34.
2. Émile Zola (1898), *Paris*, trans. E. Vizetelly, 2d ed. (London: Chatto & Windus), 486, 116.
3. Robert Millikan (1927), *Evolution in Science and Religion* (New Haven, CT: Yale University Press).
4. Mervyn Leroy's film *Madame Curie* (1943) was followed by John Glenister's TV miniseries *Marie Curie* (1977), the BBC documentary *The Genius of Marie Curie: The Woman Who Lit up the World* (2013), *Marie Curie* (2016), and the European biopic *Maria Sklodowska-Curie* (2016). After the death of Pierre Curie, Marie Curie's scientific success was marred by scandal regarding her affair with the married physicist Paul Langevin.
5. De Kruif later listed the names of the actual persons who had suggested the characters: for Arrowsmith, R. R. Hussey, later professor of pathology at Yale; for Max Gottlieb, F. G. Novy of the University of Michigan and Jacques Loeb of the Rockefeller Institute; for Terry Wickett, T. J. LeBlanc and J. H. Northrop; for Almus Pickerbaugh, William Kleine, medical director of the Red Cross. See Richard R. Lingeman (2005), *Sinclair Lewis: Rebel from Main Street* (St. Paul, MN: Borealis), 206, 222.
6. One of the moral dilemmas that confronts Arrowsmith occurs in the West Indies, where he goes to test the efficacy of his bacteriophage against an outbreak of bubonic plague. A clinical trial involves refusing inoculation to half the victims as a control, but Arrowsmith proves too humane to adhere to this requirement, and therefore the trial is inconclusive.
7. Sinclair Lewis (1961), *Arrowsmith* [1925] (New York: Signet), 267–68.
8. We hear also of Gottlieb's callousness toward his wife when his research assumes priority and of his injunction to Arrowsmith to ignore the pleas of the plague victims and inoculate only the numbers required for the statistical experiment. For an extended discussion of Loeb's dedication to quantitative method see Charles E. Rosenberg (1963), "Martin Arrowsmith: The Scientist as Hero," *American Quarterly* 15: 452.
9. "The sense one has of exquisite and wonderful rhythms—just beyond sight and sound! . . . as if the whole world was fire and crystal and a-quiver." H. G. Wells (1927), *Marriage*, 2d ed. (London: Unwin), 130.
10. For Leo Szilard's acknowledgment to Wells's novel for both information and moral guidance, see chapter 11, note 19 above.
11. Peter Kapitza, a Soviet citizen, worked with Rutherford in the Cavendish Laboratory, Cambridge, becoming a Fellow of the Royal Society before returning to Moscow as director of the Institute of Physical Problems.
12. Bertolt Brecht's "Der Mantel des Ketzers" (The heretic's coat) concerns the last months in the life of Giordano Bruno, who was executed by the church on charges of heresy that included

his cosmic theory. Bruno is presented in an entirely heroic light. Not only is the charge against him false, but he is concerned to ensure that his coat, his only possession of value, is returned to the tailor to whom he owes a debt.

13. Bertolt Brecht (1965), notes on *The Life of Galileo*, trans. Peter Tegel, in *Collected Plays*, ed. by John Willett and Ralph Manheim (London: Methuen), 1:347.

14. The plague scene is unhistorical, as the plague did not reach Tuscany until 1632, sixteen years after the period in which the play is set. Brecht included it to absolve Galileo from the charge of cowardice.

15. See "Bertolt Brecht Testifies before the House Un-American Activities Committee," www.openculture.com/2012/11/bertolt_brecht_testifies_before_the_house_un_american_activities_committee_1947.html.

16. Charles Morgan (1953), *The Burning Glass* (London: Macmillan), 11.

17. No suspicion attached to Fuchs until late in 1949, and he made his confession in 1950.

18. Heinrich Schirmbeck (1960), *Ärgert dich dein rechtes Auge*, trans. Norman Denny as *The Blinding Light* (London: Collins), 28.

19. Douglas Stewart (1973), "Rutherford," in *Selected Poems* (Sydney: Angus & Robertson), 179.

20. Shirley Hazzard (1982), *The Transit of Venus* (Harmondsworth: Penguin), 16.

21. Aldous Huxley (1962), *Island* (Harmondsworth: Penguin), 219.

22. Ursula Le Guin (1985), *The Dispossessed* (London: Panther), 9.

Conclusion • New Images of Scientists

1. Theodore Roszak (1974), "The Monster and the Titan: Science, Knowledge and Gnosis," *Daedalus* 103, no. 3: 31.

For an extended bibliography, please visit jhupbooks.press.jhu.edu/content/madman-crime-fighter.

Part 1: Works of Fiction with Scientist Characters

Akenside, Mark. "The Virtuoso" [1737]. In *A Book of Science Verse*, edited by W. Eastwood. London: Macmillan, 1961.
Aldridge, James. *The Diplomat* [1949]. London: Bodley Head, 1950.
Andres, Stefan. *Die Sintflut*. Göttingen: Wallenstein, 2007.
Asimov, Isaac. *I, Robot* [1967]. London: Collins, 1989.
———. Introduction to "I, Robot." In *Isaac Asimov Presents the Great SF Stories*. New York: Daw Books, 1979.
———."On Computers." *DP Solutions*, Apr. 1975, 2.
———. *View from a Height*. New York: Doubleday, 1963.
Atlas, Martin. *Die Befreiung: Ein Zukunftsroman*. Berlin: F. Dümmler, 1910.
Bacon, Francis. *New Atlantis* [1626]. In *The Works of Francis Bacon*, edited by James Spedding, Robert Leslie Ellis, and Douglas Denon Heath, 3:119–66. London: Longman, 1857–74. Facs. ed. Stuttgart: Friedrich Fromann, 1962–64.
Balchin, Nigel. *The Small Back Room* [1943]. London: Collins, 1949.
———. *A Sort of Traitors*. London: Collins, 1949.
Balmer, Edwin, and Philip Wylie. *When Worlds Collide and After Worlds Collide* [1932]. New York: J. B. Lippincott, 1933.
Balzac, Honoré de. *The Elixir of Life* [1830]. In *La comédie humaine*, vol. 42. London: Caxton, n.d.
———. *The Quest of the Absolute* [1834]. Translated by Ellen Marriage. London: Newnes, n.d.
Barney, J. Stewart. *L.P.M.: The End of the Great War* [1915].
Behn, Aphra. *The Emperor of the Moon* [1687]. In *The Works of Aphra Behn*, edited by M. Summers. London: Heinemann, 1915.
Bellamy, Edward. *Dr. Heidenhoff's Process* [1880]. In *Edward Bellamy: Works*. London: Frederick Warne, n.d.
———. *Looking Backward, 2000–1887* [1888]. Edited by John L. Thomas. Cambridge: Belknap, 1967.
Bellow, Saul. *Mr. Sammler's Planet* [1969]. Harmondsworth: Penguin, 1978.
Bierce, Ambrose. "Moxon's Master" [1899]. In *Science-Fiction Thinking Machines: Robots, Androids, Computers*, edited by Groff Conklin. New York: Vanguard, 1954.
Binder, Eando. "Adam Link's Vengeance." *Amazing Stories*, Feb. 1940.

———. "I, Robot." *Amazing Stories*, Jan. 1939.
Blake, William. *Blake: Complete Writings*.Edited by Geoffrey Keynes. London: Oxford University Press, 1966.
Blish, James. *A Case of Conscience* [1959]. Harmondsworth: Penguin, 1963.
Böll, Heinrich. *Fürsorgliche Belagerung* [1979]. Translated by Leila Vennewitz as *The Safety Net*. New York: Melville House, 1981.
Borchert, Wolfgang. *Lesebuchgeschichten* [1946]. In *Draussen vor der Tür und ausgewählte Enzählungen*. Reinbek bei Hamburg: Rowohlt, 1956.
Boyd, William. *Brazzaville Beach*. London: Sinclair-Stevenson, 1990.
Brecht, Bertolt. "The Experiment" [1939]. Translated by Yvonne Kapp. In *Short Stories, 1921–1946*, edited by J. Willett and R. Manheim. London: Methuen, 1983.
———. *The Life of Galileo*. Translated by Desmond I. Vesey. London: Methuen, 1963.
Brenton, Howard. *The Genius* [1983]. Royal Court Writers. London: Methuen, 1985.
Broch, Hermann. *Die unbekannte Größe* [1933]. In *Gesammelte Werke*, edited by Ernst Schönweise, vol. 10. Zurich: Rhein, 1961.
Brod, Max. *Galilei in Gefangenschaft*. Winterthur: Mondial, 1948.
Browning, Robert. "Paracelsus" [1835]. In *The Poetical Works of Robert Browning*. London: Oxford University Press, 1953.
Buck, Pearl S. *Command the Morning*. New York: John Day, 1959.
Budrys, Algis. *Who?* [1958]. London: Quartet, 1975.
Bulwer-Lytton, Edward. *A Strange Story*. London: Routledge, 1861.
———. *The Coming Race*. Charleston, SC: BiblioLife, 2008.
———. *Zanoni* [1842]. London: Routledge, 1853.
Burdick, Eugene, and Harvey Wheeler. *Fail-Safe*. London: Hutchinson, 1963.
Butler, Samuel. *Erewhon* [1872]. New York: Airmont Classics, 1967.
———. "A Satire on the Royal Society." In *The Poetical Works of Samuel Butler*, edited by Rev. George Gilfillan, 2:138–39. Edinburgh: James Nicol, 1854.
Buzzati, Dino. "Appointment with Einstein" [1958]. In *Restless Nights*, translated by Lawrence Venuti. Manchester: Carcanet, 1984.
Byatt, A. S. *A Whistling Woman*. New York: Vintage Books, 2004.
Byron, George Gordon, Lord. *Manfred* [1817]. In *Poetical Works*, edited by Frederick Page. London: Oxford University Press, 1973.
Campbell, John C. "Who Goes There?" *Astounding Stories*, Aug. 1938.
Čapek, Karel. *Krakatit* [1923]. Translated by Laurence Hyde. London: Geoffrey Bles, 1925.
———. *R.U.R. (Rossum's Universal Robots: A Fantastic Melodrama)* [1921]. Translated by Paul Selver. Garden City, NY: Doubleday, Page, 1923.
Carlyle, Thomas. "Signs of the Times" [1829]. In *English and Other Critical Essays*. London: Dent, 1915.
Cavendish, Margaret, Duchess of Newcastle. *The Description of a New World, Called The Blazing World* [1666]. Edited by Susan James. Cambridge: Cambridge University Press, 2003.
———. *Observations upon Experimental Philosophy* [1668]. Edited by Eileen O'Neill. Cambridge: Cambridge University Press, 2001.
Centlivre, Susannah. *The Basset Table* [1705]. In *Plays*. New York: Garland, 1982.
———. *A Bold Stroke for a Wife* [1718]. Edited by Thalia Stathas. London: Edward Arnold, 1969.
Chatterton, Ruth. *The Betrayers*. Boston: Houghton Mifflin, 1953.
Chaucer, Geoffrey. *The Canon's Yeoman's Prologue and Tale* [ca. 1391]. In *The Canterbury Tales*, translated by Nevil Coghill. Harmondsworth: Penguin, 1957.
Churchill, Caryl. *A Number*. New York: Theatre Communications Group, 2002.
Clarke, Arthur C. *2001: A Space Odyssey*. London: Hutchinson, 1968.
———. *2010: Odyssey Two*. London: Granada, 1982.

Cobb, Weldon J. *A Trip to Mars*. New York: Street & Smith, 1901.
Colby, Merle. *The Big Secret*. New York: Viking, 1949.
Collins, Wilkie. *Heart and Science*. London: Chatto & Windus, 1893.
Conrad, Joseph. *Heart of Darkness* [1899]. London: Penguin, 2012.
Cowley, Abraham. "To the Royal Society." Frontispiece to Thomas Sprat, *History of the Royal Society* [1667]. London: Routledge, 1959.
Cowper, William. "The Task" [1784]. In *The Poetical Works of William Cowper*, edited by H. S. Milford. London: Oxford University Press, 1959.
Cramer, Heinz von. *Aufzeichnungen eines ordentlichen Menschen*. In *Leben wie in Paradies*. Hamburg: Hoffmann & Campe, 1964.
———. *Die Konzessionen des Himmels*. Hamburg: Hoffmann & Campe, 1961.
Crichton, Michael. *The Andromeda Strain*. London: Corgi, 1976.
Cromie, Robert. *The Crack of Doom*. London: Digby, Long, 1895.
———. *A Plunge into Space* [1891]. Westport, CT: Hyperion, 1976.
Cummins, Harle Owen. "The Man Who Made a Man." In *Welsh Rarebit Tales*. Boston: Mutual Book, 1902.
Darwin, Erasmus. *Zoonomia; or, The Laws of Organic Life*. London: J. Johnson, 1794–96.
Daumann, Rudolf Heinrich. *Protuberanzen: Ein utopischen Roman*. Berlin: Schützen, 1940.
De Camp, L. Sprague. "Judgement Day." *Astounding Science Fiction* 55, no. 6 (1955).
De Kruif, Paul. *The Sweeping Wind*. London: Rupert Hart-Davis, 1962.
De l'Isle-Adam, Auguste Villiers. *Tomorrow's Eve*. Translated by Robert Martin Adams. Urbana: University of Illinois Press, 1982.
Desaguliers, J. T. *The Newtonian System of the Universe, the Best Model of Government: An Allegorical Poem*. London: John Sexex, 1729.
Dick, Philip K. *Do Androids Dream of Electric Sheep?* New York: Signet, 1968.
———. *Dr. Bloodmoney; or, How We Got Along after the Bomb* [1965]. Boston: Gregg, 1977.
———. *The Man in the High Castle*. London: Gollancz, 1975.
———. "The Preserving Machine" [1953]. In *The Preserving Machine and Other Stories*. London: Gollancz, 1971.
———. *Vulcan's Hammer*. New York: Ace Books, 1960.
Dickens, Charles. "Full Report of the First Meeting of the Mudfog Association for the Advancement of Everything." *Bentley's Miscellany* 2: 397–413 (1837).
———. *Hard Times* [1854]. Ware, HERTS: Wordsworth Classics, 1998.
———. "The Haunted Man" [1848]. In *Christmas Books*. London: Chapman & Hall, n.d.
Disraeli, Benjamin. *Tancred: or, The New Crusade*. London: Henry Colburn, 1847.
Döblin, Alfred. *Berge, Meere und Giganten: Roman*. Berlin: S. Fischer, 1924.
Dominik, Hans. *Atomgewicht 500*. Berlin: Scherl, 1935.
———. *Die Macht der Drei: Roman*. Berlin: Scherl, 1922.
Doyle, Arthur Conan. "The Disintegration Machine." In *The Complete Professor Challenger Stories*. London: John Murray & Jonathan Cape, 1976.
———. "The Final Problem" [1893]. In *A Treasury of Sherlock Holmes*, edited by Adrian Conan Doyle. Garden City, NY: Hanover House, 1955.
———. *The Land of Mist* [1926]. In *The Complete Professor Challenger Stories*. See Doyle, "Disintegration Machine."
———. *The Lost World* [1912]. London: John Murray & Jonathan Cape, 1979.
———. *The Maracot Deep* [1928]. London: Pan, 1977.
———. *The Poison Belt*. London: Hodder & Stoughton, 1913.
———. "A Study in Scarlet." In *A Treasury of Sherlock Holmes*. See Doyle, "Final Solution."
Dryden, John. "Annus Mirabilis" [1666]. In *The Poems and Fables of John Dryden*, edited by J. Kinsley. London: Oxford University Press, 1962.

Dürrenmatt, Friedrich. *Der Mitmacher: Komodie.* Basel: Reiss, 1973.

———. *Die Physiker* [1961]. Translated as *The Physicists* [1962]. London: Samuel French's Acting Edition, 1963.

Dwyer, Molly. *Requiem for the Author of Frankenstein.* Fort Bragg, CA: Lost Coast, 2008.

Eastwood, W., ed. *A Book of Science Verse: The Poetic Relations of Science and Technology.* London: Macmillan, 1961.

Edmonds, H. *The Professor's Last Experiment.* London: Rich & Cowan, 1935.

Eliot, George. *Middlemarch: A Study of Provincial Life.* 2 vols. [1871–72]. London: Dent, 1959.

Fielding, Henry. *The History of Tom Jones* [1749]. London: Dent, 1957.

Fineman, Irving. *Doctor Addams.* London: Cresset, 1939.

Flaubert, Gustave. *Madame Bovary* [1856]. Harmondsworth: Penguin, 1965.

Forster, E. M. "The Machine Stops" [1909]. In *Collected Short Stories.* Harmondsworth: Penguin, 1977.

Franklin, H. B., ed. *Future Perfect: American Science Fiction of the Nineteenth Century.* New York: Oxford University Press, 1966.

Frayn, Michael. *Copenhagen.* New York: Knopf Doubleday, 2010.

Freeman, Richard Austin. *The Famous Cases of Doctor Thorndyke.* London: Hodder & Stoughton, 1929.

Frisch, Max. *Don Juan; or, The Love of Geometry* [1962]. Translated by Michael Bullock. In *Max Frisch, Four Plays.* London: Methuen, 1969.

Gaines, Susan. *Carbon Dreams.* Berkeley, CA: Creative Arts Book, 2001.

Gaskell, Elizabeth. "Cousin Phillis" [1865]. In *Cousin Phillis and Other Tales*, edited by Angus Easson. Oxford: Oxford University Press, 1981.

———. *Wives and Daughters* [1866]. Harmondsworth: Penguin, 1969.

George, Peter. *Two Hours to Doom.* London: T. V. Boardman, 1958.

Gide, André. *The Vatican Cellars* [1914]. Translated by Dorothy Bussy. London: Cassell, 1952.

Gissing, George. *Born in Exile* [1892]. London: Gollancz, 1970.

Godfrey, Hollis. *The Man Who Ended War.* Boston: Little, Brown, 1908.

Godwin, Francis. *The Man in the Moone: or A Discourse of a Voyage thither by Domingo Gonsales*[1638]. Menston, Yorkshire: Scolar, 1971.

Godwin, William. *Caleb Williams* [1794]. London: Oxford University Press, 1970.

———. *St. Leon: A Tale of the Sixteenth Century* [1799]. New York: Garland, 1974.

Goethe, Johann Wolfgang von. *Elective Affinities.* Translated by David Constantine. Oxford: Oxford University Press, 1999.

———. *Faust* [part 1, 1808; part 2, 1832]. *Faust. Part 1.* Translated by David Luke. Oxford: Oxford University Press, 1987.

Goldschmidt, Pippa. *The Falling Sky.* Edinburgh: Freight Books, 2013.

Goldstein, Rebecca. *Properties of Light.* Boston: Houghton Mifflin, 2001.

Goodman, Allegra. *Intuition.* New York: Random House, 2006.

Gosse, Edmund. *Father and Son* [1907]. Edited by Peter Abbs. Harmondsworth: Penguin, 1986.

Graf, Oskar Maria. *Die Erben des Untergangs: Roman einer Zukunft.* Frankfurt: Nest, 1959.

Grover, Richard. "Poem on Sir Isaac Newton." In *View of Sir Isaac Newton's Philosophy*, by Henry Pemberton [1728]. New York: Johnson Reprint, 1972.

Halley, Edmund. "Ode to the Illustrious Man, Isaac Newton." In Newton, *Sir Isaac Newton's Mathematical Principles of Natural Philosophy and His System of the World*, xiii–xv.

Hamilton, Edmond. "The Metal Giants." *Weird Tales*, Dec. 1926.

Hammerström, Jan. *Die Abenteuer der Sibylle Kyberneta.* Düsseldorf: Diederichs, 1963.

Harbou, Thea von. *Metropolis.* August Scherl, 1925. Published in English, London: Readers Library, 1927.

Hardy, Thomas. *A Pair of Blue Eyes* [1873]. Edited by Roger Ebbatson. Harmondsworth: Penguin, 1986.
———. *Two on a Tower* [1882]. London: Macmillan, 1922.
———. *The Woodlanders* [1887]. London: Macmillan, 1923.
Hawthorne, Nathaniel. "The Birthmark" [1845]. In *Mosses from an Old Manse*, 12–31. Mechanicsville, VA: Electric Book, 2001.
———. "Ethan Brand" [1851]. In *The Portable Hawthorne*, edited by Malcolm Cowley, 262–81. Harmondsworth: Penguin Books, 1986.
———. "Rappaccini's Daughter" [1844]. In *The Portable Hawthorne*, 178–212. See Hawthorne, "Ethan Brand."
———. *The Scarlet Letter* [1850]. In *The Portable Hawthorne*, 337–546. See Hawthorne, "Ethan Brand."
———. *Septimius Felton; or, The Elixir of Life* [1871]. In *The Portable Hawthorne*, 589–605. See Hawthorne, "Ethan Brand." .
Hazzard, Shirley. *The Transit of Venus* [1980]. Harmondsworth: Penguin, 1982.
Heard, Gerald. *Doppelgängers* [1948]. London: Science Fiction Books, 1965..
Herbert, Frank. *Destination: Void*. New York: Berkeley Medallion, 1966.
Hilton, James. *Nothing So Strange* [1947]. Melbourne: Macmillan, 1948.
Hoffmann, E. T. A. "The Sandman" [1816]. Translated by J. T. Bealby. In *Isaac Asimov Presents the Best Science Fiction of the Nineteenth Century*, edited by I. Asimov, C. G. Waugh, and M. Greenberg. London: Gollancz, 1983.
Hogan, James P. *The Genesis Machine*. New York: Ballantine, 1978.
Houellebecq, Michel. *Les Particules élémentaires*. Paris: Flammarion, 1998.
Hoyle, Fred, and John Elliott. *A for Andromeda*. London: Corgi, 1962.
———. *Andromeda Breakthrough*. London: Corgi, 1966.
Huxley, Aldous. *After Many a Summer* [1939]. London: Chatto & Windus, 1953.
———. *Antic Hay* [1923]. London: Chatto & Windus, 1923.
———. *Ape and Essence* [1948]. London: Chatto & Windus, 1951.
———. *Brave New World* [1932]. Harmondsworth: Penguin, 1971.
———. *The Genius and the Goddess* [1955]. London: Chatto & Windus, 1955.
———. *Island* [1962]. Harmondsworth: Penguin, 1968.
———. *Jesting Pilate* [1926]. London: Chatto & Windus, 1926.
———. *Point Counter Point* [1928]. London: Chatto & Windus, 1947.
Jahnn, Hans Henny. *Trümmer des Gewissens (Der staubige Regenbogen)*. Edited by Walter Muschg. Frankfurt am Main: Europäische Verlagsanstalt, 1961.
James, Henry. *The Bostonians* [1886]. Harmondsworth: Penguin, 1971.
Jameson, Malcolm. *The Giant Atom* [1944]. New York: Bond-Charteris, 1945.
Johnson, Samuel. *The History of Rasselas, Prince of Abissinia* [1759]. In *Rasselas; Poems and Selected Prose*, edited by B. H. Bronson. New York: Holt, Rinehart & Winston, 1958.
Jonson, Ben. *The Alchemist* [1610]. In *The Complete Plays of Ben Jonson*, edited by G. A. Wilkes. Oxford: Clarendon, 1982.
Jünger, Ernst. *Gläserne Bienen*. Stuttgart: Ernst Klett, 1957.
Kaiser, Georg. *Gas I* [1918]. Translated by Hermann Scheffauer. In *Twenty-Five Modern Plays*, edited by S. M. Tucker and A. S. Downer. New York: Harper, 1953.
———. *Gas II* [1920]. Translated by Winifred Katzin. In Tucker and Downer, *Twenty-Five Modern Plays*. See Kaiser, *Gas I*.
Kellermann, Bernhard. *Der Tunnel*. Berlin: S. Fischer, 1913.
Kingsley, Charles. *Glaucus; or, The Wonders of the Shore* [1855]. London: Dent, 1949.
———. *Two Years Ago* [1857]. London: Macmillan, 1889.

———. *The Water Babies* [1863]. London: Dent, 1949.
Kingsolver, Barbara. *Flight Behavior*. New York: HarperCollins, 2012.
Kipling, Rudyard. "As Easy as A.B.C." [1912]. In *A Diversity of Creatures*. London: Macmillan, 1952.
———. "With the Night Mail: A Story of 2000 A.D." In *Actions and Reactions*. New York: Doubleday, Page, 1909.
Kipphardt, Heinar. *In the Matter of J. Robert Oppenheimer* [1964]. Translated by Ruth Speirs. London: Methuen, 1967.
Kornbluth, C. M. "Gomez" [1955]. In *Best Science Fiction Stories of C. M. Kornbluth*. London: Faber, 1968.
La Mettrie, Julien Offray de. *L'Homme machine* [1747]. In *La Mettrie's L'Homme Machine: A Study in the Origins of an Idea*, edited by Aram Vartanian. Princeton, NJ: Princeton University Press, 1960.
Lasswitz, Kurd. *Bilder aus der Zukunft*. Breslau: Verlag von S. Schottländer, 1878.
———. *Gegen das Weltgesetz: Erzählung aus dem Jahre 3877* (Against the world law: A story from the year 3877) [1878]. Available from Project Gutenberg, http://gutenberg.spiegel.de/buch/gegen-das-weltgesetz-i-3122/1.
———. "Über Zukunftsträume." Parts 1 and 2. *Die Nation*, 13 and 20 May 1899.
Lawrence, D. H. *The Rainbow* [1915]. Harmondsworth: Penguin, 1966.
Le Guin, Ursula. *The Dispossessed* [1974]. London: Panther, 1985.
Leinster, Murray. "The Man Who Put Out the Sun." *Argosy*, 14 June 1930.
———. "The Storm That Had to Be Stopped." *Argosy*, 1 Mar. 1930.
Lessing, Gotthold Ephraim. *Faust* (fragment). In *Gotthold Ephraim Lessings Sämtliche Schriften*, edited by Karl Lachmann, vol. 3. Stuttgart: G. J. Göschensche Verlagshandlung, 1924.
Levin, Ira. *The Boys from Brazil*. New York: Random House, 1976.
———. *The Stepford Wives*. New York: Random House, 1972.
Lewis, C. S. *Out of the Silent Planet* [1938]. London: John Lane, the Bodley Head, 1946.
———. *Perelandra* [1943]. New York: Macmillan, 1944.
———. *That Hideous Strength* [1945]. London: John Lane, the Bodley Head, 1946.
Lewis, Sinclair. *Arrowsmith* [1925]. New York: Signet, 1961.
MacHarg, William, and Edwin Balmer. *The Achievements of Luther Trant*. Boston: Small, Maynard, 1910.
MacNeice, Louis. "The Kingdom" [1943]. Part 6. "The Scientist." In *Collected Poems of Louis MacNeice*, edited by E. R. Dodds. London: Faber & Faber, 1979.
Mallett, David. *The Excursion: A Poem in Two Books* [1728]. In *The Works of the English Poets*, edited by Alexander Chalmers, vol. 14. London, 1810.
Marlowe, Christopher. *The Tragical History of Doctor Faustus* [1604]. In *Marlowe's Plays and Poems*, edited by M. R. Ridley. London: Dent, 1963.
Maron, Monika. *Flugasche: Roman*. Frankfurt am Main: S. Fischer, 1981.
Masters, Dexter. *The Accident*. London: Cassell, 1955.
Maugham, Somerset. *The Magician* [1908]. London: Heinemann, 1956.
McCormmach, Russell. *Night Thoughts of a Classical Physicist*. Cambridge, MA: Harvard University Press, 1982.
Meredith, George. "Melampus" [1883]. In *The Poetical Works of George Meredith*, edited by George M. Trevelyn. London: Constable, 1912.
Miller, James. *The Humours of Oxford: A Comedy* [1726]. 2nd ed. London: J. Watts, 1730.
Milton, John. *Paradise Lost*. In *Complete Poetry and Selected Prose*. New York: Modern Library, 1950.
Morgan, Charles. *The Burning Glass*. London: Macmillan, 1953.

Musil, Robert. "Eine Geschichte aus drei Jahrhunderten, 1927." In *Nachlass zu Lebzeiten*. Berlin: Ernst Rowohlt, 1981.

———. *The Man without Qualities* [1930–43]. 3 vols. Translated by Eithne Wilkins and Ernst Kaiser. London: Seeker & Warburg, 1953–60.

———. "Tonka" [1924]. Translated by Eithne Wilkins and Ernst Kaiser. In *Tonka and Other Stories*. London: Seeker & Warburg, 1965.

———. *Young Törless* [1906]. Translated by Ernst Kaiser and Eithne Wilkins. New York: Pantheon, 1978.

Muske-Dukes, Carol. *Saving St. Germ*. New York: Open Road Integrated Media, 2014.

Newcomb, Simon. *His Wisdom, the Defender*. New York: Harper, 1900.

Newman, B. *Armoured Doves*. London: Jarrolds, 1931.

Newton, Isaac [1729]. *Sir Isaac Newton's Mathematical Principles of Natural Philosophy and His System of the World*. 2d ed. Translated by Andrew Motte. Edited by Florian Cajori. Berkeley: University of California Press, 1946. *Mathematical Principles of Natural Philosophy and System of the World* [1687]. Edited by R. T. Crawford. Berkeley: University of California Press, 1934.

———. *Opticks: or, A Treatise of the Reflexions, Refractions, Inflexions and Colours of Light*. London: Sam. Smith and Benjamin Walford, Printers to the Royal Society, 1704.

Nossack, Erich. "Die Schalttafel." In *Spirale*. Frankfurt: Suhrkamp, 1956.

O'Brien, Fitz-James. "The Diamond Lens" [1858]. In *The Diamond Lens and Other Stories*. New York: AMS, 1969.

———. "The Golden Ingot" [1858]. in *The Diamond Lens and Other Stories*. See O'Brien, "The Diamond Lens."

Piercy, Marge. *Small Changes*. Robbinsdale, MN: Fawcett Crest, 1974.

Poe, Edgar Allan. "Maelzel's Chess Player" [1836]. In *Complete Tales and Poems of Edgar Allen Poe*. New York: Random House, 1938.

———. "The Murders in the Rue Morgue." In *Tales of Mystery and Imagination*. London: Nelson, 1909.

———. *Narrative of Arthur Gordon Pym of Nantucket*. In *Complete Tales and Poems of Edgar Allen Poe*. See Poe, "Maelzel's Chess Player."

———. "The Unparalleled Adventure of One Hans Pfaall" [1835]. In *Complete Tales and Poems of Edgar Allen Poe*. See Poe, "Maelzel's Chess Player."

Pope, Alexander. *The Dunciad* [1728, 1742]. In *Selected Poetry and Prose*, edited by W. K. Wimsatt Jr. New York: Holt, Rinehart & Winston, 1962.

———. *Essay on Man* [1732–34]. In *Selected Poetry and Prose*. See Pope, *Dunciad*.

Powell, G. *All Things New*. London: Hodder & Stoughton, 1926.

Priestley, J. B. *The Doomsday Men*. London: Heinemann, 1938.

Pynchon, Thomas. *The Crying of Lot 49*. New York: Bantam, 1966.

———. *Gravity's Rainbow*. New York: Viking, 1973.

Ramsay, Allan. "Ode to the Memory of Sir Isaac Newton: Inscribed to the Royal Society" [1731]. In *Poems by Allan Ramsay and Robert Fergusson*, edited by Alexander Manson Kinghorn and Alexander Law. Edinburgh: Scottish Academic Press, 1974.

Rankine, W. J. M. "The Mathematician in Love" [1874]. In *Songs and Fables*, edited by W. J. Millar. London: Griffin, 1881.

Ray, John. *The Wisdom of God* [1691]. London: W. Innys, 1743.

Rehberg, Hans. *Johannes Kepler: Schauspiel in drei Akten*. Berlin: S. Fischer, 1933.

Renard, Maurice. *Le docteur Lerne sous-dieu* [1908]. Translated as *New Bodies for Old*. New York: Macaulay, 1923.

Rhodes, William Henry. "The Case of Summerfield" [1871]. In *Caxton's Book*, edited by Daniel O' Connell. Westport, CT: Hyperion, 1974.
Rohn, Jennifer. *Experimental Heart*. New York: Cold Spring Harbor Laboratory Press, 2009.
———. *The Honest Look*. New York: Cold Spring Harbor Laboratory Press, 2010.
Rose, F. H. *The Maniac's Dream*. London: Duckworth, 1946.
Roszak, Theodore. *Memoirs of Elizabeth Frankenstein*. New York: Random House, 1995.
Sandburg, Carl. "Mr. Attila." In *Complete Poems*. New York: Harcourt Brace, 1950.
Savage, Richard. *The Wanderer*. In *Works of the English Poets*, edited by Samuel Johnson [1779–81], edited by G. B. Hill, vol. 45. Oxford: Oxford University Press, 1905.
Schirmbeck, Heinrich. *The Blinding Light* [1957]. Translated by Norman Denny. London: Collins, 1960.
Serviss, Garrett P. *Edison's Conquest of Mars* [1898]. Los Angeles: Carcosa House, 1947.
———. "The Second Deluge" [1912]. *Fantastic Novels Magazine* 2 (July 1948).
Sewell, Stephen. *Welcome the Bright World* [1982]. Sydney: Alternative Publishing Cooperative and Nimrod Theatre Press, 1983.
Shadwell, Thomas. *The Sullen Lovers, or The Impertinents* [1688]. *The Complete Works of Thomas Shadwell*, edited by Montague Summers, vol. 1. New York: B. Blom 1968.
———. *The Virtuoso: A Comedy* [1676]. http://name.umdl.umich.edu/A59463.0001.001. Ann Arbor, MI: Text Creation Partnership, 2003.
Shakespeare, William. *The Tempest* [1611]. In *William Shakespeare: The Complete Works*, edited by Alfred Harbage. London: Allen Lane, Penguin, 1969.
Shelley, Mary W. *Frankenstein, or the Modern Prometheus* [1818]. Edited by J. Kinglsey and M. K. Joseph. Oxford: Oxford University Press, 1980.
———. *The Last Man* [1826]. Ware, HERTS: Wordsworth Editions, 2004.
———. "The Mortal Immortal" [1834]. In *Isaac Asimov Presents the Best Science Fiction of the Nineteenth Century*. See Hoffmann, "Sandman."
Sinclair, Upton. *A Giant's Strength*. London: T. Werner Laurie, 1948.
Skinner, B. F. *Walden Two*. New York: Macmillan, 1948.
Smith, Martin Cruz. *Stallion Gate*. Glasgow: HarperCollins, 1992.
Smith, Michael Martin. *Spares*. New York: Bantam Books, 1997.
Snow, C. P. *The Affair* [1960]. Harmondsworth: Penguin, 1968.
———. *The Corridors of Power*. London: Macmillan, 1964.
———. *Death under Sail* [1932]. Rev. ed. London: Heinemann, 1959.
———. *The New Men*. London: Macmillan, 1954.
———. *The Search* [1934, rev. ed. 1938]. Harmondsworth: Penguin, 1979.
———. *The Sleep of Reason*. London: Macmillan, 1968.
Snyder, Alexander. "Blasphemers' Plateau." *Amazing Stories*, Oct. 1926.
Steinbeck, John. *The Log from* The Sea of Cortez. New York: Viking, 1951.
Stevenson, Robert Louis. "The Strange Case of Dr. Jekyll and Mr. Hyde" [1886]. In *Dr. Jekyll and Mr. Hyde and Other Stories*. New York: Magnum, 1968.
Stewart, Douglas. "Rutherford." In *Selected Poems*. Sydney: Angus & Robertson, 1962.
Stoker, Bram. *Dracula* [1897]. New York: Airmont, 1965.
Stuart, Don A. [John W. Campbell Jr.]. "Atomic Power." *Astounding Stories*, Dec. 1934.
Swift, Jonathan. *Gulliver's Travels* [1726]. Edited by Harold Williams. London: Dent, 1956.
———. "The Mechanical Operation of the Spirit" [1710]. In *Collected Works*, vol. 1. Oxford: Basil Blackwell, 1957.
Tasker, William. "An Ode to Curiosity." In *Poems*. London, 1779.
Tennyson, Alfred. *In Memoriam A.H.H.* [1850]. In *Tennyson's Poetry*, edited by Robert W. Hill, Jr. New York: W. W. Norton, 1971.

Thomson, James. "A Poem Sacred to the Memory of Sir Isaac Newton" [1727]. In *The Castle of Indolence and Other Poems*, edited by A. D. McKillop. Lawrence: University of Kansas Press, 1961.
———. *The Seasons* [1726–30]. In *The Poetical Works of James Thomson*, vol. 1. London: Bell & Daldy, 1860.
Thoreau, Henry David. *Walden, or Life in the Woods* [1854]. New York: W. W. Norton, 1966.
Train, Arthur, and Robert William Wood. *The Man Who Rocked the Earth*. New York: Doubleday, Page, 1915.
Verne, Jules. *Hector Servadac* [1877]. Translated by Ellen E. Frewer. London: Sampson, Low, Marston, 1878.
———. *Journey to the Centre of the Earth* [1864]. Translated by R. Baldick. Harmondsworth: Penguin, 1970.
———. *The Mysterious Island* [1875]. Translated by Lowell Blair. London: Corgi, 1976.
———. *Robur the Conqueror*. Paris: Hetzel, 1886.
———. *Twenty Thousand Leagues under the Sea* [1870]. Translated by H. Frith. London: Dent, 1968.
Verrill, A. Hyatt. "The Plague of the Living Dead." *Amazing Stories*, Apr. 1927.
———. "The Ultra-Elixir of Youth." *Amazing Stories*, Aug. 1927.
Vonnegut, Kurt, Jr. *Cat's Cradle* [1963]. London: Gollancz, 1965.
Waterloo, Stanley. *Armageddon*. Chicago: Rand McNally, 1898.
Weldon, Fay. *The Cloning of Joanna May*. London: Collins, 1989.
Wells, H. G. "Argonauts of the Air" [1895]. In *Selected Short Stories*. Harmondsworth: Penguin, 1970.
———. *Early Writings in Science and Science Fiction*. Edited by R. M. Philmus and D. Y. Hughes. Berkeley: University of California Press, 1975.
———. *Experiment in Autobiography*. London: Jonathan Cape, 1934.
———. *First and Last Things*. London: T. Fisher Unwin, 1926.
———. *The First Men in the Moon* [1901]. Atlantic ed. London: T. Fisher Unwin, 1924.
———. *The Food of the Gods, and How It Came to Earth*. London: Macmillan, 1904.
———. *The Invisible Man* [1897]. New York: Penguin Putnam, 2010.
———. *The Island of Doctor Moreau* [1896]. Harmondsworth: Penguin, 1967.
———. *Marriage*. London: Unwin, 1927.
———. *Men Like Gods*. New York: Macmillan, 1923.
———. *Star Begotten* [1937]. London: Sphere, 1975.
———. "The Time Machine" [1895]. In *Selected Short Stories*. See Wells, "Argonauts of the Air."
———. *Tono-Bungay* [1908]. London: Unwin, 1909.
———. *The War of the Worlds* [1898]. Harmondsworth: Penguin, 1971.
———. *The World Set Free* [1914]. London: Unwin, 1926.
White, Stewart Edward. *The Sign at Six*. Indianapolis: Bobbs-Merrill, 1912.
White, Stewart Edward, and Samuel Hopkins Adams. *The Mystery* [1907]. New York: Arno, 1975.
Whitman, Walt. "When I Heard the Learn'd Astronomer" [1865]. From *Leaves of Grass*. In *The Portable Walt Whitman*, edited by Mark van Doren. New York: Viking, 1958.
Wilder, Thornton. *The Skin of Our Teeth* [1942]. In *Three Plays by Thornton Wilder*. London: Longmans Green, 1958.
Wilson, Mitchell. *Meeting at a Far Meridian*. London: Seeker & Warburg, 1961.
Wolf, Christa. "Self-Experiment: Appendix to a Report" [1973]. Translated by Jeanette Clausen. *New German Critique* 13 (Winter 1978): 109–31.
———. *Störfall. Nachrichten eines Tag* [1987]. Frankfurt am Main: Luchterhand, 1988.

Wordsworth, William. *Wordsworth Poetical Works*. Edited by Thomas Hutchinson. Oxford: Oxford University Press, 1978.
Wright, Thomas. *The Female Virtuoso's: A Comedy*. London: J. Wilde, 1693.
Yeats, William Butler. *The Collected Poems of W. B. Yeats*. Ware, HERTS: Wordsworth Editions, 1994.
Zamyatin, Yevgeny. *We*. Translated by Gregory Zilborg. New York: E. P. Dutton, 1924.
Zola, Émile. *Doctor Pascal* [1893]. Translated by Vladimir Jean. London: Elek Books, 1957.
———. *Paris*. 2d ed. Trans. E. Vizetelly. London: Chatto & Windus, 1898.
Zuckmayer, Carl. *The Cold Light* [1955]. Translated by Elizabeth Montagu. Typescript, Columbia University Libraries, n.d.
Zwillinger, Frank. *Galileo Galilei: Schauspiel* [1953]. Bayreuth: Reta Baumann, 1962.

Part 2: Criticism

Arnold, Matthew

Donovan, R. A. "Mill, Arnold and Scientific Humanism." *Annals of the New York Academy of Sciences* 360 (20 Apr. 1981).

Asimov, Isaac

Asimov, Isaac. *The Early Asimov, or Eleven Years of Trying*. London: Gollancz, 1973.
Hassler, Donald M. *Isaac Asimov*. Mercer Island, WA: Starmont House, 1991.
Patrouch, Joseph F., Jr. *The Science Fiction of Isaac Asimov*. Garden City, NY: Doubleday, 1974.

Bacon, Francis

Adams, Robert P. "The Social Responsibilities of Science in *Utopia, New Atlantis*, and After." *Journal of the History of Ideas* 10 (1949).
Anderson, F. H. *The Philosophy of Francis Bacon*. Chicago: University of Chicago Press, 1948.
Aughterson, Kate. "'Strange Things So Probably Told': Gender, Sexual Difference and Knowledge in Bacon's *New Atlantis*." In *Francis Bacon's New Atlantis: New Interdisciplinary Essays*, edited by Bronwen Price, 156–97. Manchester: Manchester University Press, 2002.
Blish, James. *Doctor Mirabilis*. London: Faber & Faber, 1964.
Blodgett, E. D. "Bacon's *New Atlantis* and Campanella's *Civitas Solis*." *PMLA* 46 (1931).
Bullough, G. "Bacon and the Defense of Learning." In *Seventeenth-Century Studies Presented to Sir Herbert Grierson*, edited by J. Dover Wilson. New York: Octagon, 1967.
Haydn, Hiram C. "The Science of the Counter-Renaissance." In *The Counter-Renaissance*. New York: Charles Scribner's Sons, 1950.
Lloyd, Genevieve. "Francis Bacon: Knowledge as the Subjugation of Nature." In *The Man of Reason: "Male" and "Female" in Western Philosophy*, 1–15. London: Routledge, 1984.
Metz, Rudolph. "Bacon's Part in the Intellectual Movements of His Time." Translated by Joan Drever. In *Seventeenth-Century Studies Presented to Sir Herbert Grierson*. See Bullough, "Bacon and the Defense of Learning."
Prior, Moody E. "Bacon's Man of Science." *Journal of the History of Ideas* 15 (1954).

Balzac, Honoré de

Luce, Louise Fiber. "Honoré de Balzac and the Voyant: A Recovered Alchemical Discourse." *L'Esprit Créateur* 18 (1978).
Murard, Jean. "Balzac, la médecine et les médecins." *Histoire de la Médicine*, Aug.<N>Sept. 1971, Oct. 1971, Nov. 1971.
Pritchett, V. S. *Balzac*. London: Chatto & Windus, 1933.

Blake, William

Ault, Donald. *Visionary Physics: Blake's Response to Newton*. Chicago: University of Chicago Press, 1974.
Damon, S. Foster. *A Blake Dictionary: The Ideas and Symbols of William Blake*. Hanover, NH: Dartmouth College Press, 2013.

Brecht, Bertolt

Brecht, Bertolt. *Notes on The Life of Galileo*. Trans. Peter Tegel. In *Collected Plays*, edited by John Willett and Ralph Manheim, vol. 1. London: Methuen, 1965.
Cohen, M. A. "History and Moral in Brecht's *The Life of Galileo*." *Contemporary Literature* 11 (1970).
Demetz, P., ed. *Brecht: A Collection of Critical Essays*. Englewood Cliffs, NJ: Prentice Hall, 1962.
Jameson, Frederic. *Brecht and Method*. London: Verso, 1998.
Jevons, F. R. "Brecht's *Life of Galileo* and the Social Relations of Science." *Technology and Society* 4, no. 3 (1968).
Rütlicke, Käthe. "Bemerkungen zur Schluß-Szene." In Bertolt Brecht, *Materialen zu Brechts "Leben des Galilei"*. Frankfurt: Suhrkamp, 1963.
Spalter, M. *Brecht's Tradition*. Baltimore: Johns Hopkins University Press, 1967.
White, A. D. "Brecht's *Leben des Galilei*: Armchair Theatre?" *German Life and Letters*, n.s., 27 (Jan. 1974).

Broch, Hermann

Ziolkowski, Theodore. "Hermann Broch and Relativity in Fiction." *Wisconsin Studies in Contemporary Literature* 8, no. 3 (1967).

Browne, Sir Thomas

Chalmers, G. K. "Sir Thomas Browne, True Scientist." *Osiris* 2 (1936).

Bulwer-Lytton, Edward

Christensen, Allan Conrad. *Edward Bulwer-Lytton: The Fiction of New Regions*. Athens: University of Georgia Press, 1976.
Salmon, Richard. "The Subverting Vision of Bulwer Lytton: Bicentenary Reflections." *Victorian Studies*, 48, no. 3 (2006), 566–568.

Butler, Samuel

Bruun, S. V. "Who's Who in Samuel Butler's 'The Elephant in the Moon.'" *English Studies* 50 (1969).

Čapek, Karel

Čapek, Karel. "The Meaning of *R.U.R.*" *Saturday Review* 136 (21 July 1923).
Harkins, W. E. *Karel Čapek*. New York: Columbia University Press, 1962.

Carlyle, Thomas

Turner, Frank M. "Victorian Scientific Naturalism and Thomas Carlyle." *Victorian Studies* 18, no. 3 (1975): 325–43.

Cavendish, Margaret

Bowerbank, Sylvia, and Sara Mendelson. *Paper Bodies: A Margaret Cavendish Reader*. Peterborough, ON: Broadview, 2000.

Chaucer, Geoffrey

Baum, Paul F. "The Canon's Yeoman's Tale." *Modern Language Notes* 40 (1925).
Damon, S. Foster. "Chaucer and Alchemy." *PMLA* 39 (1924).
Duncan, Edgar H. "The Literature of Alchemy and Chaucer's *Canon's Yeoman's Tale*: Framework, Theme, and Characters." *Speculum* 43 (1968).
Gardner, John. *"The Canon's Yeoman's Prologue and Tale*: An Interpretation." *Philological Quarterly* 46 (1967).
Grennan, Joseph Edward. "Chaucer's Characterization of the Canon and His Yeoman." *Journal of the History of Ideas* 25 (1964).
———. "Chaucer's 'Secree of Secrees': An Alchemical Topic." *Philological Quarterly* 42 (1963).
Hamilton, Marie P. "The Clerical Status of Chaucer's Alchemist." *Speculum* 16 (1941).
Hartung, Albert E. "Inappropriate Pointing in *The Canon's Yeoman's Tale* G1236–1239." *PMLA* 77 (1962).
Young, Karl. "The 'Secree of Secrees' of Chaucer's Canon's Yeoman." *Modern Language Notes* 58 (1943).

Coleridge, Samuel Taylor

Barfield, Owen. *What Coleridge Thought*. Middleton, CT: Wesleyan University Press, 1971.
Coburn, Kathleen. "Coleridge, a Bridge between Science and Poetry: Reflections on the Bicentenary of His Death." *Proceedings of the Royal Institution of Great Britain* 46 (1973).
Potter, G. R. "Coleridge and the Idea of Evolution." *PMLA* 40 (1925).

Cowley, Abraham

Laprevotte, Guy. "To the Royal Society d'Abraham Cowley." *Études Anglaises* 26 (1973): 129–44.

Dick, Philip K.

Benesch, Klaus. "Technology, Art, and the Cybernetic Body: The Cyborg as Cultural Other in Fritz Lang's "Metropolis" and Philip K. Dick's 'Do Androids Dream of Electric Sheep?'" *Amerikastudien/American Studies*, 44, no. 3 (1999).
Jameson, Fredric. "After Armageddon: Character Systems in *Dr. Bloodmoney*." *Science Fiction Studies* 2, no. 1 (1975).
Lem, Stanislaw. "A Visionary among the Charlatans." *Science Fiction Studies* 2, no. 1 (1975).
Suvin, Darko. "P. K. Dick's Opus: Artifice as Refuge World View." *Science Fiction Studies* 2, no. 1 (1975).

Dickens, Charles

Metz, N. A. "Science in *Household Words*." *Victorian Periodicals Newsletter*, 1979.
Wilkinson, Ann Y. "*Bleak House*: From Faraday to Judgement Day." *English Literary History* 34 (1967).

Dominik, Hans

Fischer, William B. *The Empire Strikes Out: Kurd Lasswitz, Hans Dominik, and the Development of German Sceince Fiction*. Bowling Green, OH: Bowling Green State University Popular Press, 1984.

Donne, John

Coffin, C. M. *John Donne and the New Philosophy*. London: Routledge & Kegan Paul, 1937.

Doyle, Arthur Conan

Higham, Charles. *The Adventures of Conan Doyle: Life of the Creator of Sherlock Holmes*. London: Hamish Hamilton, 1976.
Rauber, D. F. "Sherlock Holmes and Nero Wolfe: The Role of the 'Great Detective' in Intellectual History." *Journal of Popular Culture* 6 (1972).
Rose, Phyllis. "Huxley, Holmes, and the Scientist as Aesthete." *Victorian Newsletter* 38 (1970): 22–24.

Dryden, John

Bredvold, Louis I. "Dryden, Hobbes, and the Royal Society." *Modern Philology* 25 (May 1928).
Lloyd, Claude. "John Dryden and the Royal Society." *PMLA* 45 (1930).

Dürrenmatt, Friedrich

Jenny, Urs. *Dürrenmatt: A Study of His Plays*. London: Methuen, 1978.
Matlack, Samuel. "*The Physicists* at Fifty." *New Atlantis* 36 (Summer 2012): 63–78.
Muschg, W. "Dürrenmatt und *Die Physiker*." *Moderna Sprak* 56 (1961).

Eliot, George

Adam, Ian. "A Huxley Echo in *Middlemarch*." *Notes and Queries* 2 (1964).
———, ed. *This Particular Web: Essays on Middlemarch*. Toronto: University of Toronto Press, 1975.
Briggs, Asa. "*Middlemarch* and the Doctors." *Cambridge Journal*, Sept. 1948.
Cline, C. L. "Qualifications of Medical Practitioners of *Middlemarch*." In *Nineteenth-Century Perspectives: Essays in Honor of Lionel Stevenson*, edited by C. de L. Ryals. Durham, NC: Duke University Press, 1974.
Cross, J. W., ed. *George Eliot's Life as Related in Her Letters and Journals*. 3 vols. Edinburgh: Blackwood, 1885.
Derrow, H. A. "*Middlemarch* and the Physician." *Annals of Medical History* 9 (1927).
Greenberg, R. A. "Plexuses and Ganglia: Scientific Allusion in *Middlemarch*." *Nineteenth-Century Fiction* 30 (1976).
Harvey, W. J. "The Intellectual Background of the Novel." In *Middlemarch: Critical Approaches to the Novel*, edited by Barbara Hardy. London: Athlone, 1967.
Hulme, Hilda M. "*Middlemarch* as Science-Fiction: Notes on Language and Imagery." *Novel* 2 (Fall 1968).
Kitchel, Anna, ed. "Quarry for *Middlemarch*." *Nineteenth-Century Fiction* 4 (1949–50), suppl.
Knoepflmacher, U. C. *Religious Humanism and the Victorian Novel: George Eliot, Walter Pater, and Samuel Butler*. Princeton, NJ: Princeton University Press, 1970.
Levine, George. "Determinism and Responsibility in the Works of George Eliot." *PMLA* 77 (1962).
———. "George Eliot's Hypothesis of Reality." *Nineteenth-Century Fiction* 35 (1980).
———. *Realism, Ethics and Secularism: Essays on Victorian Literature and Science*. Cambridge: Cambridge University Press, 2008.
McCarthy, P. J. "Lydgate, 'The New, Young Surgeon' of Middlemarch." *Studies in English Literature, 1500–1900* 10 (1970).

Mason, Michael York. "*Middlemarch* and Science: Problems of Life and Mind." *Review of English Studies*, n.s., 22, no. 116 (1971).
Newton, K. M. "George Eliot, George Henry Lewes, and Darwinism." *Durham University Journal* 65 (June 1974).
Paris, Bernard. *Experiments in Life: George Eliot's Quest for Values*. Detroit: Wayne State University Press, 1965.
Shuttleworth, Sally. *George Eliot and Nineteenth-Century Science*. Cambridge: Cambridge University Press, 1984.
———. "The Language of Science and Psychology in *Daniel Deronda*." *Annals of the New York Academy of Sciences* 360 (20 Apr. 1981).

Emerson, Ralph Waldo

Haugrud, Raychel A. "John Tyndall's Interest in Emerson." *American Literature* 41 (1970).

Fielding, Henry

Eales, Nellie B. "A Satire on the Royal Society Dated 1743, Attributed to Henry Fielding." *Notes and Records of the Royal Society of London* 23 (1968).

Frisch, Max

Gontrum, P. "Max Frisch's Don Juan: A New Look at a Traditional Hero." *Comparative Literary Studies* 2, no. 2 (1965).

Gaskell, Elizabeth Cleghorn

Gaskell, Elizabeth Cleghorn. *The Letters of Mrs. Gaskell*. Edited by J. A. V. Chapple and Arthur Pollard. Manchester: Manchester University Press, 1966.
Lucas, John. "Mrs. Gaskell: The Nature of Social Change." In *The Literature of Change*. Brighton, Sussex: Harvester, 1980.

Gissing, George

Korg, Jacob. "The Spiritual Theme of *Born in Exile*." In *Collected Articles on George Gissing*, edited by Pierre Coustillas. London: Frank Cass, 1968.

Goethe, Johann Wolfgang von

Cottrell, Alan P. *Goethe's View of Evil and the Search for the New Image of Man in Our Time*. Edinburgh: Edinburgh University Press, 1982.
Fink, Karl J. *Goethe's History of Science*. Cambridge: Cambridge University Press, 1992.
Gray, Ronald D. *Goethe, the Alchemist: A Study of Alchemical Symbolism in Goethe's Literary and Scientific Work*. Cambridge: Cambridge University Press, 1952.
———. *Goethe: A Critical Introduction*. Cambridge: Cambridge University Press, 1967.
Heller, Otto. *Faust and Faustus: A Study of Goethe's Relation to Marlowe*. New York: Cooper Square, 1972.
Smeed, J. W. *Faust in Literature*. New York: Oxford, University Press, 1975.

Hardy, Thomas

Hardy, F. E. *The Early Life of Thomas Hardy*. London: Macmillan, 1928.
Ingham, Patricia. "Hardy and *The Wonders of Geology*." *Review of English Studies*, n.s., 31 (Feb. 1980), 59–64.
Millgate, Michael. *Thomas Hardy: His Career as a Novelist*. London: Bodley Head, 1971.

Wickens, G. G. "Literature and Science: Hardy's Response to Mill, Huxley, and Darwin." *Mosaic* 14, no. 3 (1981), 63.

Hawthorne, Nathaniel

Bales, Kent. "Hawthorne's Prefaces and Romantic Perspectivism." *Emerson Society Quarterly* 23 (1977): 55–69.
———. "Sexual Exploitation and the Fall from Natural Virtue in Rappaccini's Garden." *Emerson Society Quarterly* 24 (1978): 133–44.
Crews, Frederick C. *The Sins of the Fathers: Hawthorne's Psychological Themes*. London: Oxford University Press, 1966.
Fogle, Richard Harter. *Hawthorne's Fiction: The Light and the Dark*. Norman: University of Oklahoma Press, 1964.
Fryer, Judith. *The Faces of Eve: Women in the Nineteenth-Century American Novel*. New York: Oxford University Press, 1978.
Gollin, Rita. *Nathaniel Hawthorne and the Truth of Dreams*. Baton Rouge: Louisiana State University Press, 1979.
Heilman, R. B. "'The Birthmark': Science as Religion." *South Atlantic Quarterly* 48 (1949).
Kaul, A. N., ed. *Hawthorne: A Collection of Critical Essays*. Englewood Cliffs, NJ: Prentice Hall, 1966.
Laser, Marvin. "Head, Heart, and Will in Hawthorne's Psychology." *Nineteenth-Century Fiction* 10 (1955).
Martin, Terence. *Nathaniel Hawthorne*. Boston: Twayne, 1983.
Noble, David W. "The Analysis of Alienation by Twentieth-Century Social Scientists and Nineteenth-Century Novelists: The Example of Hawthorne's *The Scarlet Letter*." *Festschriften* 112 (1972).
Pollin, Burton R. "'Rappaccini's Daughter'—Sources and Names." *Names* 14, no. 1 (1966).
Ringe, Donald A. "Hawthorne's Psychology and the Head and Heart." *PMLA* 65 (1950).
Schroeder, John W. "That Inward Sphere: Notes on Hawthorne's Heart Imagery and Symbolism." *PMLA* 65 (1950).
Stein, W. B. *Hawthorne's Faust: A Study of the Devil Archetype*. Gainesville: University of Florida Press, 1953.
Stoehr, Taylor. *Hawthorne's Mad Scientists: Pseudoscience and Social Science in Nineteenth-Century Life and Letters*. Hamden, CT: Archon, 1978.
Thompson, W. R. "Aminadab in Hawthorne's 'The Birthmark.'" *Modern Language Notes* 52 (1955).
Turner, Arlin. *Nathaniel Hawthorne: A Biography*. New York: Oxford University Press, 1980.
Wentersdorf, Karl P. "The Elements of Witchcraft in *The Scarlet Letter*." *Folklore* 83 (1972).

Hoffmann, E. T. A.

Brantly, Susan. "A Thermographic Reading of E. T. A. Hoffmann's *Der Sandmann*." *German Quarterly* 55, no. 3 (1982).
Ellis, J. M. "Clara, Nathanael, and the Narrator: Interpreting Hoffmann's *Der Sandmann*." *German Quarterly* 54, no. 1 (1981).
Lawson, Ursula D. "Pathological Time in E. T. A. Hoffmann's *Der Sandmann*." *Monatshefte* 60 (1968).
Prawer, S. S. "Hoffmann's Uncanny Guest: A Reading of *Der Sandmann*." *German Life and Letters* 18 (1965).
Von Matt, Peter. *Die Augen der Automaten: E. T. A. Hoffmanns Imaginationslehre als Prinzip seiner Einzählkunst*. Tübingen: Niemeyer, 1971.

Huxley, Aldous

Birnbaum, Milton. *Aldous Huxley's Quest for Values*. Knoxville: University of Tennessee Press, 1971.
Brander, Lawrence. *Aldous Huxley: A Critical Study*. London: Rupert Hart-Davis, 1969.
Clareson, Thomas D. "The Classic: Aldous Huxley's *Brave New World*." *Extrapolation* 2 (1961).

James, Henry

Long, R. E. "A Source for Dr. Mary Prance in *The Bostonians*." *Nineteenth-Century Fiction* 19 (1964).
Purdy, S. B. *The Hole in the Fabric: Science, Contemporary Literature, and Henry James*. Pittsburgh: Pittsburgh University Press, 1977.

Johnson, Samuel

Boswell, James. *The Life of Samuel Johnson, L.L.D.* [1791]. New York: Oxford University Press, 1961.
Eberwein, Robert. "The Astronomer in [Samuel] Johnson's Rasselas." *Michigan Academician* 5, no. 1 (1973).
Philip, J. R. "Samuel Johnson as Anti-scientist." *Notes and Records of the Royal Society of London* 29 (1975).
Schwartz, Richard B. *Samuel Johnson and the New Science*. Madison: University of Wisconsin Press, 1971.

Jünger, Ernst

Kochhar-Lindgren, Gray. "Ethics, Automation, and the Ear: Capitalism, Technology, and the Suspension of Animation in Ernst Jünger's *The Glass Bees*." Ctheory.net. https://journals.uvic.ca/index.php/ctheory/article/view/14623/5489.

Kingsley, Charles

Chitty, Susan. *The Beast and the Monk: A Life of Charles Kingsley*. London: Hodder & Stoughton, 1974.
Gillespie, H. R., Jr. "George Eliot's Tertius Lydgate and Charles Kingsley's Tom Thurnall." *Notes and Queries* 2, no. 1 (1964).
Kingsley, Charles. *Charles Kingsley: His Letters and Memories of His Life*. Edited by Frances Eliza Grenfell Kingsley. London: Henry S. King, 1877.
Thorp, Margaret F. *Charles Kingsley, 1819–1875*. New York: Octagon, 1969.

Kipphardt, Heiner

Zipes, Jack D. "Documentary Drama in Germany: Mending the Circuit." *Germanic Review* 62 (Jan. 1967).

La Mettrie, Julien Offray de

Vartanian, Aram. *La Mettrie's "L'Homme Machine": A Study in the Origins of an Idea*. Princeton, NJ: Princeton University Press, 1960.

Lasswitz, Kurd

Fischer, William B. *The Empire Strikes Out: Kurd Lasswitz, Hans Dominik, and the Development of German Sceince Fiction*. Wisconsin: Popular Press, 1984.
Hillegas, Mark R. "The First Invasion from Mars." *Michigan Alumnus Quarterly Review* 66 (1959).

Le Guin, Ursula

Bierman, Judah. "Ambiguity in Utopia: *The Dispossessed*." *Science Fiction Studies* 2, no. 3 (1975): 249.
Bittner, James W. *Approaches to the Fiction of Ursula K. Le Guin*. Ann Arbor: UMI Research Press, 1984.
Bucknall, Barbara J. *Ursula K. Le Guin*. New York: Frederick Ungar, 1981.
Cummins, Elizabeth. *Understanding Ursula K. Le Guin*. Columbia: University of South Carolina Press, 1990.
Moylan, Tom. "Ursula K. Le Guin, *The Dispossessed*." In *Demand the Impossible*. London: Methuen, 1986.
Selinger, Bernard. *Le Guin and Identity in Contemporary Fiction*. Ann Arbor: UMI Research Press, 1958.
Spivak, Charlotte. *Ursula K. Le Guin*. Boston: Twayne, 1984.
Tavormina, M. Teresa. "Physics as Metaphor: The General Temporal Theory in *The Dispossessed*." *Mosaic* 13 (1980).

Lem, Stanislaw

Jarzebski, Jerzy. "Stanislaw Lem: Rationalist and Sensualist." *Science Fiction Studies* 4, no. 2 (July 1977).
Kandel, Michael. "Lem in Review (June 2238)." *Science Fiction Studies* 4, no. 1 (March 1977).
———. "Stanislaw Lem on Man and Robots." *Extrapolation* 14 (1972).
Philmus, Robert. "*Futurological Congress* as Metageneric Text." *Science Fiction Studies* 13, no. 3 (1986).
Rodnianskaia, Irina. "Two Faces of Stanislaw Lem: On *His Master's Voice*." *Science Fiction Studies* 13, no. 3 (1986).
Ziegfield, Richard E. *Stanislaw Lem*. New York: Frederick Ungar, 1985.

Lewis, C. S.

Sammons, Martha C. *A Guide through C. S. Lewis' Space Trilogy*. Westchester, IL: Cornerstone Books, 1980.

Lewis, Sinclair

Geismar, M. "Sinclair Lewis." In *The Last of the Provincials: The American Novel, 1915–1925*. London: Seeker & Warburg, 1947.
Kazin, Alfred. "The New Realism: Sherwood Anderson and Sinclair Lewis." In *On Native Grounds: An Interpretation of Modern American Prose Literature*. New York: Harcourt, Brace & World, 1942.
Lingeman, Richard R. *Sinclair Lewis: Rebel from Main Street*. St. Paul, MN: Borealis, 2005.
Rosenberg, Charles E. "Martin Arrowsmith: The Scientist as Hero." *American Quarterly* 15 (1963).
Schorer, Mark, ed. *Sinclair Lewis: A Collection of Critical Essays*. Englewood Cliffs, NJ: Prentice Hall, 1962.
———. *Sinclair Lewis: An American Life*. New York: McGraw-Hill, 1961.

Mann, Thomas

Prusok, Rudi. "Science in Mann's *Zauberberg*: The Concept of Space." *PMLA* 88 (1973).

Marlowe, Christopher

Brown, Beatrice D. "Marlowe, Faustus, and Simon Magus." *PMLA* 54 (1939).

McAlindon, T. "The Ironic Vision: Diction and Theme in Marlowe's *Doctor Faustus*." *Review of English Studies* 32, no. 126 (1981).

Szőnyi, Gye. "The Quest for Omniscience: The Intellectual Background of Marlowe's Doctor Faustus." *Papers in English and American Studies* 1 (1980).

Meredith, George

Grabar, Terry H. "Scientific Education and Richard Feverel." *Victorian Studies* 14, no. 2 (1970).

Musil, Robert

Goldgar, Harry. "Freud and Robert Musil's Törless." *Comparative Literature* 17 (1965).

Kirchberger, Lida. "Musil's Trilogy: An Approach to *Drei Frauen*." *Monatshefte* 55 (1963)

Luft, David S. *Robert Musil and the Crisis of European Culture, 1880–1942*. Berkeley: University of California Press, 1980.

Peters, Frederick G. *Robert Musil, Master of the Hovering Life: A Study of the Major Fiction*. New York: Columbia University Press, 1978.

Sokel, W. H. "Kleist's Marquise of O, Kierkegaard's Abraham, and Musil's Tonka: Three Stages of the Absurd as the Touchstones of Faith." *Wisconsin Studies in Contemporary Literature* 8, no. 4 (1967).

O'Brien, Fitz-James

Franklin, H. Bruce. "O'Brien and Science Fiction." In *Future Perfect: American Science Fiction of the Nineteenth Century*. New York: Oxford University Press, 1966.

Poe, Edgar Allan

Howarth, W., ed. *Twentieth-Century Interpretations of Poe's Tales*. Englewood Cliffs, NJ: Prentice Hall, 1971.

Ketterer, D. "The Science Fiction Element in the Work of Poe." *Science Fiction Studies* 1 (1974).

Pope, Alexander

Nicolson, Marjorie Hope, and George S. Rousseau. *This Long Disease, My Life: Alexander Pope and the Sciences*. Princeton, NJ: Princeton University Press, 1968.

Prior, Matthew

Spears, Monroe K. "Matthew Prior's Attitude toward Natural Science." *PMLA* 63 (1948).

Proust, Marcel

Bottiger, L. E. "Remembrance of Disease Lifelong: Marcel Proust and Medicine." *British Medical Journal* 287 (3 Dec. 1983).

Pynchon, Thomas

Bloom, Harold, ed. *Thomas Pynchon*. New York: Chelsea House, 1986.

Friedman, Alan J. "Science and Technology." In *Approaches to Gravity's Rainbow*, edited by Charles Clerc. Columbus: Ohio State University Press, 1983.

Friedman, Alan J., and Manfred Puetz. "Science as Metaphor: Thomas Pynchon and *Gravity's Rainbow*." *Contemporary Literature* 15 (1974).

Morrison, Philip. Review of *Gravity's Rainbow*, by Thomas Pynchon. *Scientific American* 229 (Oct. 1973).

Tanner, Tony. *Thomas Pynchon*. London: Methuen, 1982.

Ritter, Johann Wilhelm

Rehm, Else. "Johann Wilhelm Ritter und die Universität Jena." *Jahrbuch des Freien Deutschen Hochstifts.* Tübingen: Niemeyer, 1973.

——. "Über den Tod und den letzten Verfügungen des Physikers Johann Wilhelm Ritter." *Jahrbuch des Freien Deutschen Hochstifts.* Tübingen: Niemeyer, 1974.

Shadwell, Thomas

Gilde, Joseph M. "Thomas Shadwell and the Royal Society: Satire in *The Virtuoso.*" *Studies in English Literature* 10 (1970).

Lloyd, Claude. "Shadwell and the Virtuosi." *PMLA* 44 (1929).

Shelley, Mary

Awad, Louis. "The Alchemist in English Literature, Part 1: Frankenstein." *Bulletin of the Faculty of Arts / Fouad University* 13 (May 1951).

Bennett, Betty T., and Charles E. Robinson, eds. *The Mary Shelley Reader.* New York: Oxford University Press, 1990.

Bloom, Harold. "*Frankenstein,* or the New Prometheus." *Partisan Review* 32, no. 4 (1965).

——. *Mary Shelley's Frankenstein.* New York: Chelsea House, 2007.

Brooks, P. "Godlike Science / Unhallowed Arts: Language and Monstrosity in *Frankenstein.*" *New Literary History* 9 (1978).

Buchen, Irving H. "*Frankenstein* and the Alchemy of Creation and Evolution." *Wordsworth Circle* 8 (1977).

Butler, Marilyn. "The First *Frankenstein* and Radical Science." *Times Literary Supplement,* 9 (Apr. 1993).

Callahan, P. J. "*Frankenstein,* Bacon, and the Two Truths." *Extrapolation* 14 (1972).

Cude, Wilfred. "Mary Shelley's Modern Prometheus: A Study in the Ethics of Scientific Creativity." *Dalhousie Review* 52 (1972).

Fleck, P. D. "Mary Shelley's Notes to Shelley's Poems and *Frankenstein.*" *Studies in Romanticism* 6, no. 4 (1967).

Florescu, Radu. *In Search of Frankenstein.* Boston: New York Graphic Society, 1975.

Goldberg, M. P. "Moral and Myth in Mrs. Shelley's *Frankenstein.*" *Keats-Shelley Journal* 8 (Winter 1959).

Hume, Robert D. "Gothic versus Romantic: A Revaluation of the Gothic Novel." *PMLA* 84 (1969).

Ketterer, David. *Frankenstein's Creation: The Book, the Monster, and Human Reality.* English Literary Studies 16. Victoria, BC: University of Victoria, 1979.

Kiely, Robert. "Frankenstein." In *The Romantic Novel in England.* Cambridge, MA: Harvard University Press, 1973.

Kreutz, C. "Mary Wollstonecraft Shelleys Prometheusbild." in *Das Prometheussymbol in der Dichtung der Englischen Romantik, Palaestra* 236 (1963).

Levine, George. "*Frankenstein* and the Tradition of Realism." *Novel* 7 (Fall 1973).

Levine, George, and U. C. Knoepflmacher, eds. *The Endurance of Frankenstein: Essays on Mary Shelley's Novel.* Berkeley: University of California Press, 1979.

Lovell, Ernest J., Jr. "Byron and Mary Shelley." *Keats-Shelley Journal* 2 (Winter 1953).

Lund, Mary Graham. "Mary Godwin Shelley and the Monster." *University of Kansas City Review* 28 (June 1962).

——. "Shelley as Frankenstein." *Forum* 4, no. 2 (1963).

McInerney, Peter. "Frankenstein and the Godlike Science of Letters." *Genre* 13, no. 4 (1980): 455–75.

McLeod, Patrick G. "Frankenstein: Unbound and Otherwise." *Extrapolation* 21 (1980).
Mays, Milton A. "*Frankenstein*, Mary Shelley's Black Theodicy." *Southern Humanities Review* 3 (1969).
Millhauser, M. "The Noble Savage in *Frankenstein*." *Notes and Queries* 190 (1946).
Moers, Ellen. "Female Gothic." In *Literary Women*. London: Allen, 1977.
———. "Female Gothic: The Monster's Mother." *New York Review of Books*, 21 Mar. 1974.
Nelson, Lowry, Jr. "Night Thoughts on the Gothic Novel." *Yale Review* 52 (1962).
Nitchie, Elizabeth. *Mary Shelley, Author of "Frankenstein*. New Brunswick, NJ: Rutgers University Press, 1953.
Palmer, D. J., and R. E. Dowse. "*Frankenstein*: A Moral Fable." *Listener*, 23 Aug. 1962.
Pollin, B. R. "Philosophical and Literary Sources of *Frankenstein*." *Comparative Literature* 17 (1965).
Railo, Eino. *The Haunted Castle*. New York: Humanities Press, 1964.
Reed, John R. "Will and Fate in *Frankenstein*." *Bulletin of Research in Humanities* 83 (1980).
Rieger, James. "Dr. Polidori and the Genesis of *Frankenstein*." *Studies in English Literature* 3 (1963).
Roszak, Theodore. "The Monster and the Titan: Science, Knowledge, and Gnosis." *Daedalus* 103, no. 3 (1974).
Scholes, Robert, and E. S. Rabkin. *Science Fiction*. New York: Oxford University Press, 1977.
Scott, Sir Walter. "Remarks on *Frankenstein*." *Blackwood's Edinburgh Magazine* 2, no. 12 (1818).
Seed, David. "*Frankenstein*: Parable or Spectacle?" *Criticism* 24, no. 4 (1982).
Shelley, Percy Bysshe. "On *Frankenstein*." In *The Complete Works of Percy Bysshe Shelley*, edited by Roger Ingpen and Walter E. Peck, vol. 6. New York: Benn, 1926–29.
———. "Preface to *Frankenstein; or, the Modern Prometheus*, 1818." In *The Complete Works of Percy Bysshe Shelley*, vol. 6. See Shelley, "On *Frankenstein*."
Small, Christopher. *Ariel like a Harpy: Shelley, Mary, and Frankenstein*. London: Gollancz, 1972.
———. *Mary Shelley's Frankenstein: Tracing the Myth*. Pittsburgh: University of Pittsburgh Press, 1973.
Spark, Muriel. *Child of Light: A Reassessment of Mary Shelley*. Hadleigh, Essex: Tower Bridge, 1951.
———. "Mary Shelley: A Prophetic Novelist." *Listener*, 22 Feb. 1951.
Sterrenburg, Lee. "Mary Shelley's Monster: Politics and Psyche in *Frankenstein*." In Levine and Knoepflmacher, *Endurance of Frankenstein*, 143–71.
Swingle, L. J. "*Frankenstein*'s Monster and Its Romantic Relatives: Problems of Knowledge in English Romanticism." *Texas Studies in Literature and Language* 15, no. 1 (1973).
Turney, Jon. *Frankenstein's Footsteps: Science, Genetics and Popular Culture*. New Haven, CT: Yale University Press, 1998.
Vasbinder, S. H. *Scientific Attitudes in Mary Shelley's Frankenstein*. Ann Arbor, MI: UMI Research Press, 1984.
Walling, W. A. *Mary Shelley*. New York: Twayne, 1972.
Ziolkowski, T. "Science, Frankenstein, and Myth." *Sewanee Review* 89, no. 1 (1981).

Shelley, Percy Bysshe

Grabo, Carl. *A Newton among Poets: Shelley's Use of Science in Prometheus Unbound*. New York: Cooper Square, 1968.
———. *Prometheus Unbound: An Interpretation*. Chapel Hill: University of North Carolina Press, 1935.

Smart, Christopher

Greene, D. J. "Smart, Berkeley, the Scientists, and the Poets." *Journal of the History of Ideas* 14 (1953).

Snow, C. P.

Bergonzi, B. "The World of Lewis Eliot." *Twentieth Century* 167 (Mar. 1960).
Bezel, Nail. "Autobiography and the 'Two Cultures' in the Novels of C. P. Snow." *Annals of Science* 32 (1975).
Greacen, Robert. *The World of C. P. Snow*. London: Scorpion, 1962.
Karl, F. R. *C. P. Snow: The Politics of Conscience*. Carbondale: Southern Illinois University Press, 1963.
Ramanathan, Suguna. *The Novels of C. P. Snow: A Critical Introduction*. London: Macmillan, 1978.
Shusterman, David. *C. P. Snow*. Boston: Twayne, 1975.
Snow, C. P. "Interview." *Review of English Literature* 3 (July 1962).
Thale, Jerome. *C. P. Snow*. Edinburgh: Oliver & Boyd, 1964.
Waring, A. G. "Science, Love, and the Establishment in the Novels of D. A. Granin and C. P. Snow." *Forum for Modern Language Studies* 14, no. 1 (1978).

Stevenson, Robert Louis

Chesterton, G. K. *Robert Louis Stevenson*. London: Houghton & Stoughton, 1927.

Stoker, Bram

Hennelly, Mark M., Jr. "Dracula: The Gnostic Quest and Victorian Wasteland." *English Literature in Transition* 20 (1977).

Swift, Jonathan

Case, Arthur, E. *Four Essays on Gulliver's Travels*. Princeton, NJ: Princeton University Press, 1945.
Crane, Ronald S. "The Houyhnhnms, the Yahoos, and the History of Ideas. "In *Reason and the Imagination*, edited by J. A. Mazzeo. New York: Columbia University Press, 1962.
Eddy, W, A. *Gulliver's Travels: A Critical Study*. Gloucester, MA: Peter Smith, 1963.
Foster, M. P., ed. *A Casebook on Gulliver among the Houyhnhnms*. New York: Crowell, 1961.
Jeffares, A. Norman, ed. *Fair Liberty Was All His Cry: A Tercentenary Tribute to Jonathan Swift, 1667–1745*. London: Macmillan, 1967.
Kiernan, Colin. "Swift and Science." *Historical Journal* 14, no. 4 (1971).
Korshin, Paul J. "The Intellectual Context of Swift's Flying Island." *Philological Quarterly* 50 (1971).
Merton, Robert C. "The 'Motionless' Motion of Swift's Flying Island." *Journal of the History of Ideas* 27 (1966).
Monro, John M. "Book III of *Gulliver's Travels* Once More." *English Studies* 49 (1968).
Nicolson, Marjorie Hope. "The Scientific Background of Swift's *Voyage to Laputa*." In *Science and Imagination*. Ithaca, NY: Cornell University Press, 1956.
Nicolson, Marjorie Hope, and Nora Monier. "Swift's 'Flying Island' in the *Voyage to Laputa*." *Annals of Science* 2 (1937).
Ross, Angus. *Swift: Gulliver's Travels*. London: Edward Arnold, 1972.
Sutherland, John H. "A Reconsideration of Gulliver's Third Voyage." *Studies in Philology* 54 (1957).

Tuveson, Ernest, ed. *Swift: A Collection of Critical Essays*. Englewood Cliffs, NJ: Prentice Hall, 1964.
Williams, Kathleen. "Gulliver in Laputa." In *Twentieth-Century Interpretations of Gulliver's Travels*, edited by F. Brady. Englewood Cliffs, NJ: Prentice Hall, 1968.

Tennyson, Alfred, Lord

Gibson, Walker. "Behind the Veil: A Distinction between Poetic and Scientific Language in Tennyson, Lyell, and Darwin." *Victorian Studies* 2, no. 1 (1958).
Gliserman, Susan M. "Early Victorian Science Writers and Tennyson's 'In Memoriam': A Study in Cultural Exchange." *Victorian Studies* 18, nos. 3 and 4 (1975).
Millhauser, Michael. *Fire and Ice: The Influence of Science on Tennyson's Poetry*. Lincoln: Tennyson Society, 1971.
Wickens, G. G. "The Two Sides of Early Victorian Science and the Unity of 'The Princess.'" *Victorian Studies* 23, no. 3 (1980).

Thomson, James

Drennon, Herbert. "James Thomson's Contact with Newtonianism and His Interest in Natural Philosophy." *PMLA* 49 (1934).
———. "Newtonianism in James Thomson's Poetry." *Englische Studien* 70 (1936).
———. "Scientific Rationalism and James Thomson's Poetic Art." *Studies in Philology* 31 (1934).
Ketcham, Michael G. "Scientific and Poetic Imagination in James Thomson's 'Poem Sacred to the Memory of Sir Isaac Newton.'" *Philological Quarterly* 61 (1982).
McKillop, Alan D. *The Background of Thomson's The Seasons*. Hamden, CT: Archon, 1961.
———, ed. *The Castle of Indolence, and Other Poems*. Lawrence: University of Kansas Press, 1961.

Thoreau, Henry David

Baym, Nina. "Thoreau's View of Science." *Journal of the History of Ideas* 26 (1965).
Griffin, David. "The Science of Henry David Thoreau." *Synthesis* 1, no. 4 (1973).
Harding, Walter. "Walden's Man of Science." *Virginia Quarterly* 57 (1981).

Verne, Jules

Barthes, Roland. "The *Nautilus* and the Drunken Boat." In *Mythologies*, translated by Annette Lavers. London: Jonathan Cape, 1972.
Butor, Michel. "Le Point suprême et l'Age d'Or: à travers quelques oeuvres de Jules Verne." *Répertoire* 1 (1960).
Costello, Peter. *Jules Verne, Inventor of Science Fiction*. London: Hodder & Stoughton, 1978.
De la Fuye, Marguerite Allotte. *Jules Verne*. Translated by Erik de Maury. London: Staples, 1954.
Evans, Arthur B. *Jules Verne Rediscovered: Didacticism and the Scientific Novel*. New York: Greenwood, 1988.
Kylstura, Peter H. "Some Backgrounds of Jules Verne." *Janus* 57 (1970).
Suvin, Darko. "Communication in Quantified Space: Utopian Liberalism of Jules Verne's Science Fiction." *Clio* 4, no. 1 (1974).
Winandry, André, "The Twilight Zone: Imagination and Reality in Jules Verne's *Strange Journeys*." *Yale French Studies* 43 (1969).

Voltaire

Wade, Ira O. *Voltaire's "Micromégas."* Princeton, NJ: Princeton University Press, 1950.

Von Arnim, Achim

Riley, Helene M. "Scientist, Sorcerer, or Servant of Humanity: The Many Faces of Faust in the Work of Achim von Arnim." *Journal of Germanic Studies* (University of British Columbia) 13 (1977).

Vonnegut, Kurt

Klinkowitz, Jerome. *Kurt Vonnegut.* London: Methuen, 1982.

McNelly, Willis E. "Kurt Vonnegut as Science Fiction Writer." in *Vonnegut in America: An Introduction to the Life and Work of Kurt Vonnegut*, edited by Jerome Klinkowitz and Donald L. Lawler. New York: 1977.

Zins, Daniel L. "Rescuing Science from Technocracy: *Cat's Cradle* and the Play of the Apocalypse." *Science Fiction Studies* 13, no. 2 (1986).

Wells, H. G.

Bergonzi, Bernard. "Another Early Wells Item." *Nineteenth-Century Fiction* 13 (1958).

———. *The Early H. G. Wells: A Study of the Scientific Romances.* Manchester: Manchester University Press, 1961.

———. *H. G. Wells: A Collection of Critical Essays.* Englewood Cliffs, NJ: Prentice Hall, 1976.

Bower, R. "Science, Myth, and Fiction in H. G. Wells's *Island of Dr. Moreau.*" *Studies in the Novel* 8, no. 3 (1976).

Eisenstein, A. "Very Early Wells: Origins of Some Major Physical Motifs in *The Time Machine* and *The War of the Worlds.*" *Extrapolation* 13, no. 2 (1972), 119–126.

Harris, Wilson, ed. *Arnold Bennett and H. G. Wells: A Record of a Personal and a Literary Friendship.* London: Rupert Hart-Davis, 1960.

Haynes, Roslynn D. *H. G. Wells, Discoverer of the Future: The Influence of Science on His Thought.* London: Macmillan, 1980.

———. "The Unholy Alliance of Science in *The Island of Doctor Moreau.*" *Wellsian* 11 (1988).

———. "Wells's Debt to Huxley and the Myth of Doctor Moreau." *Cahiers Victoriens et Édouardiens* 13 (Apr. 1981).

Hennelly, Mark M., Jr. "*The Time Machine*: A Romance of 'The Human Heart.'" *Extrapolation* 20 (1979).

Hillegas, Mark R. "Cosmic Pessimism in H. G. Wells's Scientific Romances." *Papers of the Michigan Academy of Science, Arts and Letters* 46 (1961).

Kagarlitski, J. *The Life and Thought of H. G. Wells.* Translated by Moura Budberg. London: Sidgwick & Jackson, 1966.

Le Mire, E. D. "H. G. Wells and the World of Science Fiction." *University of Windsor Review* 2, no. 2 (1967).

Lindsay, C. B. "A Formalist Approach to Wells' Science Fantasies." *Rocky Mountain Review of Language and Literature* 34, no. 3 (1980).

Locke, G. "Wells in Three Volumes?" *Science Fiction Studies* 3, no. 3 (1976).

McConnell, Frank. *The Science Fiction of H. G. Wells.* Oxford: Oxford University Press, 1981.

Newell, K. B. *Structure in Four Novels by H. G. Wells.* The Hague: Mouton, 1968.

Parrinder, Patrick. "*The First Men in the Moon*: H. G. Wells and the Fictional Strategy of His 'Scientific Romances.'" *Science Fiction Studies* 7, no. 2 (1980).

———. *H. G. Wells.* Edinburgh: Oliver & Boyd, 1970.

———. "*News from Nowhere*, *The Time Machine*, and the Break-up of Classical Realism." *Science Fiction Studies* 3, no. 3 (1976).

Parrinder, Patrick, ed. *H. G. Wells: The Critical Heritage.* London: Routledge, 1972.

Philmus, Robert. "'The Time Machine': Or, The Fourth Dimension as Prophecy." *PMLA* 84 (1969).
Platzner, Robert L. "H. G. Wells's 'Jungle Book': The Influence of Kipling on *The Island of Dr. Moreau*." *Victorian Newsletter* 36 (1969).
Raknem, Ingvald. *H. G. Wells and His Critics*. Trondheim, Norway: Universitets-forlaget / Allen & Unwin, 1962.
Roemer, Kenneth M. "H. G. Wells and the 'Momentary Voices' of a Modern Utopia." *Extrapolation* 23 (1982).
Stiles, Anne. "Literature in Mind: H. G. Wells and the Evolution of the Mad Scientist." *Journal of the History of Ideas* 70, no. 2 (2009).
Suvin, Darko. "*The Time Machine* versus *Utopia* as a Structural Model for Science Fiction." *Comparative Literature Studies* 10, no. 4 (1973).
Suvin, Darko, and Robert M. Philmus, eds. *H. G. Wells and Modern Science Fiction*. London: Associated University Presses, 1977.
Wagar, Warren. *H. G. Wells and the World State*. New Haven, CT: Yale University Press, 1963.
Williamson, Jack. *H. G. Wells: Critic of Progress*. Baltimore: Mirage, 1973.
Zamyatin, Yevgeny. "H. G Wells." In *A Soviet Heretic: Essays by Yevgeny Zamyatin*, edited and translated by Mirra Ginsburg, 259–90. Chicago: University of Chicago Press, 1970.

Whitman, Walt

Lindfors, Berndt. "Whitman's 'When I Heard the Learn'd Astronomer.'" *Walt Whitman Revue* 10 (1964).

Wolf, Christa

Buehler, George. *The Death of Socialist Realism in the Novels of Christa Wolf*. Frankfurt am Main: Peter Lang, 1984.
Eigler, Friederike. "Rereading Christa Wolf's 'Selbstversuch': Cyborgs and Feminist Critique of Scientific Discourse." *German Quarterly* 73, no. 4 (2000): 401–415.
Kuhn, Anna K. *Christa Wolf's Utopian Vision: From Marxism to Feminism*. Cambridge: Cambridge University Press, 1988.

Wordsworth, William

Gaull, M. "From Wordsworth to Darwin: 'On to the Fields of Praise.'" *Wordsworth Circle* 1 (1979).

Yeats, William Butler

Spivey, Ted R. "W. B. Yeats and the 'Children of Fire': Science, Poetry, and Visions of the New Age." *Studies in the Literary Imagination* 14, no. 1 (1981).

Zamyatin, Yevgeny

Shane, Alex M. *The Life and Works of Evgenyi Zamyatin*. Berkeley: University of California Press, 1968.
Warrick, Patricia. "The Sources of Zamiatin's *We* in Dostoevsky's *Notes from the Underground*." *Extrapolation* 17 (1975).

Zola, Émile

Schmidt, Günther. *Die literarische Rezeption des Darwinismus: Das Problem der Vergebung bei Emile Zola und in Drama des deutschen Naturalismus*. Berlin: Akademie-Verlag, 1974.

Zuckmayer, Carl

Glade, Henry. "Carl Zuckmayer's Theory of Aesthetics." *Monatshefte* 52 (1960).
Peppard, Murray B. "Carl Zuckmayer: Cold Light in a Divided World." *Monatshefte* 49 (1957).

INDEX

2001: A Space Odyssey, 272

A for Andromeda, 306–7
Albertus Magnus, 15, 265
alchemists, 2, 4–5, 12, 15
alchemy, 43, 344n5; Arabic, 14; Chinese, 14, 342n2; origins, 12, 13
Aldridge, James, *The Diplomat*, 328
Amazing Stories, 165
Andrade, E.N. da C., 62
Andres, Stefan, *Die Sintflut*, 259–60
androids, 261, 263, 269–72
Aristotle, 13, 31, 34
Arnold, Thomas, 112
artificial intelligence, 269–70
Asimov, Isaac, 266–69, 306, 332; "Evidence," 267; "I, Robot," 267; *View from a Height*, 268
Astounding Stories, 165, 175
Atlas Martin, *Die Befreiung*, 181
atomic power/weapons, 174–75, 179, 195–99, 239–40, 243–57, 284, 367n23
Atomised, 274
atom spies, 184–85, 232–33, 249, 292, 328
Atwood, Margaret, *Oryx and Crake*, 280
Auburn, David, *Proof*, 338
Aughterson, Kate, 40
Avatar, 167, 310, 338

B——, Madame, *La femme endormie*, 147–48
Bacon, Roger, 15, 17
Bacon, Sir Francis, Lord Verulam, 5, 9, 25, 29–34, 35, 36, 38–39, 43, 60; and nature, 39–40, 100–101, 345n3, 346n35; *New Atlantis*, 7, 30–34, 38, 39, 40, 41–42, 345n17

Balchin, Nigel: *The Small Back Room*, 286; *A Sort of Traitors*, 286–87
"Ballad of Gresham Colledge," 47
Balmer, Edwin, and Philp Wylie: *After Worlds Collide*, 174; *When Worlds Collide*, 173–74
Balzac, Honoré de, *La recherché de l'absolu*, 86
Barney, J. Stewart, L.P.M.: *The End of the Great War*, 171
Barthes, Roland, 130–31, 134
Baruch Plan, 245, 367n20
Batman and Robin, 305, 317
Battlestar Galactica, 262
Beautiful Mind, A, 338
Becker, Carl, 161
Behn, Aphra, *The Emperor of the Moon*, 52
Bell, Alexander Graham, 130, 162
Bellow, Saul, *Mr. Sammler's Planet*, 257
Bergman, Kerstin, 310, 311
Bichat, Marie-François-Xavier, 114, 116
Bierce, Ambrose, "Moxon's Master," 144–45
Binder, Eando, "I, Robot," 266
Blade Runner, 270
Blaikley, Alexander, 107
Blake, William, 65, 74, 80–82; *Newton*, 81–82
Blish, James, *A Case of Conscience*, 176
Blueprint, 277
Bohr, Niels, 291
Böll, Heinrich, *Fürsorgliche Belagerung*, 260
Bones, 311
Borchert, Wolfgang, *Lesebuchgeschichten*, 243
Boyd, William, *Brazzaville Beach*, 307–8, 318
Boyle, Robert, 18, 38, 43, 51
Boys from Brazil, The, 208
Braun, Wernher von, 218, 255

Breaking the Code, 338
Brecht, Bertolt: *Das Experiment*, 325; *Leben des Galilei*, 241–42, 324–26
Brenton, Howard, *The Genius*, 297–98
British Association for the Advancement of Science, 106
Broad, William, *Star Warriors*, 174, 257
Broch, *Die unbekannte Große*, 226–27
Brod, Max, *Galilei in Gefangenschaft*, 285, 324
Browne, Sir Thomas, 64
Browning, Robert, *Paracelsus*, 85
Buck, Pearl, *Command the Morning*, 242, 250–51
Budrys, Algis, *Who?*, 217
Butler, Samuel, 48–50, 143–44; *The Elephant in the Moon*, 48–49; *Erewhon*, 143–44; *Hudibras*, 48; "A Satire on the Royal Society," 49–50
Buzzati, Dino, "Appointment with Einstein," 288–89
Byatt, A. S., *A Whistling Woman*, 315

Campbell, John W., *Who Goes There?*, 283
Cavendish, Margaret, Duchess of Newcastle, 46, 52, 305; *Description of a New World, Called the Blazing World*, 46–47
chain reaction, 309
Chatterton, Ruth, *The Betrayers*, 329
Chaucer, Geoffrey, 344n16; *Canon's Yeoman's Tale*, 15, 18–19
Cobb, Weldon, *Trip to Mars*, 164
Colatrella, Carol, 317
Colby, Merle, *The Big Secret*, 327–28
Contact, 308, 311–12, 313, 317
Cotes, Roger, 64
Cowley, Abraham, "To the Royal Society," 36, 66
Cramer, Heinz Von: "Aufzeichnungen eines ordentlichen Menschen," 225; *Die Konzessionen des Himmels*, 248
Crichton, Michael: *Jurassic Park*, 204, 273, 276; *The Lost World: Jurassic Park*, 204; *Next*, 280
CRISPR technology, 281
Cromie, Robert: *The Crack of Doom*, 193, 196; *A Plunge into Space*, 164
Crookes, Sir William, 189, 193
Curie, Marie, 189, 303, 321
Curie, Pierre, 189, 321
cybernetics, 247–48, 258
cyborgs, 261

Damnation de Faust, 187
Darwin, Charles, 105, 112, 121; *Descent of Man*, 108, 142, 154; *Origin of the Species*, 105, 108, 119, 121
Darwin, Erasmus, 92; *Zoonomia*, 108
Darwinism, 108, 119–21, 154–56, 192–93
Daumann, Rudolf, *Protuberanzen*, 174–75
Day after Trinity, The, 236, 250
De Camp, L. Sprague, "Judgement Day," 197
Dee, John, 16
Desaguliers, J. T., 61
Descartes, René, 17, 42, 48, 57, 58, 60, 66, 75
Dick, Philip: "The Android and the Human," 269; "Do Androids Dream of Electric Sheep?", 269–70; *Dr Bloodmoney*, 198–99
Dickens, Charles, 85, 261
Diderot, Denis, 38; *Encyclopédie*, 38
Disraeli, Benjamin, 149, 357n39, 360n16; *Tancred*, 119–20
Dobbs, Betty Jo Teeter, 57
Döblin, Alfred, *Berge, Meere und Giganten*, 203–4
Doctor Who?, 262
Dolly the sheep, 277
Dominik, Hans: *Atomgewicht 500*, 182; *Die Macht der Drei*, 181–82
Doyle, Arthur Conan, 138–41; "The Disintegration Machine," 193; *The Land of Mist*, 140; *The Lost World*, 138–39; *The Maracot Deep*, 140; *The Poison Belt*, 139; "A Study in Scarlet," 177
Dr. Cyclops, 195
Dr. Renault's Secret, 202
Dr. Strangelove, 197–98, 264
Dr. X, 202
Dürrenmatt, Friedrich: *Der Mitmacher*, 295; *Die Physiker*, 242, 289–91, 330
Dyson, Freeman, 191, 236, 239
dystopias, 144, 205, 215–16, 219, 268, 280

Easlea, Brian, 238
ecofeminists, 40
Eden. See Garden of Eden
Edison, Thomas Alva, 130, 163, 361n4
Edison Studios, 187
Egyptian metal workers, 13
Einstein, Albert, 176, 237, 239, 249, 363n17
Eliot, George, 113, 115–18; "The Lifted Veil," 115; *Middlemarch*, 113–18
Elizabeth I of England, 15
Ellul, Jacques, 261

Enigma, 338
Enlightenment, the, 61, 76, 93, 102
Enthiran, 261
environmental issues, 259–60, 300, 337–38
Erin Brokovich, 299
Evolution. *See* Darwinism
Ex Machina, 270

Fail-Safe, 198
Fall, the. *See* Garden of Eden
Faraday, Michael, 106–7
Fat Man and Little Boy, 254
Faust, Dr. Johann, 20
Faust character, 20–22; in German Romanticism, 21, 77–79
Fermi, Enrico, 236, 250
Feyerabend, Paul, 39
Figala, Karin, 57
Fineman, Irving, *Doctor Addams*, 223–24
Flammarion, Camille, 163
Flicker, Eva, 302, 305–9
Fly, The, 203
Forster, E. M., "The Machine Stops," 144, 205, 259
France, Anatole, *Penguin Island*, 194
"Frankenfoods," 2, 26
Frankenstein, 2, 91–104, 153, 191, 263; ballet, 104; films of, 102–4, 202; as a myth of modernity, 2, 25, 91
Franklin, Benjamin, 76, 162
Fray, Michael, *Copenhagen*, 291
Frisch, Max, *Don Juan; or, The Love of Geometry*, 224–25
Fuseli, Henry, *The Nightmare*, 100

Gaines, Susan, *Carbon Dreams*, 307, 310, 313, 317–18
Galileo Galilei, 57
Galton, Francis, *Natural Inheritance*, 142
Galvani, Luigi, 76
Garden of Eden, 2, 14, 28
Gaskell, Elizabeth, 10; "Cousin Phillis," 125; *Wives and Daughters*, 10, 111–13
Gattaca, 273, 277–78
Geber, 14
Genetic engineering, 272–73, 275–77, 279–81
Gernsback, Hugo, 165
Gide, André, *Les Caves du Vatican*, 201
Gissing, George, *Born in Exile*, 126
Godsend, 277

Godwin, Francis, 34; *The Man in the Moone*, 34–35
Godwin, William, 93, 98
Godzilla, King of Monsters, 283
Goethe, Johann Wolfgang von, 77–79; *Faust*, 78–79; theory of light, 77, 352n6
Goldschmidt, Pippa, *The Falling Sky*, 309–10
Goldstein, Rebecca, *Properties of Light*, 338
golem legend, 17, 344n12
Goodman, Allegra, *Intuition*, 310, 313
Gosse, Edmund, *Father and Son*, 120
Gosse, Philip, 110, 120; *Omphalos*, 121
Graf, Oskar Maria, *Die Erben des Untergangs*, 176
Grass, Günter, *Die Rättin*, 260, 300

Haldane, J.B.S., 206
Halley, Edmund, 58, 66, 132, 347n7, 350n35
Hammerström, Jan, *Die Abenteur der Sibylle Kyberneta*, 258–59
Haraway, Donna, 40; "A Cyborg Manifesto," 266, 272
Hardy, Thomas, 122–25; *A Pair of Blue Eyes*, 122–23; *Two on a Tower*, 122, 123–25, 192; *The Woodlanders*, 126
Harry Potter, 23
Hartlib, Samuel, 35
Harvey, William, 42
Harvey, W. J., 117
Hawking, Stephen, 261, 338
Hawthorne, Nathaniel, 87–89; "The Birthmark," 88; "Rappaccini's Daughter," 88–89; *The Scarlet Letter*, 89
Hazzard, Shirley, *The Transit of Venus*, 333
Heisenberg, Werner, 289, 291
Henderson, Thomas, 107
Herbert, Frank, *Destination Void*, 268
Hermes Trismegistus, 13
Herschel, Caroline, 302
Herschel, William, 9
Hetzel, Pierre-Jules, 130
Hevelius, Johannes, 45
Hobbes, Thomas, 17, 43
Hoffmann, E.T.A., 86; "Der Sandmann," 87, 264
Hogan, James P., 176, 332; *The Genesis Machine*, 176
Hogarth, William, 73–74
Holst, T., 97
homunculus, 17
horror fiction and film, 3, 26
Houellebecq, Michel, *Les Particulaires élémentaires*, 274

Hoyle, Fred, and John Elliot, *A for Andromeda*, 306–7
Humans, 272
Huxley, Aldous, 204; *After Many a Summer Dies the Swan*, 220; *Antic Hay*, 220; *Brave New World*, 144, 182, 205–6, 216, 219, 273; *The Genius and the Goddess*, 220–21; *Island*, 204, 333–34; *Point Counter Point*, 220
Huxley, Thomas Henry, 110, 112, 119, 153, 161, 180, 200; *Evolution and Ethics*, 151, 153–54

I, Robot, 262, 268–69
ideological treason, 293
Imitation Game, The, 338
Imperato Ferrante, 45
Industrial Revolution, 79, 142, 261
internationalism in science, 31, 32, 180, 184, 238, 324, 327–28
Invisible Ray, The, 195
Ishiguro, Katzuo, *Never Let Me Go*, 279
Island, The, 279

Jahnn, Hans Henny, *Die Trümmer des Gewissens*, 287–88
Johnson, Samuel, 61, 68, 73; *Rasselas*, 73
Jonson, Ben, 22–23; *The Alchemist*, 22–23
Jünger, Ernst, 266; *Gläserne Bienen*, 265–66
Jungk, Robert, *Brighter Than a Thousand Suns*, 289
Jurassic Park, 204, 273, 276, 338; *Jurassic Park III*, 204; *Jurassic World*, 204

Kaiser, Georg: *Gas I*, 194–95; *Gas II*, 195
Kellermann, Bernhard, *Der Tunnel*, 171–72
Kepler, Johannes, 15, 57
Keynes, John Maynard, 56–57
Kingsley, Charles, 110, 121; *Glaucus*, 110–11; *Two Years Ago*, 109–10
Kingsolver, Barbara, *Flight Behavior*, 299
Kipphardt, Heiner, *In the Matter of J. Robert Oppenheimer*, 252, 292–95
Kirby, David, 278, 281
Klaes, Larry, 312
Klinger, Friedrich von, 78
Koestler, Arthur, 242
Kornbluth, C. M., "Gomez," 330
Kuhn, Thomas, 39, 309–10

La Mettrie, Julien Offray de, 75, 142; *L'Homme machine*, 75, 273
Laplace, Pierre-Simon, 58, 350n2
Lasswitz, Kurd: *Gegen das Weltgesetz*, 179; "Über Zukunftsträume I u. II," 178
Lavoisier, Antoine, 76
Lawrence, D. H., *The Rainbow*, 214–15
Le cabinet de Méphistophélès, 187
Le Guin, Ursula, 167; *The Dispossessed*, 334–35
Leinster, Murray: "The Man Who Put Out the Sun," 194; "The Storm That Had to Be Stopped," 194
Levin, Ira: *The Boys from Brazil*, 208, 273, 274–75; *The Stepford Wives*, 270
Le Voyage dans la lune, 187
Lewis, C. S., 206; *Out of the Silent Planet*, 206–7; *Perelandra*, 207–8; *That Hideous Strength*, 208
Lewis, Sinclair, *Arrowsmith*, 322
Locke, John, 75, 80–81
Lovelace, Ada, 302, 371n3
Lowell, Percival, 163
L'uomo meccanico, 264
Lyell, Sir Charles, *Principles of Geology*, 108

Macneice, Louis, "The Kingdom," 214
Madame Curie, 303, 321
Mallett, David, *The Excursion*, 66
Man Who Knew Infinity, The, 338
Marlowe, Christopher, *The Tragical History of Dr. Faustus*, 20–21
Maron, Monika, *Flugasche*, 260, 300
Mars/Martians, 163–66
Martian, The, 167
Marvel, 168
Masters, Dexter, *The Accident*, 251–52
mathematicians, 338–39
Matrix, The, 262
"Matthew effect," 302–3
Maugham, Somerset, *The Magician*, 23, 201
McConnell, William, 129
McCormmach, Russell, *Night Thoughts of a Classical Physicist*, 238–39, 296
Mechanical Man, The, 262
mechanization, 75, 77, 79–80, 90, 142–43, 200, 214–15, 261, 322–23
Medicine Man, 316–17, 338
Méliès, Georges, 187

Meredith, George, *Melampus*, 111
Merton, Robert K., 32, 42, 238, 366n6
Metropolis, 264–65
Miller, James, *The Humours of Oxford*, 53
Milner, H. M., *Frankenstein; or the Man and the Monster*, 102
Moleschott, Jacob, 108
Morgan, Charles: *The Burning Glass*, 326–27; *The Flashing Stream*, 326
Morris, William, *News from Nowhere*, 142
Muller, Friedrich, 78
Mumford, Lewis, 129–30
Musil, Robert: *Der Mann ohne Eigenschaften*, 229–30; "Tonka," 228–29; *Die Verwirrungen des Zöglings Törleß*, 228
Muske-Duke, Carol, *Saving St. Germ*, 304–5, 314
myths, 2

natural philosophers, 107
natural philosophy, 9, 70, 154
natural theology, 64
Nature, 161
Naturphilosophie, 77
NCIS, 7, 303, 311
Newcomb, Simon, 170–71; *His Wisdom, the Defender*, 170–71
Newton, Sir Isaac, 9, 18, 38, 54–67; as alchemist, 56–57; *Opticks*, 58, 64, 81; *Principia*, 56, 58–61, 64, 80, 82
Nossack, Erich, *Die Schalttafel*, 225
Novalis, 77
nuclear weapons, 170–74, 183–85, 186

O'Brien, Fitz-James, 85–86
Oldenburg, Henry, 35
Olson, R. G., 61
Oppenheimer, J. Robert, 184, 190, 236, 244, 249, 252–54, 292–95
Orphan Black, 275
orrery, 63, 350n22
Orwell, George, *1984*, 205
Outbreak, 338

Paracelsus, 16, 17
Perée, Jacques Louis, 16
Perkowitz, Sidney, 303
Picoult, Jodi, *My Sister's Keeper*, 277, 279–80
Piercy, Marge, *Small Changes*, 316

Poe, Edgar Allan: "Maelzel's Chess Player," 10, 144; *The Murders in the Rue Morgue*, 202
Pope, Alexander, 68, 69–70; *The Dunciad*, 69; *Essay on Man*, 69
Popper, Karl, 11, 39
Powers, Richard, 281
Priestley, J. B., *The Doomsday Men*, 195–96
Priestley, Joseph, 76
progress, 128, 161
Puritans, 42, 53
Pynchon, Thomas, 217–19; *The Crying of Lot 49*, 218; *Gravity's Rainbow*, 218, 254–55

radiation-produced monsters, 283–84
radioactivity, 189–90, 193–94
Ramsay, Alan, "Ode to the Memory of Sir Isaac Newton," 65
Rankine, W.J.M., "The Mathematician in Love," 85
Raspail, François-Vincent, 117
Ray, John, 43; *The Wisdom of God Manifested in the Word of the Creation*, 43, 65
Raymond, Alex, *Flash Gordon*, 173, 190
Rehberg, Hans, *Johannes Kepler*, 285, 324
Reich, Kathy, *Devil Bones*, 311
religion and science, 105–6, 108–9, 118–21, 153–55, 159–60, 169, 176, 201–2, 206, 275, 320, 350n26, 350n28, 351n16
Renard, Maurice, *Le docteur Lerne, sous-dieu*, 200
Restoration stage, 42, 53
Rhodes, W. H., "The Case of Summerfield," 192
"risk society," 1, 339
Ritter, J. W., 77
Roberts, Robin, 306
Robida, Albert, 238
RoboCop, 262
robots, 261–69, 271–72, 368n3; ethical treatment of, 271–72
Rohn, Jennifer: *Experimental Heart*, 307, 309, 310, 315, 317; *The Honest Look*, 307, 309, 313, 314, 316
Romantic stereotype of the scientist, 76–90, 125, 127
Romantic writers, 76–90; English poets, 79–83; German writers, 76–79; novelists, 84–90
Röntgen, Wilhelm, 189
Rose, F. W., *The Maniac's Dream*, 196
Rosicrucians, 52, 57
Ross, Sydney, 9, 10

Roszac, Theodore, 337; *Memoirs of Elizabeth Frankenstein*, 104
Rotblat, Joseph, 240, 366n10
Rowlandson, Thomas, 49
Royal Society, The, 35, 36–37, 41–42, 43, 44, 47, 50, 52; *Philosophical Transactions of*, 35
Rudolph II of Prague, 15
Runaway, 262
Rutherford, Sir Ernest, 190, 193

Sala, George Augustus, 129
Sambourne, Linley, 120
Sandburg, Carl, "Mr. Atilla," 241
Savage, Richard, *The Wanderer*, 66–67
Schelling, Friedrich, 76–77
Schiaparelli, Giovanni, 163
Schirmbeck, Heinrich, *Ärgert dich dein rechtes Auge*, 243, 246–48, 330–31
Schlegel, Friedrich, 77
scientific materialism, 80, 107, 108, 142, 143, 205, 212, 214, 219, 263, 335n6
scientism, 61, 152, 156, 206, 216
"scientist," 10–11, 343n10
scientist: as absent-minded professor, 5, 44, 135, 187; as adventurer, 6, 129–41; as alchemist, 5, 18–27, 138, 153, 200; as amoral, 158, 235–60; as detective, 177–78; as helpless, 6, 145, 284–98, 300–301; as hero, 161–62; as idealist, 5, 153, 319–36; as inventor, 162–68; as mad, bad, dangerous, 3, 6, 152–56, 186–210, 280, 283; stereotypes, 4, 5, 337; as unemotional, 5–6, 90, 95, 125–27, 158, 177, 211–34, 322–23; as virtuoso, 5, 41, 43–45, 50–53; as woman, 302–18; as world savior / utopian ruler, 168–77, 178–83
Serviss, Garrett P., 163; *Edison's Conquest of Mars*, 163, 164; *The Second Deluge*, 171
Seven Days to Noon, 197
Sewell, Stephen, *Welcome the Bright World*, 227–28
Shadwell, Thomas, 50–52; *The Sullen Lovers*, 52; *The Virtuoso*, 50–52, 135
Shakespeare, William, 22; *The Tempest*, 22
Shelley, Mary, 2, 91–104, 127; *Frankenstein*, 2, 91–104, 145, 273; *The Last Man*, 104
Shelley, Percy, 84, 92, 93
Simon Magus, 17
Sinclair, Upton: *A Giant's Strength*, 287; *The Millennium*, 193
Sixth Day, The, 273

Skinner, B. F., *Walden Two*, 182, 248
Smeed, J. W., 79
Smith, Martin Cruz, *Stallion Gate*, 253–54
Smith, Michael Marshall, *Spares*, 279
Snow, C. P., 7, 221, 365n16; *Death under Sail*, 221–22; *The New Men*, 222, 243–46; *The Search*, 222–23; *In Their Wisdom*, 296
Snyder, Alexander, "Blasphemer's Plateau," 202
Social Darwinism, 107, 127
Soddy, Frederick, 190
spiritualism, 140–41
Sprat, Thomas, 36; *History of the Royal Society of London*, 36, 37, 42, 65
Stapledon, Olaf, 207
Star Gate, 167, 262
Star Trek, 167, 303
Star Trek: Voyager, 272
Star Wars, 268
Steinke, Jocelyn, 303, 305, 308, 313
Stevenson, Robert Louis, 145; *Dr. Jekyll and Mr. Hyde*, 94, 145–46, 147, 202
Stewart, Douglas, "Rutherford," 331–32
Stoker, Bram, *Dracula*, 169
Surrogates, 271
Sussman, H. L., 144
Swift, Jonathan, 68, 70–73; *Gulliver's Travels*, 70–73, 155–56
Symmes, John Cleves, 132
Szilard, Leo, 175–76

Taoism, 13
Tasker, William, 60
Teller, Edward, 246, 252–53
Tennyson, Alfred, *In Memoriam A.H.H.*, 118–19, 153
Tesla, Nikola, 130, 163, 164
Thing, The, 283
Thomson, James, "Poem Sacred to the Memory of Sir Isaac Newton," 59–60, 66
three laws of robotics, 267–69, 368n16
Train, Arthur, and Robert Williams Wood, *The Man Who Rocked the Earth*, 172–73
Transcendence, 201
Tudor, Andrew, 3, 186
Turing, Alan, 270; Turing test, 369n19
Tyndall, John, 143, 200

Ussher, Bishop, 108
utopias, 30–33, 142, 170–72, 178–83, 205

Verne, Jules, 130–38, 141, 147–49; *The Amazing Adventure of the Barsac Mission*, 148; *The Begum's Fortune*, 148; *Five Weeks in a Balloon*, 132; *The Floating Island*, 148; *For the Flag*, 148; *Journey to the Centre of the Earth*, 132–33, 135–36; *Master of the World*, 148; *Mysterious Island*, 134–35; *Robur the Conqueror*, 148; *Topsy Turvy*, 148; *Twenty Thousand Leagues under the Sea*, 136–37
Verrill, A. Hyatt, "The Plague of the Living Dead," 201–2
Villiers de l'Isle-Adam, Auguste, 146; *L'Eve future*, 146
virtuosi, 5, 41, 43–45, 347n1
Voltaire, 59
Vonnegut, Kurt, *Cat's Cradle*, 243, 255–56

Waterloo, Stanley, *Armageddon*, 170
Weart, Spencer, 190–91, 276
Weingart, Peter, 3, 186, 302, 308
Weir, Andy, *The Martian*, 136
Weldon, Fay, *The Cloning of Joanna May*, 273, 275
Wells, H. G., 9, 39, 40, 99, 138, 149–60, 164–65, 171, 172; *Anticipations*, 179–80; "The Chronic Argonauts," 138; films of *The Island of Doctor Moreau*, 156; *The First Men in the Moon*, 157–58, 180; *The Food of the Gods*, 180, 204; *The Invisible Man*, 156–57; *The Island of Doctor Moreau*, 152–56; "The Land Ironclads," 180; "The Lord of the Dynamos," 152; *Marriage*, 323; *A Modern Utopia*, 180–81; "The Moth," 151–52; "The Time Machine," 149–51; *Tono-Bungay*, 159–60; *War of the Worlds*, 164–65, 190; *When the Sleeper Awakes*, 158; *The World Set Free*, 172, 190, 323–24, 361n19
Whewell, William, 10
Whitehead, Alfred North, 128
Whitman, Walt, 84
Wiener, Norbert, 258–59
Wilberforce, Bishop Samuel, 119
Wilder, Thornton, *The Skin of Our Teeth*, 298–99
Wilson, Mitchell, *Meeting at a Far Meridian*, 329
Wöhler, Friedrich, 143
Wolf, Christa, 209–10, 232; "Selbstversuch," 230–31; *Störfall*, 231–32, 257, 260, 299–300
Wordsworth, William, 82–83
World Is Not Enough, The, 311
Wouk, Herman, *The Traitor*, 329
Wright, Joseph, 24, 63, 83
Wright, Thomas, *The Female Virtuoso's*, 52

Yeats, W. B., 84

Zamyatin, Yevgeny, *We*, 144, 205, 215–16
Zola, Émile, 319; *Doctor Pascal*, 320; *Paris*, 320
Zuchmayer, Carl, *Das kalte Licht*, 232–34, 330
Zwillinger, Frank, *Galileo Galilei, Schauspiel*, 285–86